GENE FUNCTION IN PROKARYOTES

GENE FUNCTION IN PROKARYOTES

Edited by

Jon Beckwith
Harvard Medical School

Julian Davies
Biogen S.A.

Jonathan A. Gallant
University of Washington

Cold Spring Harbor Laboratory
1983

**COLD SPRING HARBOR
MONOGRAPH SERIES**

The Lactose Operon
The Bacteriophage Lambda
The Molecular Biology of Tumour Viruses
Ribosomes
RNA Phages
RNA Polymerase
The Operon
The Single-Stranded DNA Phages
Transfer RNA:
    Structure, Properties, and Recognition
    Biological Aspects
Molecular Biology of Tumor Viruses, Second Edition:
    DNA Tumor Viruses
    RNA Tumor Viruses
The Molecular Biology of the Yeast Saccharomyces:
    Life Cycle and Inheritance
    Metabolism and Gene Expression
Mitochondrial Genes
Lambda II
Nucleases
Gene Function in Prokaryotes

GENE FUNCTION IN PROKARYOTES

MONOGRAPH 15

© 1983 by Cold Spring Harbor Laboratory

Printed in the United States of America

Book design by Emily Harste

Library of Congress Cataloging in Publication Data

Main entry under title

Gene function in prokaryotes.
    (Cold Spring Harbor monograph series; 15)
    Papers presented at a symposium held at the Banbury
Center of the Cold Spring Harbor Laboratory, June 1982.
    Includes index.
    1. Bacterial genetics—Congresses.   2. Gene expression
—Congresses.   I. Beckwith, Jon R.   II. Davies, Julian E.
III. Gallant, Jonathan A.   IV. Cold Spring Harbor
Laboratory.   V. Series.
QH434.P76   1983   589.9'015   83-15229
ISBN 0-87969-164-6

All Cold Spring Harbor Laboratory publications are available through booksellers or may be
ordered directly from Cold Spring Harbor Laboratory, Box 100, Cold Spring Harbor, New York
11724.

SAN 203-6185

Contents

Preface

In June of 1982, a meeting was held at the Banbury Center of Cold Spring Harbor Laboratory in honor of the memory of Luigi Gorini. The participants heard speakers whose research was in the areas that Luigi had pioneered during his career. Out of this meeting has evolved this volume, which covers not only these research areas but several others at the forefront of bacterial genetics.

Luigi was an imaginative scientist, a great humanitarian, and a good friend. His great joy in research was to ask questions about the way in which bacteria "work" and to try to answer these questions by the clever use of genetics. Luigi isolated many different kinds of mutants, a number of which remain uncharacterized. What particularly intrigued him were the "funny" mutants—those derivative microorganisms that did not behave in the predicted manner. It was the analysis of "funny" mutants that provided Luigi with information that began many of the studies mentioned in this book. His insight and curiosity led to the isolation and analysis of mutants that affected protein secretion, gene regulation, the fidelity of translation, transcription-translation coupling, and other important biological processes. Luigi Gorini's contribution to these areas is amply described in the Biographical Memoir by Jon Beckwith and Dan Fraenkel and in appropriate scientific reviews.

The development of understanding of the topics to which Luigi contributed and which are presented in this book owes much to the application of bacterial genetics. The intelligent isolation and characterization of mutants have provided most of the framework for our understanding of bacterial structure, metabolism, regulation, and cell division. In addition, these concepts have provided the framework for approaches to the study of eukaryotic cell function. More recently, basic knowledge of transposition and protein secretion have owed much to genetic studies in bacteria. These studies, in turn, have influenced the direction of research in eukaryotic organisms.

The future seems equally bright for the key role of prokaryotic genetics in the development and understanding of a variety of biological research problems. Burgeoning areas include the study of the complex machinery involved in chemotaxis and other proteins that pass signals through membranes, the mechanism of cell division, and the nature of interactions between cells. Many of the speakers emphasized that this meeting was not a memorial but a celebration—a celebration to acknowledge Luigi Gorini's contributions to studies of microorganisms and to point the way to new applications of genetics in unraveling these more complicated bacterial functions that provide overlap between prokaryotic and eukaryotic cell function.

We are grateful for financial support for the meeting provided by the following: Bayer AG/Cutter/Miles; Biogen S.A.; Bristol-Myers Co. Fermentation Research and Development; Bristol-Myers Co. Pharmaceutical Research and Development Division; Cetus Corp.; E.I. du Pont de Nemours & Co. Inc.; Hoffmann-La Roche Inc.; Eli Lilly and Co.; Merck, Sharp and Dohme Research Laboratories; Schering Corp.; Searle Research and Development; The Squibb Institute for Medical Research; Stauffer Chemical Company; The Upjohn Company. In organizing this meeting we were ably assisted by Michael Shodell and Beatrice Toliver of the Banbury Conference Center.

We also wish to thank Annamaria Torriani-Gorini for her continuing interest and assistance with this project and Nancy Ford, Nadine Dumser, Mary Cozza, Joan Ebert, and Adrienne Guerra of the Publications staff of Cold Spring Harbor Laboratory for their efforts in making the publication of this book a reality.

J. Beckwith
J. Davies
J.A. Gallant

GENE FUNCTION
IN PROKARYOTES

The following article first appeared as *A Biographical Memoir* published in 1980 by the National Academy Press and has been reproduced here in its entirety. We have foot-noted the text with updated or corrected information where appropriate. Following the article are photographs of Luigi Gorini at work and at leisure in various stages of his life.

The photographs were graciously provided by Annamaria Torriani-Gorini. We are also grateful to her for providing us with her insights and personal remembrances (both humorous and profound), which accompany the photographs.

LUIGI GORINI

November 13, 1903–August 13, 1976

BY JONATHAN BECKWITH
AND
DAN FRAENKEL

L UIGI GORINI, professor in the Department of Microbiology and Molecular Genetics at Harvard Medical
School and a member of the National Academy of Sciences,
died August 13, 1976. He was born on November 13, 1903 in
Milan, Italy. His father was a microbiologist. Luigi obtained
his first degree from the University of Pavia in 1925; his
thesis (1925) was in organic chemistry, but his interest was in
biology. He continued his studies in organic chemistry, but he
was to publish only four papers in the next twenty years.

In 1931 the Italian government moved to control the
universities by requiring a Fascist oath. Luigi described this
period in a speech at Montana State University on February
10, 1970.

The first uproar was *no* unanimously—we will never do that. But then came
second thoughts, the rationalization: we scientists should not be involved in
politics, we should not permit that others, worse than us, would take our
responsibilities, etc. At the end, we were about one hundred *no's* out of
about 10,000 university people. And so we quit. It was not an easy thing to
do, not only materially but especially for the spirit. We, the one percent,
started a double life, political underground for our soul and professional
marginal for our belly. I discovered very quickly that the ability to convey
opinions, to convince others, was not a gift that I had, so I did my underground work which may look romantically wonderful in retrospect, but
seen from inside was a day by day realization of inefficiency.

3

The next ten years were spent in Turin in a succession of small pharmaceutical houses where his politics, which were Socialist, were tolerated. The work was research, development, and quality control. In these years he was married and had two children. His son from this marriage, Jan,[1] is now following a career of research in immunology in the Laboratory of Radiation Pathology, Casaccia-Rome, and his daughter, Isa, is now a biochemist in Milan. The external circumstances of his life were relatively comfortable.

When the war came, Luigi refused induction and went partially underground with the assumed name Carlo Cattaneo. Cattaneo was a nineteenth century Italian patriot and opponent of the monarchy who edited a journal of science and politics. Luigi avoided arrest when the police came for him in 1942 and escaped to Milan, where he found work in a very small research institute (Istituto Giuliana Ronzoni) owned by an anti-Fascist industrialist. There he met Annamaria Torriani, who had just finished her studies. She was to be his colleague in the laboratory and in the resistance, and later his wife. They had one son, Daniel, who is now eighteen and a student at Rhode Island School of Design.[2]

In the resistance, Luigi was involved in the collection and distribution of news among several cities. He also carried food, medicines, and documents to the partisans in the mountains above Milan. Although a pacifist and nominally unarmed, one of his occasional duties was to collect money. This meant going to the prospective contributor, taking out a gun, and explaining the advantages of supporting the cause.

When Milan was liberated (April 25, 1945), the Socialist party gave Luigi the task of taking over a property in the mountains at Selvino which had been a summer camp for children of Fascists. The most needy at the time were Jewish children from the liberated concentration camps who had

[1] *Gion (correct spelling) is presently at the Istituto di Immunologia Centro Nazionale della Ricerche in Rome.*

[2] *Daniel, an artist, graduated from the Rhode Island School of Design in 1981 and is now living in Boston.*

begun to appear in Milan. Luigi and Annamaria decided to use Selvino for them. In the next three years it served as a rehabilitation center for about a thousand children. They were from several countries of origin, and ranged from three-year-olds to teenagers. Selvino was to help rebuild their confidence prior to their emigration to Palestine. Luigi was nominally the administrator, but mainly a friend and counsellor. At the same time he was doing scientific work at the Institute in Milan. In 1976 Luigi and Annamaria were honored by the government of Israel for their work at Selvino, and an account of these activities was placed in the Martyrs and Heroes Archives at Yad-Vashem, Israel.

The last group of children left for Israel in 1948. Meanwhile, Luigi's academic title had been restored, but only at its former level as beginning assistant. Annamaria went to the Pasteur Institute in Paris. Her work there together with Melvin Cohn and Jacques Monod is well known (she is now a professor of biology at Massachusetts Institute of Technology). Luigi joined the laboratory of Claude Fromageot at the Sorbonne as a member of the CNRS (Centre National Recherche Scientifique), and he soon was independent.

Over the next seven years there were seventeen papers published dealing with aspects of bacterial proteolysis and the biochemistry of extracellular enzymes. Much of this work was on the mechanism of protection of various bacterial proteases by ions such as calcium and manganese. He and his co-workers were able to show that the metal ions protected these enzymes against autodigestion by stabilizing particular protein conformations. This work had wide impact in that it provided a strong suggestion that proteins do not have unique folding patterns, but can exist in several different stable states. This work was a continuation of his earlier interests in microbiology, and its quality was recognized early by the award of the Kronauer Prize (1949, University of Paris).

The work on physiology of proteolysis led to the discovery in 1954 of an unusual bacterial growth factor, catechol. Bernard Davis, who was then interested in aromatic biosynthesis, invited Luigi to his Department of Pharmacology at New York University. In 1957, Luigi joined the Department of Bacteriology and Immunology at Harvard Medical School, of which Dr. Davis had become head.

Soon after arriving in New York, Luigi, working together with Werner Maas, made a fundamental discovery in bacterial regulation. It was known at the time that some bacterial enzymes in sugar degradative pathways were inducible. There were also indications of regulation of enzyme synthesis in biosynthetic pathways, since the level of such enzymes was somewhat lower when the end-product was available than when it had to be made. Gorini and Maas showed that if partial starvation of the end-product of the pathway was arranged—they used an arginine-limited chemostat—the rate of synthesis of an enzyme in the arginine pathway became high (derepression). This phenomenon, "bacteria in overdrive," showed that enzyme synthesis in biosynthetic pathways was variable over a wide range, somehow responding to the endogenous level of end-product. This finding had a profound impact on thinking about regulation of gene expression and played a major role in the development of the concept of the repressor. Kenneth Schaffner, who has reviewed the early history of this field, puts it this way:

> Arthur Pardee recalls that the short paper by Gorini and Maas particularly "attracted attention" because it was "simply presented." . . . The demonstration, particularly striking in the case of Gorini's and Maas' experiment, that elimination of the repressing metabolite could result in a rapid and continued rate of constitutive enzyme synthesis, suggested . . . that inducible systems might perhaps be analyzed by a similar mechanism of negative control. . . .*

*K. Schaffner, "Logic of Discovery and Justification in Regulatory Genetics," *Studies in the History and Philosophy of Science*, 4 (4) (1974):349–85.

At Harvard the arginine pathway was Luigi's main research for some years. His group was concerned early with sorting out the physiological role of the derepression phenomenon from the other mechanism controlling flow in the pathway, end-product inhibition of the first enzyme. Luigi was interested in whether the system might really function by a combination of induction and repression and eventually established that apparent strain differences in regulation reflected differences in repressor protein only. Luigi and his co-workers continued to publish work on the regulation of the arginine biosynthetic genes until his death.

In 1964 Luigi and his colleagues published the first of a long series of papers on bacterial ribosomes that were to dramatically change the thinking of biologists about the function of the ribosomes. Up until that time, it was thought that all the specificity of translation of the genetic code lay in the interaction between transfer RNA and messenger RNA. Ribosomes were seen as passive templates upon which this process took place. In 1961 Gorini, Gundersen, and Berger noticed the peculiarity that an arginine auxotroph in the presence of a streptomycin-resistant mutation could be restored to prototrophy by the addition of streptomycin to the growth medium. Rather than ignoring this finding as one often does with peculiar observations, Luigi followed it up, and in 1964 he and Eva Kataja presented evidence that streptomycin was altering the specificity of translation via an interaction with the ribosome. (There already existed evidence that streptomycin acted on the ribosome.) From this they suggested that "the ribosomal structure could include the accuracy of the reading of the code during translation."* There quickly followed work in collaboration with Drs. Julian Davies and Walter Gilbert providing direct *in vitro* confirmation of this proposal.

* Luigi Gorini and E. Kataja, "Phenotypic Repair by Streptomycin of Defective Genotypes in *E. coli, Proceedings of the National Academy of Sciences* (USA), 51:487–93.

Luigi proceeded over the next twelve years to develop a new field: the study of factors influencing the fidelity of translation of the genetic code. The influence of ribosomal mutations was extensively studied. Certain mutations to drug resistance, which affected a ribosomal protein, were found to decrease drug-dependent misreading. Other mutations in the same protein caused total dependence on streptomycin for growth in any medium. It appeared that the ribosome was then so distorted as to function usefully only in the presence of an agent causing translational ambiguity. A new type of ribosome mutation, "ram" (ribosomal ambiguity), was discovered which increased misreading even in the absence of antibiotics.

Much work followed on the types of mutations corrected by misreading. While initially it appeared that chain-terminating (nonsense) mutations were the only ones affected, work from Luigi's laboratory subsequently showed that the translation of missense and even frame-shift mutations could be changed by alteration of the ribosome. Further, altered transfer RNA molecules appeared particularly sensitive to ribosomal mutations.

Luigi also had characteristically original ideas about other aspects of antibiotic action, such as the possibility that streptomycin might bind to RNA directly and affect ribosome assembly. In some of his last work, evidence was obtained for a link between mutations affecting the ribosome and mutations in RNA polymerase, suggesting that there may be unexplored levels of interaction between transcription and translation.

All this work, of course, was done with a long succession of collaborators—graduate and medical students, postdoctoral fellows, and other visitors. But Luigi always worked in the laboratory himself. He arrived first in the morning and was not above looking at his colleagues' experiments before they came in themselves. He was blessed with a remarkable vitality. The whole story of ribosomal suppression was discovered when he was in his sixties, and even after his formal retirement at seventy the work continued with fifteen papers. Luigi's science was well recognized. He became an American Cancer Society Professor (1964), received Harvard's Ledlie Prize (1965), and was elected to membership in the National Academy of Sciences (1971).

But it was not only science that he discussed with his colleagues; it was more often politics or literature. He slept little and was extraordinarily well organized. He read the local papers and the *New York Times, The New York Review, The Guardian, Le Monde,* and *Jerusalem Post* weeklies as well as books they mentioned, and that is what he talked about, often indignantly, passionately, always interestingly.

He had an unusually genuine and strong sense of outrage over injustice and inequality. He was particularly concerned about the plight of minority groups in this country and of third world peoples in general. Luigi accepted many invitations to speak at black southern colleges, taking these opportunities to actively oppose the pseudoscientific theories that were used to support racism. For instance, in a talk at Southern University on February 21, 1974, referring to genetic theories of inequality:

All this nonsense could be disregarded as no more than science fiction in bad taste if it were not the fact that in this way science is dangerously and irresponsibly misused to justify the right to power and wealth for the benefit of only a few racial groups, or families, or individuals, no matter what were the means these groups or their ancestors used to acquire their present dominant position in society.

Luigi was heavily involved in anti-Vietnam War activities, and when Henry Kissinger was awarded the Nobel Prize for Peace in 1973, Luigi organized a petition protesting the award. The petition was sent to the Nobel Committee and received publicity in this country.

His attitudes toward science and the role of scientists in society influenced many around him. This influence is exemplified by a paragraph from the Ph.D. thesis acknowledgement of one of his students, Dirk Elseviers.

> Luigi Gorini directed my work in Boston. His creativity, enthusiasm and energy are a constant stimulus for everybody around him. He has taught me that the satisfaction in doing science lies in doing it and in nothing else. [But] above all that it is of capital importance to keep in touch with reality; our lives are in the hands of politicians and not of Science. I really like him.

And in Luigi's own words, again from his speech at Montana State University:

> My job here tonight is to make you realize that for me, like for hundreds of us scientists, my own scientific interest means a lot intellectually but, morally speaking, science alone does not satisfy entirely my conscience. I will try to be the most unequivocal radical possible and at the same time constructive, so that when I quit, your opinion about me should not be similar to that expressed a long time ago by the fascist Italian police about someone whom I know after his first confrontation with them. He was very happy to be released, for a time at least, but a few years later he discovered by chance the written motivation for letting him out and he was really not satisfied. The police file sounds like the following: "Lonely anarchist; he is not dangerous."

When he "quit," Luigi left behind him a spirit of rigorous scientific curiosity and social conscience which has affected many of those who were close to him.

HONORS AND DISTINCTIONS

ACADEMIC POSITIONS

1946–1949	In charge of Department of Biochemistry, Instituto Scientifico di Chimica e Biochimica Giuliana Ronzoni, Milan, Italy
1949–1951	Attaché de Recherches, Centre National Recherche Scientifique, Laboratoire de Chimie Biologique, Sorbonne, Paris, France
1951–1954	Chargé de Recherches, Centre National Recherche Scientifique, Laboratoire de Chimie Biologique, Sorbonne, Paris, France
1954–1955	Maître de Recherches, Centre National Recherche Scientifique, Laboratoire de Chimie Biologique, Sorbonne, Paris, France
1955–1957	Visiting Researcher, Department of Pharmacology, College of Medicine, New York University, New York
1957–1962	Lecturer, Department of Bacteriology and Immunology, Harvard Medical School, Boston, Massachusetts
1962–1964	American Cancer Society Associate Professor, Department of Bacteriology and Immunology, Harvard Medical School
1964–[3] June 30, 1974	American Cancer Society Professor, Department of Microbiology and Molecular Genetics, Harvard Medical School
July 1, 1974	Professor Emeritus, Department of Microbiology and Molecular Genetics, Harvard Medical School

HONORS

1925	Highest *cum laude* honors awarded by the University of Pavia
1927	Prize for Advancement in Organic Chemistry awarded by the Politecnico of Milan
1949	Prize Kronauer awarded by Faculté des Sciences, Sorbonne, Paris
1963	Elected to the American Academy of Arts and Sciences

[3]*Elected Affiliate of the Royal Society of Medicine, London.*

1965   Ledlie Prize awarded by Harvard University
1971   Elected to the National Academy of Sciences

PROFESSIONAL AND HONORARY SOCIETIES

American Society for Microbiology
Federation of American Societies for Experimental Biology
Society of General Physiologists
American Society of Biological Chemists
American Association for the Advancement of Science

BIBLIOGRAPHY

1924

Analogia di Costituzione tra il fenantrene ed il 2-N-fenil-α, β-Naftotriazolchinone. Ph.D. thesis, Ist Chimica Generale Universita di Pavia.

With A. Dansi. Intorno all'azione delle sostanze coloranti sulla sensibilita della gelatina-bromuro d'argento. I) Rivista Fotografica Italiana (April): 1–36.

1925

With A. Dansi. Intorno all'azione delle sostanze coloranti sulla sensibilita della gelatina-bromuro d'argento. II) Rivista Fotografica Italiana (June):3–8.

With G. Charier and A. Manfredi. Sul 2-Nofenil [α,β] nafto-1-2-3-triazolchinone. Gazz. Chim. Ital., 56:196–207.

1933

Azionone[4] della trimetilamina su l'esametidiaminoisopropanolo dioduro. Gazz. Chim. Ital., 63:751–56.

1935

With C. Gorini. Ulteriori richerche sulle proteasi degli acidoproteolitici. Rend. R. Ist. Lomb. Sci. Lett., 68:115–25.

1938

Ancora sul sistema proteasico degli acidoproteolitici. Rend. R. Ist. Lomb. Sci. Lett., 72:133–46.

Sulle proteasi degli acidoproteolitici. Enzymologia, 10:192–202.

1946

With A. Torriani. Sulla purificazione della penicillinase da Bacterium Coli. Boll. Soc. Ital. Biol. Sper., 22:1.

1947

With A. Torriani. Biochemistry of *Escherichia coli* and the production of penicillinase. Nature, 160:332–33.

[4]*Azione.*

1948

With A. Torriani. Action de la penicilline sur l'activité protéolytique des bactéries acido-protéolytiques. Biochim. Biophys. Acta, 2:226–38.

1949

With Cl. Fromageot. Une protéinase bactérienne (*Micrococcus lysodeikticus*) nécessitant l'ion calcium pour son fonctionnement. C. R. Acad. Sci., 229:559–61.

1950

With Cl. Fromageot. Les facteurs physiologiques conditionnant la présence de protéinase dans les cultures de *Micrococcus lysodeikticus*. Biochim. Biophys. Acta, 5:524–34.

Le rôle du calcium dans l'activité et la stabilité de quelques protéinases bactériennes. Biochim. Biophys. Acta, 6:237–55.

1951

Rôle du calcium dans le systeme trypsine-serumalbumine. Biochim. Biophys. Acta, 7:318–34.

With L. Audrain. Nécessité du calcium dans la croissance de bactéries lorsque la source d'azote est une protéine pure. Biochim. Biophys. Acta, 6:477–86.

With M. Grevier. Le comportement de la protéinase endocellulaire de *Micrococcus lysodeikticus* au cours de la lyse de cet organisme par le lysozyme. Biochim. Biophys. Acta, 7:291–95.

1952

With L. Audrain. Influence du calcium sur la stabilité du complex trypsine-ovomucoide. Biochim. Biophys. Acta, 8:702–3.

With L. Audrain. Action de quelques métaux bivalents sur la sensibilité de la serumalbumine à l'action de la trypsine. Biochim. Biophys. Acta, 9:180–92.

With L. Audrain. Influence du zinc sur la stabilité de la plasmine. Biochim. Biophys. Acta, 9:337–38.

1953

With F. Felix. Influence du manganèse sur la stabilité du lysozyme. I. Influence du manganèse sur la vitesse d'inactivation irrever-

sible du lysozyme par la chaleur. Biochim. Biophys. Acta, 10:128–35.
With L. Audrain. Le complexe ovomucoide-trypsine. Son activité protéolytique et le role de quelques métaux dans la stabilité de ses constituents. Biochim. Biophys. Acta, 10:570–79.
With F. Felix. Sur le mécanisme de protection de la trypsine par Ca^{++} ou Mn^{++}. Biochim. Biophys. Acta, 11:535–42.
With F. Felix and Cl. Fromageot. Influence du manganèse sur la stabilité du lysozyme. II. Role protecteur du manganèse lors de l'hydrolyse du lysozyme par la trypsine. Biochim. Biophys. Acta, 12:283–88.

1954

With J. Labouesse-Mercouroff. Sur les facteurs conditionnant l'activité enzymatique de la carboxypeptidase. Biochim. Biophys. Acta, 13:291–93.
With G. Lanzavecchia. Recherches sur le mécanisme du production d'une protéinase bactérienne. I. Nouvelle technique de détermination d'une protéinase par la coagulation du lait. Biochim. Biophys. Acta, 14:407–14.
With G. Lanzavecchia. Recherches sur le mécanisme de production d'une protéinase bactérienne. II. Mise en evidence d'un zymogène précurseur de la protéinase de Coccus P. Biochim. Biophys. Acta, 15:399–410.

1956

With R. Lord. Necessité des orthodiphenols pour la croissance de Coccus P (Sarcina Sp.). Biochim. Biophys. Acta, 19:84–90.
With L. Audrain. Relations entre degré de dénaturation et sensibilité à la trypsine de la serumalbumine. Influence de Ca^{++} et de Mn^{++} et rôle des ponts disulfure. Biochim. Biophys. Acta, 19:289–96.

1957

With W. K. Maas. End-product control for the formation of a biosynthetic enzyme. Fed. Proc. Fed. Am. Soc. Exp. Biol., 16:215.
With W. K. Maas. The potential for the formation of a biosynthetic enzyme in Escherichia coli. Biochim. Biophys. Acta, 25:208–9.

1958

With W. K. Maas. Negative feedback control of the formation of biosynthetic enzymes. In: *Physiological Adaptation*, pp. 151–58. Wash., D.C.: American Physiological Society.

With W. K. Maas. Feedback control of the formation of biosynthetic enzymes. In: *A Symposium on the Chemical Basis of Development*, ed. W. D. McElroy and B. Glass, pp. 469–78. Baltimore: Johns Hopkins Univ. Press.

Regulation en retour (feedback control) de la synthèse de l'arginine chez *Escherichia coli*. Bull. Soc. Chim. Biol., 40:1939–52.

1959

With H. L. Ennis. Feedback control of the synthesis of enzyme and end product in arginine biosynthesis in *Escherichia coli*. Fed. Proc. Fed. Am. Soc. Exp. Biol., 18:222.

1960

With H. Kaufman. Selecting bacterial mutants by the penicillin method. Science, 131:604–5.

Antagonism between substrate and repressor in controlling the formation of a biosynthetic enzyme. Proc. Natl. Acad. Sci. USA, 46:682–90.

1961

With H. L. Ennis. Control of arginine biosynthesis in strains of *Escherichia coli* not repressible by arginine. J. Mol. Biol., 3:439–46.

With W. Gundersen. Repressor and modulator, two cellular tools for controlling synthesis of biosynthetic enzymes. In: *Proceedings of the 5th International Congress of Biochemistry*, vol. 1, pp. 155–59. Oxford: Pergamon Press.

With W. Gundersen. Induction by arginine of enzymes of arginine biosynthesis in *Escherichia coli* B. Proc. Natl. Acad. Sci. USA, 47:961–71.

With M. Berger and W. Gundersen. Coordinate repression and genetic sequence of the arginine biosynthetic enzymes in *Escherichia coli*. Communication at the 5th International Congress of Biochemistry, Moscow.

Effect of L-cystine on initiation of anaerobic growth of *Escherichia coli* and *Aerobacter aerogenes*. J. Bacteriol., 82:305–12.

With W. Gundersen and M. Berger. Genetics of regulation of enzyme synthesis in the arginine biosynthetic pathway of *Escherichia coli*. Cold Spring Harbor Symp. Quant. Biol., 26:173–82.

1963

With S. M. Kalman. Control by uracil of carbamyl phosphate synthesis in *Escherichia coli*. Biochim. Biophys. Acta, 69:355–60.

Control by repression of a biochemical pathway. Bacteriol. Rev., 27:182–90.

1964

With E. E. Sercarz. Different contributions of exogenous and endogenous arginine to repressor formation. J. Mol. Biol., 8:254–62.

Conditional streptomycin dependent mutants and control mechanisms. Communication presented at 6th International Congress of Biochemistry, New York.

With E. Kataja. Phenotypic repair by streptomycin of defective genotypes in *E. coli*. Proc. Natl. Acad. Sci. USA, 51:487–93.

With E. B. Horowitz. Coordination between repression and retroinhibition in control of a biosynthetic pathway. In: *Comparative Biochemistry of Arginine and Derivatives* (Ciba Foundation, Study Group No. 19), pp. 64–81. Boston: Little, Brown.

With J. E. Davies and W. Gilbert. Streptomycin suppression and the code. Proc. Natl. Acad. Sci. USA, 51:883–90.

With E. Kataja. Streptomycin-induced oversuppression in *E. coli*. Proc. Natl. Acad. Sci. USA, 51:995–1001.

Streptomycin and the ambiguity of the genetic code. New Sci., 24:776–79.

1965

With J. Yashphe. Phosphorylation of carbamate *in vivo* and *in vitro*. J. Biol. Chem., 240:1681–86.

With E. Kataja. Suppression activated by streptomycin and related antibiotics in drug sensitive strains. Biochem. Biophys. Res. Commun., 18:656–63.

With J. Davies and B. D. Davis. Misreading of RNA codewords induced by aminoglycoside antibiotics. Mol. Pharmacol., 1:93–106.

With W. F. Anderson and L. Breckenridge. Role of ribosomes in streptomycin-activated suppression. Proc. Natl. Acad. Sci. USA, 54:1076–83.

With D. Old. Amino acid changes provoked by streptomycin in a polypeptide synthesized *in vitro*. Science, 150: 1290–92.

1966

Antibiotics and the genetic code. Sci. Am., 214:102–9.

The action of streptomycin on protein synthesis *in vivo*. Bull. N.Y. Acad. Med., 42:633–37.

With J. R. Beckwith. Suppression. Annu. Rev. Microbiol., 20:401–22.

With G. Jacoby and L. Breckenridge. Ribosomal ambiguity. Cold Spring Harbor Symp. Quant. Biol., 31:657–64.

1967

Induction of code ambiguity by aminoglycoside antibiotics. Fed. Proc. Fed. Am. Soc. Exp. Biol., 26:5–8.

With G. A. Jacoby. Genetics of control of the arginine pathway in *Escherichia coli* B and K. J. Mol. Biol., 24:41–50.

With G. A. Jacoby. The effect of streptomycin and other aminoglycoside antibiotics on protein synthesis. In: *Mechanism of Action and Biosynthesis of Antibiotics*, ed. D. Gottlieb and P. Shaw, vol. I, pp. 726–47. Berlin, Heidelberg, and N.Y.: Springer-Verlag.

With R. Rosset and R. A. Zimmermann. Phenotypic masking and streptomycin dependence. Science, 157:1314–17.

Ambiguity in the translation of the genetic code into proteins, induced by aminoglycoside antibiotics. In: *Immunity, Cancer and Chemotherapy*, pp. 167–75. N.Y.: Academic Press.

1968

With J. Davies. The effect of streptomycin on ribosomal function. Curr. Top. Microbiol. Immunol., 44:100–122.

1969

With G. A. Jacoby. A unitary account of the repression mechanism of arginine biosynthesis in *Escherichia coli*. I. The genetic evidence. J. Mol. Biol., 39:73–87.

With O. Karlstrom. A unitary account of the repression mechanism of arginine biosynthesis in *Escherichia coli*. II. Application to the physiological evidence. J. Mol. Biol., 39:89–94.

With R. Rosset. A ribosomal ambiguity mutation. J. Mol. Biol., 39:95–112.

With L. Breckenridge. The dominance of streptomycin sensitivity re-examined. Proc. Natl. Acad. Sci. USA, 62: 979–85.

The contrasting role of *strA* and *ram* gene products in ribosomal functioning. Cold Spring Harbor Symp. Quant. Biol., 34:101–11.

1970

With P. Strigini. Ribosomal mutations affecting efficiency of amber suppression. J. Mol. Biol., 47:517–30.

With L. Breckenridge. Genetic analysis of streptomycin resistance in *Escherichia coli*. Genetics, 65:9–25.

Informational suppression. Annu. Rev. Genet., 4:107–34.

1971

With N. Z. Sarner, M. J. Bissell, and M. DiGirolamo. Mechanism of excretion of a bacterial proteinase. I. Demonstration of two proteolytic enzymes produced by a *Sarcina* strain *(Coccus P)*. J. Bacteriol., 105:3, 1090–98.

With M. J. Bissell and R. Tosi. Mechanism of excretion of a bacterial proteinase. II. Factors controlling accumulation of the extracellular proteinase of a *Sarcina* strain *(Coccus P)*. J. Bacteriol., 105:3, 1099–1109.

With U. Bjare. Drug dependence reversed by a ribosomal ambiguity mutation, *ram*, in *Escherichia coli*. J. Mol. Biol., 57:423–35.

With H. Momose. Genetic analysis of streptomycin dependence in *Escherichia coli*. Genetics, 67:19–38.

With R. A. Zimmermann and R. Rosset. Nature of phenotypic masking exhibited by drug-dependent streptomycin A mutants of *Escherichia coli*. J. Mol. Biol., 57:403–22.

With R. A. Zimmermann and R. T. Garvin. Alteration of a 30s ribosomal protein accompanying the *ram* mutation in *E. coli*. Proc. Natl. Acad. Sci. USA, 68:2263.

Ribosomal discrimination of tRNA's. Nature (London) New Biol., 234:52, 261–64.

1972

With D. K. Biswas. Restriction, de-restriction and mistranslation in missense suppression. Ribosomal discrimination of tRNA's. J. Mol. Biol., 64:119–34.

With P. J. Piggott and M. D. Sklar. Ribosomal alterations controlling alkaline phosphatase isozymes in *E. coli*. J. Bacteriol., 110:291–99.

With J. F. Atkins and D. Elseviers. Low level activity in β-galactosidase frameshift mutants of *E. coli*. Proc. Natl. Acad. Sci. USA, 69:1192–95.

With D. K. Biswas. The attachment site of streptomycin to the 30s ribosomal subunit. Proc. Natl. Acad. Sci. USA, 69:2141–44.

1973

I Ribosomi. In: *Enciclopedia Della Scienza e della Tecnica Mondadori*, ed. Arnaldo Mondadori, pp. 305–12. Milan: Edizioni Scientifiche e Techniche.

With R. T. Garvin and R. Rosset. Ribosomal assembly influenced by growth in the presence of streptomycin. Proc. Natl. Acad. Sci. USA, 70:2762–66.

1974

With D. Elseviers. Direct selection of ribosomal mutants with altered translation efficiency in *E. coli* B. Fed. Proc. Fed. Am. Soc. Exp. Biol., 33:1335.

With R. T. Garvin and D. K. Biswas. The effects of streptomycin or dihydrostreptomycin binding to 16S RNA or to 30S ribosomal subunits. Proc. Natl. Acad. Sci. USA, 71:3814–18.

Streptomycin and misreading of the genetic code. In: *Ribosomes*, ed. P. Lengyel, M. Nomura, and A. Tissieres, pp. 791–803. N.Y.: Cold Spring Harbor Laboratory.

1975

With S. Chakrabarti. Growth of bacteriophages MS2 and T7 on streptomycin-resistant mutants of *Escherichia coli*. J. Bacteriol., 121:670–74.
With D. Elseviers. Misreading and the mode of action of streptomycin. In: *Drug Action and Drug Resistance in Bacteria*, vol. 2, *Aminoglycosidic Antibiotics*, ed. H. Umezawa, pp. 147–75. Tokyo: Univ. of Tokyo Press.
With R. T. Garvin. A new gene for ribosomal restriction in *Escherichia coli*. Mol. Gen. Genet., 137:73–78.
With M. Duncan. A ribonucleoprotein precursor of both the 30S and 50S ribosomal subunits of *E. coli*. Proc. Natl. Acad. Sci. USA, 72:1533–37.
With A. Kikuchi and D. Elseviers. Isolation and characterization of lambda transducing bacteriophages for *argF*, *argI* and adjacent genes. J. Bacteriol., 122:727–42.
With D. Elseviers. Direct selection of mutants restricting efficiency of suppression and misreading levels in *E. coli* B. Mol. Gen. Genet., 137:277–87.
With S. L. Chakrabarti. A link between streptomycin and rifampicin mutation. Proc. Natl. Acad. Sci. USA, 72:2084–87.
With A. Kikuchi. Similarity of genes *argF* and *argI*. Nature (London), 256:621–24.

1976

With A. Kikuchi. Studies of the DNA carrying genes *valS*, *argI*, *pyrB*, and *argF* by electron microscopy and by site specific endonuclease. J. Microsc. Biol. Cell., 27:1–10.

1977

With S. Chakrabarti. Interaction between mutations of ribosomes and RNA polymerase: A pair of *strA* and *rif* mutations individually temperature-insensitive but temperature-sensitive in combination. Proc. Natl. Acad. Sci. USA, 74:1157–61.

In 1978, the following papers were published posthumously:

With M. Dohi, H. Inouye, and A. Kikuchi. *In vitro* synthesis of ornithine transcarbamilase. J. Biochem. Tokyo, 84:1389–1399.
With M. Dohi and A. Kikuchi. Some regulation profiles of ornithine transcarbamilase synthesis *in vitro*. J. Biochem. Tokyo, 84:1401–1409.

The world in all its aspects (science, only one of them) interested Luigi, who had the culture and mentality of a Renaissance man. Traveling was of special enjoyment to him and to those with him who could profit from his knowledge of art and history. He also loved mountain climbing and managed to hike in the Alps or elsewhere at least once a year. Trips to Machu Picchu in 1973 and the refuge at Monte Rosa (Macugnaga, Italy) in 1974 marked the end of his climbing as leukemia began to sap his strength.

Besides bicycling (at which he was well practiced since it was the choice transport of underground fighters), Luigi liked rowing. He was proud of coming in second in a race in Pavia ... when only two boats were competing! At the First International Meeting of Biochemistry in Cambridge, England (1949), he and a group of friends went boating on the Cam. Luigi is shown here with Lou Siminovitch and Germaine Bazir-Stanier. While punting vigorously, he became entangled in a low-lying branch and was suspended in midair, much to the amusement of his crew, as they were propelled forward in the unmanned canoe. He landed in the canoe directly behind (far left in photo).

A tour of Provence remained memorable to all of the participants in every detail. A stop at Vaucluse was ''de rigueur'' to Luigi, who knew many of Petrarch's and Dante's poems by heart.

Luigi actively participated in the underground fight against fascism by bringing medicine and food to the partisans and by helping to organize the Movimento Federalista Europeo, an international movement which he believed to be the best solution in preventing new wars. When he was invited to the United States in 1955, Luigi became attracted by the renewed democratic strength (after McCarthy) of this country. Shown above (second row, second from left), Luigi with a group of colleagues of A. Pappenheimer at N.Y.U. Medical Center.

When Luigi was offered a position in the Department of Bacteriology and Immunology at Harvard Medical School in 1957, he accepted. He felt that he could contribute openly and directly to the organization of a society in which he could merge, without being rejected as in fascist Italy or fully accepted as in chauvinist France. (*Above*) At Harvard (1966) Luigi counting streptomycin-resistant mutants; (*right*) a typical pose of Luigi doing precision pipetting (1967); (*below*) Luigi discussing the repression mechanisms of arginine biosynthesis with Olee Karlstrom, Mina Bissel, and George Jacoby (1969).

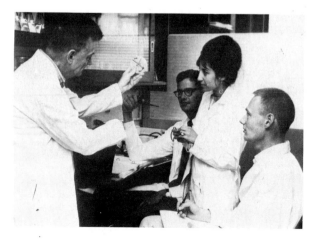

Luigi's almost religious crusade for freedom of the spirit made him too independent to ever participate in the necessities of an organized political party or religion. (*Left*) Luigi, with a political poster of Angela Davis as a backdrop. (*Below left*) Luigi at the Weizmann Institute of Science in Rehovot, Israel, at 70 years of age. (*Below right*) Luigi's father Constantino at age 80 at the University of Milan.

In 1976, Luigi wrote, "One of these days I will be confronted with my last experience. What will happen? Everything is so mysterious and I spend my time in waiting.... I can see the immediate future, the first two or three days, but after that is a mystery. For me it will be peace and even that is not the exact word. To feel peace you have to feel nonpeace, but you do not feel nonexistence."

Precision of Protein Biosynthesis

Michael Yarus and Robert C. Thompson
Department of Molecular, Cellular, and Developmental Biology
University of Colorado
Boulder, Colorado 80309

INTRODUCTION

In translation, the cyclic selection of amino acids to become associated with cognate tRNAs, and of aminoacyl-tRNAs (aa-tRNAs) to donate their amino acids to the growing peptide chain, must be carried out rapidly to yield a biologically functional protein product. Despite these demanding requirements, the fraction of the product that is accurately formed—the "yield," in chemical jargon—is probably greater than 90%. Later in this paper we discuss the value of this number in more detail. This extraordinary performance has attracted much experimentation, which was foreseen in the pioneering studies of Luigi Gorini.

For brevity and concreteness, we have concentrated on the accuracy of coding; i.e., the accuracy with which aa-tRNAs are matched to codons. We further confine discussion to chain extension. On related topics, there is an excellent recent review of chain initiation (Gold et al. 1981). Similarly, there are other recent sources of information about the accuracy of synthesis of aa-tRNAs (Yarus 1979a), and two independent and roughly contemporary surveys of translational accuracy can be compared with this one (see Buckingham and Grosjean 1983; Kurland and Gallant 1983). A survey of the earlier literature, and a different point of view, may be found in the text of Yarus (1979b). Finally, the reader desiring the broad view should not forget that translational error is always seen against a background of transcriptional errors (Rosenberger and Foskett 1981), still largely uncharacterized.

ACCURACY AND KINETICS

Accuracy is a matter of the relative rates of reaction of two similar substrates (e.g., aa-tRNAs at the ribosomal A site) or the relative rates of two reactions of the same substrate (e.g., the translocating ribosome). The evolution of an accurate translation apparatus is a problem of the maximization of these ratios.

The error rate (relative rate of reaction) when two substrates compete for reaction with the same active site (Fersht 1981) is

$$Vn/Vc = (Rn/Rc)(N/C)$$

in which Vn and Vc are the velocities of production of the noncognate and cognate products, respectively; Rn and Rc express the intrinsic accuracy of the process—they are collections of elementary rate constants, as determined by the detailed reaction scheme; and N and C are the concentrations of the noncognate and cognate substrates, respectively.

This error rate expression, which calculates the relative rate of production of the erroneous product, is extremely general and simple. It nevertheless has implications that can be antiintuitive. For example, however precise a reaction scheme may be, it can be forced to any fraction of errors by disparities in the concentrations of the substrates with which it must react. Conversely, however imprecise it is, the reaction will perform well if the cognate substrate is abundant. This sort of effect was first pointed out for tRNAs and aa-tRNA synthetases (Yarus 1972) but is quite general. More complex reaction schemes make the Rs more complex. This may decrease their ratio and increase accuracy but rarely changes the nature of the expression above. Therefore, even proofreading of the usual sort (Hopfield 1974; Ninio 1975) can only reduce, and will not prevent, mistakes forced by "pool bias" (disproportionate substrate concentrations). Discussion of one possible exception to the form of the expression above may be found in the literature (Hopfield 1980; Savageau 1981). The relative concentrations of aa-tRNAs (substrates) are therefore as basic to the accuracy of translation as is the structure of the ribosome and the message.

The error rate equation when a single reactant may choose two alternate (first-order) fates is simplified to $Vn/Vc = Rn/Rc$. The decision between the normal fate of the complex (e.g., normal ribosomal translocation) and an aberrant result (e.g., a frameshift) is made, in this case, entirely on the basis of Rn/Rc. This ratio represents the intrinsic tendency to error, built into the system by physical constraints and evolution.

These simple ideas confer order on a variety of experiments on translational errors. We begin with a detailed treatment of the amino acid incorporation cycle. Though we have tried to present the following subjects in a plausible and helpful sequence, the rest of the text was written as a series of independent sections. We hope that it will be intelligible in any part or order, and we recommend a corresponding reading strategy, on the basis of interest.

EXPERIMENTAL SYSTEMS USED TO STUDY ACCURACY

Because the answers to scientific questions often depend on the experimental system used, we have included a brief survey of the most popular techniques for investigating the accuracy of protein biosynthesis before considering the results of these studies. The most fundamental division between these techniques is between in vivo and in vitro experiments.

In Vivo

There are several ways to detect missense errors during translation in vivo, which are discussed in more detail in the section on in vivo accuracy. Whereas the detection of errors in vivo is often straightforward, calculating error rates involves assigning these errors to the ribosome, to a particular codon, and to a particular tRNA, which is far more difficult. This is a serious problem because these estimates of accuracy of translation in vivo form the basis for judging the relevance of most in vitro work on accuracy.

The most straightforward way of assigning an error to a ribosome-related reaction is to measure its frequency in strains with ribosomes of intrinsically different accuracy. The properties of these ribosomes are discussed below in Factors That Perturb Accuracy.

One way to assign the error to the codons for a single amino acid is to determine how the error frequency responds to starvation for that amino acid in rel^- strains of *Escherichia coli*. These cells lack one of the mechanisms for preventing misreading of a codon in the absence of the appropriate aa-tRNA. The frequency of errors associated with translation of a codon can be expected to rise during starvation for the appropriate amino acid. However, the only unambiguous way to determine the particular codon responsible for an error is to partially sequence the erroneous protein and the gene. Identification of the codon responsible for an error is much simpler if the error involves misreading of a termination codon, which accounts in part for the popularity of this type of investigation.

Preliminary information about which tRNA is involved in the error is usually available from a knowledge of the amino acid inserted. There is, unfortunately, no general way to define the tRNA responsible if this information and the experimenter's intuition are insufficient to do so.

In Vitro

In vitro protein synthesis using natural and synthetic mRNA has been widely used to study the mechanism of protein synthesis and its accuracy. Errors are generally detected by the incorporation of amino acids into peptides that should not contain them. The former systems are more obviously relevant to the situation in vivo but share many of the disadvantages of the in vivo systems discussed above. Specifically, the complexity of the mRNA and the translation system employed make errors more difficult to detect and assign. For this reason, the failure of Lagerkvist and his co-workers (Elias et al. 1979) and Goldman et al. (1979) to detect misreading of AAA/G and UUU/C/A/G codons of MS2 RNA in vitro should not be taken to indicate that these errors do not occur at a significant level in vivo. Although not sensitive enough to detect true errors, these systems are useful for studies of the rules of codon-anticodon pairing that underlie accuracy, since these groups found codons (CUX, Goldman et al. 1979; GUX, Mitra et al. 1979; GCX, Samuelsson et al. 1980) that are deciphered using rules for base-pairing in the "wobble" position other than those predicted by Crick (1966).

Synthetic mRNAs, in which one or a few codons are repeated, allow errors to be detected and identified because the error is amplified by repetition. These systems have often been used at Mg^{++} concentrations that make their error rates much higher than those observed in vivo, which has detracted from the significance of the results obtained. The recent work of Kurland and his co-workers, indicating that poly(U) can be translated under conditions similar to those thought to occur in vivo, has emphasized the attractiveness of this system for studies of the relationship between various experimental parameters and translational accuracy (Jelenc and Kurland 1979; Wagner et al. 1980).

With the advent of good in vitro systems to study the mechanism and accuracy of protein biosynthesis, it is useful to draw some parallels between these experimental systems and those of enzymologists. The present tendency is to study full in vitro translation systems only after they have reached steady-state. As in analogous studies of enzyme mechanisms at steady state, only two kinetic constants can be obtained. Neither of these is always simply related to the elementary rate constants for the translational process. Obtaining new information about the mechanism by which accuracy is achieved from the kinetics of these systems is therefore very difficult.

By analogy with other enzymatic processes, information about the mechanism of protein synthesis and the way in which accuracy is achieved should be obtained more easily through studying the pre-steady-state kinetics of the process. These studies are more difficult than steady-state studies because they require more reagents and occur on a faster time scale. Pre-steady-state studies of translation have been carried out on partial protein biosynthesis systems, and the results (discussed at length in the sections on the limits, mechanism, and perturbation of accuracy) are sufficiently enlightening that we look forward to pre-steady-state studies of full systems with great interest.

The experimental results obtained in pre-steady-state systems are relatively easy to interpret because each reactant, even if normally a catalyst, participates only once. This "single-turnover" condition is usually achieved by depriving the system of a component it needs to cycle, and for this reason the significance of the results obtained has been questioned (Kurland 1982). Although the physiological relevance of these studies can never be completely proved, the objection does not appear to be a serious one in the case of the aa-tRNA binding reaction, since the rate and accuracy of this reaction is similar in single-turnover and in more complex systems (Thompson et al. 1979; Thompson and Dix 1982).

HOW ACCURATE IS TRANSLATION IN VIVO?

An error rate in vivo is the relative frequency of a specific error at a known site. Determining the nature of the error and the site is usually far more difficult than detecting the occurrence of an error.

Two classes of in vivo experiment to determine missense error frequencies can be distinguished based on the method used to detect the usually low level of errors. In particularly favorable cases, where a protein or peptide should not contain a particular amino acid, an error can be inferred from its presence. The presence of valine in some tryptic peptides of hemoglobin, which should not contain this amino acid, allowed Loftfield and Vanderjagt (1972) to determine that the error rate of protein synthesis in erythrocytes is between 1×10^{-4} and 5×10^{-4}. A similar study carried out in *E. coli* by Edelmann and Gallant (1977) took advantage of the fact that flagellin should contain no cysteine. The presence of cysteine was attributed to the misreading of arginine codons at a frequency of 6×10^{-4}. A valuable addition to this type of experimental methodology, which focuses on a particular codon, has recently been published by F. Bouadloun and C.G. Kurland (in prep.). These workers took advantage of the amino acid specificity of some enzymatic and chemical methods of breaking peptide bonds to determine the frequency with which an amino acid is incorporated into a protein through an error in protein synthesis. The frequency of replacement of tryptophan by cysteine in ribosomal protein S6 is 3×10^{-3}, and the frequency of replacement of arginine by cysteine in ribosomal protein L7/L12 is close to 1×10^{-3} as measured by this technique.

A second system of somewhat broader applicability relies on the fact that some errors will result in a polypeptide of different net charge, which can be resolved from the correct product by the isoelectric focusing step of two-dimensional polyacrylamide electrophoresis (O'Farrell 1978; Parker et al. 1978). Using this technique, Parker and his colleagues (1980) have determined that at least one type of error in the translation of the coat protein gene of coliphage MS2 occurs with a surprisingly high frequency. Variants of this protein having a net charge of $+1$ relative to the wild-type form 1% or more of the protein made in vivo. By measuring the lysine content of satellite spots to the coat protein on two-dimensional gels, they find an erroneous incorporation of this amino acid. They attribute this error to misreading of asparagine codons because it is more common in *rel*$^-$ cells starved for asparagine. If the error is distributed equally between the ten Asn codons of the gene, the error rate would be between 1×10^{-3} and 2×10^{-3}. However, because this error does not seem to occur to the same extent in genes that do not use the Asn codon AAU, Parker and his colleagues attribute the error to misreading of the four AAU codons at a frequency of 5×10^{-3}.

The range of missense error frequencies reported is obviously quite wide. The factor distinguishing the highest and lowest estimates is probably the type of error involved. Errors seem to be more frequent when misreading involves the third base pair of the codon, as might be expected from the known ambiguity of base-pairing of this base in cognate codon-anticodon pairs.

A wide range of error frequencies has also been observed for translation of the three termination codons in vivo. Strigini and Gorini (1970) found levels of ornithine transcarbamoylase in an Arg F40 (amber) strain of *E. coli* consistent with a readthrough error of 1×10^{-3} of the amber codon, and Andersson et al. (1982) have observed misreading of all three termination codons in genes resulting from the fusion of *lacI* and *lacZ* at frequencies ranging from 1×10^{-4}

to 1×10^{-2}. The UGA codon seems much more prone to this type of error than UAA or UAG in this study, which implies an abnormally poor fidelity of tRNA^{Trp}, for which there is independent evidence (see below, Suppression of Natural Terminators).

The picture that emerges from these studies is of protein synthesis as a process whose accuracy demands are flexible rather than fixed. Despite a high overall accuracy, it can have accuracies at specific codons that would be too low to be generally acceptable. This view is supported by the results of other studies, discussed below, in which the accuracy of translation in cells subjected to various insults is shown to be much lower than normal. The result is slower growth, and not a catastrophic collapse of vital functions. Of course, this represents the biochemist's view of the relationship between accuracy and physiology. Outside the laboratory the cell faces a much more competitive environment in which a 1% loss in growth rate due to inaccurate protein synthesis could spell virtual extinction over a short period.

LIMITS TO ACCURACY: THE MECHANISM OF aa-tRNA SELECTION BY RIBOSOMES

The virtues of accuracy in an information transfer process like protein synthesis are self-evident. The reduced growth rate of some strains that have reduced accuracy reinforces this point in a practical fashion. However, the existence of mutants that have accuracies of protein synthesis higher than the wild type implies that the optimal level of accuracy is not necessarily the highest. Many investigators have tried to determine what limits the accuracy of translation. Most attention has been given to limits arising from thermodynamic, energy efficiency, and kinetic factors.

Thermodynamics

Although the accuracy of aa-tRNA selection by ribosomes is clearly influenced by regions of the tRNA outside the codon and anticodon (Yarus 1982), the base-pairing between these two RNAs is still the principal factor allowing the ribosome to determine whether the codon has been translated correctly. However, the fidelity of base-pairing in aqueous solution falls far short of explaining the fidelity of aa-tRNA selection by ribosomes. The noncognate G-U pair, e.g., has sufficient stability to be observed in double-helical regions of tRNA (Ladner et al. 1975). Its strength has been estimated from the melting temperatures of synthetic double-helical RNAs (Uhlenbeck et al. 1971), and the dissociation rate constants of anticodon-anticodon complexes of tRNAs (Grosjean et al. 1978), to be only $2-3$ kcal/mole less than that of the A-U pair. A free energy difference of 2 kcal/mole between the correct and incorrect tRNAs allows only a 40-fold discrimination between them in a simple reversible selection process. The great discrimination observed in translation arises because base-pairing between the codon and anticodon on the ribosome is not like the pairing occurring in solution, and the binding process is not a simple reversible one.

The essential features of a nonequilibrium binding process that would allow the ribosome to amplify the discrimination available between the cognate and noncognate codon-anticodon pairs were proposed almost simultaneously by Hopfield (1974) and by Ninio (1975). The mechanism proposed by Hopfield is closer to what is presently known about the mechanism of aa-tRNA selection by ribosomes and so will form the substance of our discussion.

$$RS + TC \overset{k_1}{\underset{k_{-1}}{\longleftrightarrow}} RS \cdot TC \overset{k_2}{\longrightarrow} RS \cdot aa\text{-}tRNA \cdot EF\text{-}Tu \cdot GDP \cdot P \overset{k_3}{\longrightarrow} RS \cdot pep\text{-}tRNA$$
$$\downarrow k_4 \qquad + EF\text{-}Tu \cdot GDP + P_i$$
$$RS + aa\text{-}tRNA$$

The reaction involves reversible binding of a ternary complex (TC) of aa-tRNA, EF-Tu, and GTP to the ribosome·mRNA complex (RS) (reactions k_1 and k_{-1}) and hydrolysis of the GTP to GDP (reaction k_2). This initial selection was earlier thought to be followed directly by incorporation of the amino acid into nascent protein (reaction k_3). Hopfield showed on theoretical grounds that this process could give error frequencies much lower than those possible in a simple one-step process if (1) it was possible for the ribosome to discard the aa-tRNA from ternary complexes after GTP hydrolysis (reaction k_4), and (2) the energy released by GTP hydrolysis was used to favor the k_1, k_2, k_3 path for aa-tRNA binding over the theoretically possible alternative path in which the aa-tRNA binds to the ribosome directly by the reversal of reaction k_4.

Hopfield's ideas were experimentally confirmed by Thompson and Stone (1977) who showed that poly(U)-programmed ribosomes could bind and hydrolyze the GTP of ternary complexes containing the noncognate aa-tRNAs of leucine and isoleucine but could not subsequently bind the aa-tRNAs. These findings have since been confirmed by Yates (1979), by Thompson et al. (1981b), and by Ruusala et al. (1982) in a full protein biosynthesis system. The selection of aa-tRNAs is therefore a two-step process; the first consisting of initial recognition and hydrolysis of a ternary complex, and the second consisting of the decision to bind or reject the aa-tRNA. This latter step has come to be known as proofreading (Hopfield 1974). By substituting two discrimination steps where there was one, this complex mechanism removes any theoretical objection to basing the fidelity of protein biosynthesis on simple base-pairing between the codon and anticodon.

Having said this, we feel it important to point out that the existence of proofreading does not justify, a posteriori, the view that thermodynamic factors would otherwise limit the fidelity of protein biosynthesis. Recent measurements of the free energy difference between a cognate and a noncognate ternary complex binding to the ribosome indicate that the ribosome is amplifying the free energy difference expected from solution studies and that the thermodynamics of discrimination should not limit the accuracy of selection even in the absence of the complex mechanism outlined above. Thompson and Karim (1982) have determined that the dissociation equilibrium constant of a Phe ternary complex from poly(U)-programmed ribosomes is 8×10^{-10} M, whereas the analogous constant for the Leu2 ternary complex, which is noncognate by a single G-U base pair, is greater than 1.6×10^{-5} M (Thompson and Dix 1982; Thompson and Karim 1982). The codon-anticodon pairing of these two species is GAA:UUU and GAG:UUU, respectively. The 20,000-fold discrimination implied by these results indicates that the ribosome is destabilizing the G-U pair found only in the noncognate complex (not in the wobble position) about 200-fold. This could be accomplished by destabilizing the nonstandard geometry of base-pairing characteristic of this pair in double-helical RNA (Ladner et al. 1975) and possibly in solution. In addition to raising the question of the true purpose of proofreading, these results cast serious doubts on the relevance of solution measurements of the pairing properties of RNAs to the accuracy of codon-anticodon pairing on the ribosome.

Energy Efficiency

Since the proofreading process for enhancing fidelity requires energy, it is possible that the real limit on accuracy comes not from thermodynamic concerns but from a concern for energy efficiency. If the proofreading process rejected the cognate aa-tRNA with any significant frequency, or the initial recognition of ternary complexes was very nonselective, a significant amount of energy would be wasted. The costs of adding an amino acid to a protein could then be significantly greater than the four pyrophosphate bonds expended to charge the tRNA, bind its ternary complex to the ribosome, and translocate the peptidyl tRNA from the A to the P site. In vitro measurements of the frequency of rejection of tRNA^{Phe} from poly(U)-programmed ribosomes following the hydrolysis of GTP show that rejection of the cognate aa-tRNA is an infrequent event (Thompson et al. 1980) and so is unlikely to cause a significant waste of energy. On the other hand, the mistaken selection of near-cognate ternary complexes by the poly(U)-programmed ribosome does occur with a significant frequency (Thompson and Stone 1977; Yates 1979; Thompson and Dix 1982) and could perceptibly increase the energy cost of protein synthesis in vivo. However, protein synthesis without this extra cost is probably not the major energy-consuming activity of a rapidly dividing *E. coli,* so it is not likely that the energy wasted by proofreading is the chief factor limiting the accuracy of protein synthesis in this organism.

Kinetics

The free energy difference between the complexes of cognate and noncognate aa-tRNAs with the ribosome dictates the maximal and not the actual discrimination between these species. Attaining the maximal discrimination requires a particular set of kinetic parameters that may conflict with the need to carry out protein synthesis rapidly. Recent studies of the kinetic parameters for selection of aa-tRNAs by poly(U)-programmed ribosomes in vitro indicate that the requirements for speed and accuracy in protein synthesis may indeed conflict and that this could be an important factor limiting the accuracy of this process in vivo.

The rate constants for the reaction of Phe and Leu ternary complexes with poly(U)-programmed ribosomes have been determined in single-turnover experiments. The values of k_1, k_2, and k_3 do not depend greatly on whether the aa-tRNA is Phe or Leu2 (Thompson et al. 1980; Thompson and Dix 1982). In contrast, the values of k_{-1} and k_4 differ for these species by at least 3000-fold and 80-fold, respectively (Thompson and Dix 1982; Thompson and Karim 1982). Under these circumstances the maximum discrimination can be attained only if k_2 is much less than k_{-1} and k_3 is much less than k_4 for the cognate species. This is clearly not the case. k_2 is 25 sec^{-1} (J.F. Eccleston et al., unpubl.), k_{-1} is 0.002 sec^{-1} (Thompson and Karim 1982), k_3 is 0.4 sec^{-1} (Thompson et al. 1979), and k_4 is less than 0.08 sec^{-1} (Thompson et al. 1980; Thompson and Dix 1982).

These values seem to be chosen to attain maximum speed rather than maximum accuracy, suggesting that the speed of protein biosynthesis is the major factor limiting accuracy. This implies that the ribosome has accuracy to spare but is sorely pressed for time. The idea that the ribosome has an ample reserve of accuracy seems quite reasonable: If the discrimination between Phe and Leu2 observed in binding the two ternary complexes was duplicated in the proofreading process, the ribosome would have a maximum ability to discriminate between these two aa-tRNAs of 4×10^8-fold, a number far in excess of that necessary to make the overall error frequency insignificantly greater than that due to transcription. The sacrifice of speed of protein synthesis is, on the other hand, likely to be serious. A rapidly dividing *E. coli* already devotes half of its dry weight to components involved in the synthesis of new protein to avoid this becoming the rate-limiting step in its growth. The value of k_1 obtained in vitro (Thompson et al. 1980) indicates that selection of aa-tRNAs may be close to being the rate-determining step of protein synthesis for some of the rarer aa-tRNAs, so an increase in accuracy that requires that this reaction be slowed might, on balance, prove detrimental to the organism.

FACTORS THAT PERTURB ACCURACY

Besides their intrinsic interest, factors that perturb the accuracy of translation are useful as tools to study how the accuracy of protein biosynthesis affects other cell functions and how the individual steps of the elongation cycle affect the accuracy of the overall process. Factors that influence accuracy include ribosomal protein mutations, antibiotics, the concentrations of polycations, and the nucleotide ppGpp. The action of several of these factors is now fairly well understood. When considered in light of the error expression given in the Introduction, most of these factors change the ratio of the rate constants. The effects of ppGpp, which are not yet understood at the molecular level, have been attributed to effects on the ratio of rate constants (Gallant and Foley 1979) and to changes in the ratio of reactant concentrations (Wagner et al. 1982).

Ribosomal Mutations

Gorini was the first to show that mutations in ribosomal protein genes could influence the accuracy of protein biosynthesis. The structure of five ribosomal proteins is known to influence accuracy; variations in the structure of S12 (Gorini 1970), S17 (Bollen et al. 1975), and L6 (Kuhlberger et al. 1979) increase accuracy, and variations in S4 (Rosset and Gorini 1969) and S5 (Piepersberg et al. 1975; Cabezon et al. 1976) decrease it. A study of the effects of these mutations on accuracy is potentially a very powerful tool because the phenomenon can be studied both in vivo and in vitro.

Most in vivo measurements of the effects of ribosomal mutations involve the efficiency of nonsense suppression. Strigini and Gorini (1970) showed that the *strA*1 mutation in protein S12 reduced erroneous readthrough of an amber mutation in the *lacZ* gene 40-fold, and Piepersberg et al. (1979b) showed that mutations in S4 and S5 increase the readthrough of an amber codon in the ornithine transcarbamoylase gene 6- and 23-fold, respectively. By these criteria, strains with two or more ribosomal mutations have the translational accuracy expected if the mutations acted independently. One exception to this rule is that strains having two mutations that reduce accuracy seem to be only a little less accurate than the least accurate of the single mutants (Piepersberg et al. 1979b).

The effects of ribosomal mutations on accuracy depend on codon, on codon context (see below, Accuracy and Message Content), and on the tRNA. As was pointed out by Gorini (1974), hyperaccurate ribosomes do not reduce all errors equally. For example, they restrict the activity of some (but not all) suppressor tRNAs. If we take this restriction as an index of the ribosome's contributions to the accuracy of tRNA action, this may be used to localize the part of the tRNA structure used by the ribosome to determine accuracy. Bradley et al. (1981) isolated tRNA mutants that overcame this restriction and showed, by sequencing the tRNAs, that they were changes in the next nearest 3' neighbor of the anticodon. These studies have been extended by interchanging the sections of the tRNA genes coding for the anticodon loops and helices (Yarus et al. 1980). If the action of the suppressors is restricted to different degrees by hyperaccurate (*strA*) ribosomes, this quality transplants with the anticodon loop and stem (D. Bradley et al., unpubl.). It appears that the more accurate ribosome perceives the anticodon loop region and is largely indifferent to the rest of the tRNA sequence. Thus, the accuracy of coding may be determined by interactions with this region, in agreement with other data (Yarus 1982).

The effects of ribosomal mutations on accuracy can be reproduced in vitro in protein synthesis systems translating poly(U). However, these studies have contributed relatively little to our understanding of why the mutations affect accuracy. The complexity of the mechanism of translation of even this simple mRNA is undoubtedly responsible for this fact. Recent work in simplified systems, in which the selection of aa-tRNAs by poly(U)-programmed ribosomes is studied in isolation, has been more instructive in this respect. Yates (1979) and Thompson et al. (1981a) have shown that mutations in ribosomal proteins S12 and S4 affect the rate at which near-cognate ternary complexes are bound to the ribosome and undergo GTP hydrolysis and the probability that the aa-tRNAs will be retained by the ribosome during proofreading. The mutations do not affect these parameters for the cognate aa-tRNA (Yates 1979; Thompson et al. 1981a; R.C. Thompson et al., unpubl.), which is why they influence accuracy.

However, a closer examination of the effects of mutations on the ribosome's interaction with the cognate species is instructive of the way accuracy in the selection of aa-tRNAs is achieved. Although the mutations do not affect the observed rate of GTP hydrolysis or the proofreading of the cognate species, they affect the elementary rate constants that determine these complex parameters in a way similar to the analogous rate constants of the near-cognate species (R.C. Thompson et al., unpubl.). It now seems likely that the effects of the mutations on the elementary rate constants of the cognate and noncognate reactions are identical and that their radically different effects on the overall rates of these two processes arise because the contributions of the elementary rate constants to the outcome of the two processes differ.

The rate constant for dissociation of the cognate ribosome·ternary complex (k_{-1}) is reduced about twofold for ribosomes from *ram*1 cells and increased about twofold for ribosomes from *strA*1 cells (R.C. Thompson et al., unpubl.). The failure of the mutations to affect the rate constant for GTP hydrolysis $(k_1 \times k_2/[k_{-1}+k_2])$, is a consequence of the fact that k_2 is much greater than k_{-1} for this species. Although the changes in k_{-1} for the noncognate species have not yet been measured, the changes observed in the rate constant for GTP hydrolysis suggest that changes in k_{-1} similar to those described for the cognate species would be observed at the level of the elementary rate constants.

Similarly, the rate constant for peptide formation by the cognate tRNA[Phe] decreased 2-fold for ribosomes from *strA*1 cells and increased 2.3- to 3-fold for ribosomes from *ram*1 cells (R.C. Thompson et al., unpubl.). The apparent absence of an effect on the proofreading ratio $(k_3/[k_3+k_4])$ for this species is due to the rate constant for peptide formation (k_3) being so much greater than that for rejection (k_4) that small changes in the former hardly affect the fraction of aa-tRNA retained. Very similar effects of the mutations are seen on the rate constant for peptide formation by the near-cognate tRNA$_2^{Leu}$ (R.C. Thompson et al., unpubl.), but the fact that the rate constant for rejection is much greater than that for peptide formation means that in this case there is also a discernible effect on the fraction of aa-tRNA retained.

The striking result applicable to both the GTPase and the proofreading steps of recognition is that changes in accuracy are correlated with nonspecific changes in elementary rate constants, i.e., changes that apply equally to the cognate and noncognate reactions. That changes in specificity could arise from nonspecific changes in elementary rate constants was first proposed by Ninio (1974). However, the observation that only some accuracy-increasing muta-tions (Galas and Branscombe 1976; Gupta and Schlessinger 1976; Zengel et al. 1977; Piepersberg et al. 1979b) and no accuracy-reducing mutations (Piepers-berg et al. 1979b; Andersson et al. 1982) affect the rate of protein biosynthesis in vivo seemed to argue against this idea. Ninio's proposal and these experi-mental findings can now be reconciled in light of the differences found between the mutant ribosomes in the single-turnover system. The mutations affect the rate and, therefore, the accuracy of the elementary steps of protein biosynthesis, but only the most restrictive of accuracy mutations reduces the rate of peptide formation to the point where that step becomes rate-limiting and slows the overall process. The finding that the rate of the reaction by which amino acids are added to the nascent chain influences the accuracy of protein synthesis is further evidence that the rate and not the intrinsic accuracy of the translational apparatus limits the accuracy of the process (Thompson and Karim 1982).

The change in accuracy due to mutations will, of course, depend on the codon and on its context in the mRNA. However, it seems reasonable to expect at least a fivefold to tenfold difference in accuracy on going from a *strA*1 to a wild-type strain and about the same going from the wild-type to a *ram*1 strain. The physiological consequences of these changes are not as large as one might expect. *ram*1 strains are known to revert readily, acquiring an uncharacterized mutation that restores the accuracy of protein synthesis (E.R. Dabbs, pers. comm.). However, the growth rates of these strains are not very different from the wild type (Piepersberg et al. 1979b). *strA*1 strains, on the other hand, do not have any perceptible advantage over wild type as a consequence of their greater accuracy in translation.

Antibiotics

A second factor that has been demonstrated to affect the accuracy of protein biosynthesis in vivo and in vitro is the presence of aminoglycoside antibiotics. The effect was first demonstrated with streptomycin in Gorini's laboratory (Davies et al. 1964) but has since been shown to be a property of neomycin, kanamycin, hygromycin B, and gentamicin (Davies et al. 1965). These aminoglycosides contain either a streptamine or a 2-deoxystreptamine group, which is probably the direct cause of the misreading (Tanaka et al. 1967). Streptomycin retains a primary role in investigations of antibiotic-induced misreading because, unlike the other antibiotics (Davies and Davis 1968), it has a single strong binding site on the ribosome and a correspondingly simpler mode of action.

In vivo, streptomycin has been shown to cause a modest decrease in the accuracy of chain termination at an amber codon in *lacZ* mRNA (Strigini and Gorini 1970) and to increase the incorporation of cysteine into flagellin sixfold (Edelmann and Gallant 1977). Unfortunately, neither of these observations allows us to state quantitatively the effect of streptomycin on misreading since it is not known what proportion of the ribosomes in these studies was complexed by the antibiotic. Concentrations of streptomycin sufficient to complex all the ribosomes would probably have killed the cells.

In vitro, streptomycin has been shown to increase errors in the translation of natural (Tai et al. 1978) and synthetic mRNA. The incorporation of leucine, isoleucine, serine, and tyrosine by poly(U)-programmed ribosomes (Davies et al. 1965) led to the conclusion that the antibiotic allows the misreading of only one base at a time. This has been substantiated by experiments with a range of defined heteropolymers as mRNA (Davies et al. 1966). These experiments do not allow an estimate of the maximal effect of streptomycin on accuracy, because both the antibiotic experiment and the control were conducted at concentrations of Mg^{++} that are now known to produce artificially high error frequencies.

A more precise definition of how streptomycin affects the accuracy of ribosomes has been provided by experiments on the reaction of poly(U)-programmed ribosomes with noncognate aa-tRNAs. Yates (1979) and Thompson et al. (1981a), working with $tRNA_2^{Leu}$ and $tRNA^{Ile}$, and Campuzano et al. (1979), working with $tRNA^{Tyr}$, have shown that the major effect of aminoglycosides is on the proofreading step of aa-tRNA binding. After reacting with the near-cognate ternary complex, the ribosomes incorporate the near-cognate aa-tRNA into peptides with about the same efficiency as the cognate species. The molecular basis for this result is shown by a recent experiment of D.B. Dix (unpubl.), which indicates that streptomycin reduces the rate of the aa-tRNA rejection reaction (k_4) at least tenfold.

Whether streptomycin also affects the initial recognition of ternary complexes is not yet clear. D.B. Dix and R.C. Thompson (unpubl.) have confirmed the earlier report of Thompson et al. (1981a) that the rate of reaction of poly(U)-programmed ribosomes with ternary complexes of $tRNA_2^{Leu}$ is unaffected by the antibiotic. However, the results of Yates (1979) with $tRNA^{Ile}$ and streptomycin, and those of Campuzano et al. (1979) with $tRNA^{Tyr}$ and neomycin, indicate that recognition may also be adversely affected at this step of reaction. One possible interpretation of these seemingly contradictory results is that the antibiotics can hinder the rejection of noncognate ternary complexes but that this effect is only apparent with tRNAs that are easily distinguished as being noncognate in the absence of the antibiotic. Despite this disagreement there is common agreement that streptomycin affects accuracy in protein biosynthesis primarily by affecting the proofreading of tRNAs. This result is in fact the best evidence for the importance of proofreading in vivo (Yates 1979; Thompson et al. 1981a).

The physiological effects of streptomycin are dramatic. Within 5 minutes of exposure to the drug, the viability of a culture drops by five logs (Gorini 1970). It is by no means clear, however, that the bactericidal effects of the antibiotic result from its effect on the accuracy of protein synthesis. Streptomycin affects other steps of protein synthesis (Cundliffe 1981), and it is possible that these effects may account for the killing. The evidence relating translational errors to the mechanism of growth inhibition is, on the other hand, much stronger and has been discussed thoroughly by Gorini (1970). The effects of sublethal concentrations of streptomycin on the accuracy of protein synthesis have been used by Gallant and Palmer (1979) to test the hypothesis that errors in translation may initiate an "error catastrophe" (Orgel 1963) through a positive feedback loop. The results of this experiment indicate that *E. coli* subjected to this abuse does not undergo a catastrophe but rather adapts to a new growth rate consistent with the burden of an increased concentration of inactive protein.

Polycations

The presence of high concentrations of polycations has long been known to reduce the accuracy of protein synthesis in vitro. It is not clear whether a similar effect ever operates in vivo, but a recent study has shown that cells with abnormally low concentrations of polyamines probably have difficulty in the synthesis of protein and, in the presence of the normally innocuous *strA*1 mutation, cannot make protein at all (Tabor et al. 1981). A full evaluation of the effect of increased polycation concentrations in vitro is difficult. Most in vitro protein synthesis systems incorporate an initiation step that is much more sensitive to polycations than elongation. Furthermore, the optimal concentrations of the relevant polycations, Mg^{++}, spermidine, and putrescine are undoubtedly interdependent. The concentrations of these three ions in the "polymix" of Jelenc and Kurland (1979) are 5 mM, 1 mM, and 8 mM, respectively, and increasing the Mg^{++} concentration in this mixture increases the error frequency. Unpublished experiments by D.B. Dix and R.C. Thompson in a single-turnover system studying only the aa-tRNA binding and peptidyl transferase steps of protein synthesis indicate that the Mg^{++} can be reduced even further, to 3.5 mM in the absence of other polycations, and to 1 mM in their presence without significantly affecting the GTPase and proofreading reactions of ribosomes with a cognate ternary complex. However, these more stringent conditions reduce the rate of the GTPase reaction and improve the proofreading of noncognate complexes. It follows that they increase the accuracy of protein synthesis (D.B. Dix and R.C. Thompson, in prep.).

Increased concentrations of Mg^{++} generally decrease the accuracy of both ternary complex recognition and proofreading. In a system in which Mg^{++} is the only polycation, increasing its concentration from 3.5 mM to 15 mM raises the rate of reaction of poly(U)-programmed ribosomes with a Leu2 ternary complex 10-fold and increases the fraction of Leu incorporated into peptide 2.5-fold. The same change increases the rate of reaction with the Phe ternary complexes only 60% and does not change the fraction of Phe subsequently incorporated into peptide. These results suffice to explain the loss of accuracy at high concentrations of polycations. As in the case of ribosomal mutations, we should beware of interpreting the failure to find effects of polycations on the observable rate constants of the cognate species as indicating the absence of an effect on the elementary rate constants k_1 through k_4. It seems probable that the deleterious effects of polycations result from a slowing of the reactions with rate constants k_{-1} and k_4, and that at this level the cognate and near-cognate reactions are affected equally.

ppGpp

Amino acid starvation of *E. coli* is known to reduce the intracellular concentration of the corresponding aa-tRNA (Bock et al. 1966; Yegian and Stent 1969; Piepersberg et al. 1979b). If, as seems likely, this causes a decreased concentration of the ternary complex, an increased error frequency would be expected, because (as shown by the equation above in Accuracy and Kinetics) the error frequency is directly related to the ratio of concentrations of the correct and incorrect substrates. In good accord with this prediction, an increased error frequency is observed in vivo during amino acid starvation of mutants unable to make the unusual guanosine nucleotide, ppGpp (O'Farrell 1978; Parker et al. 1978). However, an increased error frequency is not observed to any appreciable extent in wild-type cells, although there is precisely the same drop in the relative concentration of aa-tRNA. The way in which ppGpp allows amino-acid-starved *E. coli* to escape errors in protein synthesis is one of the more interesting and perplexing aspects of accuracy in vivo.

The error fraction, as defined previously, involves the ratio of two rate constants and the ratio of two substrate concentrations. If the error fraction is to remain constant, either these ratios must remain unchanged or they must change in opposite directions. Three independent measurements of the levels of charged tRNAs in cells indicate that the degree of aminoacylation is grossly perturbed by amino acid starvation; the species acylated by the limiting amino acid is charged to less than one-quarter the extent of the others (Yegian and Stent 1969; Piepersberg et al. 1979b), which argues strongly against the idea that the substrate levels for protein synthesis are unaffected by amino acid starvation. There is one caveat, because aa-tRNAs are not the immediate substrates for protein biosynthesis. Wagner and his colleagues (1982) have argued that, as a consequence of the reduction in EF-Tu·GTP concentration following the synthesis of ppGpp, the relative levels of the true substrates, EF-Tu·GTP-bound aa-tRNAs, are unchanged by amino acid starvation. According to their hypothesis the limiting and nonlimiting aa-tRNAs are distributed differently between the free and EF-Tu·GTP-bound forms, the limiting species being found predominantly in the bound, and the nonlimiting species being found predominantly in the free form. There are no experiments that bear directly on this question, but arguments that this distribution is different for different aa-tRNAs are unlikely to be correct on theoretical grounds.

Consider the following simplified pathway for the incorporation of aa-tRNAs into protein:

$$\text{aa-tRNA} + \text{EF-Tu} \cdot \text{GTP} \xrightarrow{k_f} \text{TC} \xrightarrow{k_b(RS)} \text{peptide} + \text{tRNA}$$

where RS is a ribosome. At steady state,

$$k_f(\text{aa-tRNA})(\text{EF-Tu}\cdot\text{GTP}) = k_b(\text{TC})(\text{RS}) \text{ or,}$$

$$\frac{(\text{TC})}{\text{aa-tRNA})} = \frac{k_{-f}}{k_b} \cdot \frac{(\text{EF-Tu}\cdot\text{GTP})}{(\text{RS})}$$

Of the four terms governing the proportion of charged tRNA found in ternary complexes, two, the concentrations of EF-Tu·GTP and ribosomes, are the same for all species. The other two terms are rate constants that will be similar for all species, or at least are unlikely to be chosen to anticipate a particular kind of amino acid starvation. We conclude, unlike Wagner and colleagues (1982), that the proportion of charged aa-tRNAs found in ternary complexes will be similar for all species. The observed decrease in the relative level of charging of an aa-tRNA following amino acid starvation translates directly into a drop in the relative concentration of its ternary complex and must be accompanied by a change in the rate constants for substrate selection by the ribosome if the accuracy of protein biosynthesis is to be preserved.

The most reasonable explanation for the effect of ppGpp on the accuracy of translation seems to be that the ratio of rate constants for the cognate and noncognate reactions changes so as to compensate for the imbalance in ternary complex concentrations. However, there is experimental evidence that the rate constants for the reaction of cognate and noncognate ($\text{tRNA}_2^{\text{Leu}}$) ternary complexes with poly(U)-programmed ribosomes are unaffected by physiologically significant levels of ppGpp in a purified, single-turnover system (Dix et al. 1983). Any effect of this nucleotide on the ribosomal rate constants that serves to maintain the accuracy of translation in the face of an imbalance in substrate concentration must therefore be indirect. The open-ended nature of the problem makes it very difficult to study in vitro.

In one in vitro study, ppGpp was shown to increase the accuracy of a system translating poly(U) (Wagner and Kurland 1980) but, as pointed out by the investigators, this effect was a consequence of the nucleotide slowing the overall rate of substrate usage and allowing the imbalance in substrate concentrations to be corrected. The result is therefore not relevant to the direct effects of ppGpp on the accuracy of translation in vivo where this correction is not observed. In another study (Wagner et al. 1982) an increased accuracy was observed on reducing the amount of EF-Tu available and, in this case, the imbalance in the relative concentrations of the correct and incorrect substrates remained unchanged. This finding might be relevant, since one effect of the accumulation of ppGpp might be to reduce the amount of EF-Tu·GTP available for protein synthesis. However, the imbalance in aa-tRNA levels used in this study (500−1000-fold) is much greater than that observed in vivo (usually 4−6-fold). Furthermore, the error rate was not sensitive to EF-Tu concentration when the EF-Tu:aa-tRNA ratio was varied from 1 to 0.3, which is the probable range of variation during amino acid starvation in vivo. The relevance of these experiments to the effects of ppGpp on the accuracy of protein biosynthesis in vivo remains to be demonstrated.

INHOMOGENEITIES IN TRANSLATION

Ribosomal Progress Is Intrinsically Uneven

Translation is unlikely to be a smooth progression from the first codon to the last. Rather, it is marked by surges and pauses. There are two distinct sources of these inhomogeneities, both relevant to accuracy and both suppressed in the usual averaged picture of translation.

Hesitation is built into translation even in the case in which the translation of every codon proceeds at the same rate. This follows because a ribosomal step is a stochastic process and therefore occurs with a predictable probability over a range of times. Each step contributes a significant variation to the total range of transit times for the mRNA. Because the overall transit time is the sum of times of the individual codon transits, the overall variance of the transit time will be the sum of all codon variances.

Assume that the rate-limiting step is first order, as if it depended on rearrangement of a ribosomal complex. The result is that the growing points of proteins begun simultaneously will spread out along the message, exhibiting a Poisson distribution of lengths. After a few tens of steps, the length distribution becomes, pretty accurately, a familiar Gaussian distribution, centered on the mean length and mean step time. Thus, the time of arrival at any chain length, including the time of completion of the protein, will also be distributed approximately as Gaussian. The inaccuracy of the double approximation (in going from chain length to arrival time) will rarely be significant, unless the time of the first or last completed chain is the quantity of interest. Because chain length goes up linearly with the number of amino acids (time), and variance increases as the square root, the distribution gets broader in absolute terms as the chains grow but shrinks as a fraction of the length achieved. This will tend to make the distribution of relative transit times sharper for large than for small proteins. The reader may wish to pause and recall that all codons in the protein being discussed were translated with identical rate constants.

We now theorize that codons are translated with different rates. Perhaps some tRNAs translate their codons slowly, either because of intrinsically low rate constants or low aa-tRNA concentration. Radical depression of $tRNA^{Trp}$ levels does pause translation at the cognate codons in the MS2 coat cistron in vitro (Goldman 1982) and probably in vivo as well (S.W. Koontz et al., unpubl.). Pauses probably also occur when aa-tRNA concentrations are normal. These data are reviewed below. Alternatively, it has been suggested that structural features of the message may temporarily halt translation (von Heijne et al. 1978).

This new complication alters our previous discussion because the variance of a first-order process increases as the square of the average time required for the reaction. The variance of the transit time will not only be increased (with respect to synthesis in the same average time using equivalent steps), but the broadening of the transit time distribution will be dominated by the slow steps. The transition to the symmetrical Gaussian transit time distribution, from the skewed Poisson, will be less complete. Pauses therefore are synonymous with a more variable and less symmetric message transit time distribution. There is a considerable body of evidence that protein synthesis is inhomogeneous in this way.

Evidence for Pausing during Normal Translation

Slowly translated regions of the message should accumulate ribosomes, and the distribution cf nascent chain lengths will show peaks at these positions. Nascent rabbit globin chains have been isolated from intact reticulocytes and fractionated by gel permeation under denaturing conditions. They show just such a set of peaks, instead of the smooth distribution expected of a random distribution of growing points (Protzel and Morris 1974). When the same fractionation technique is applied to nascent bacteriophage MS2 coat protein from infected cells, the same result is found (Chaney and Morris 1979). In vitro translation of the MS2 coat cistron also gives a clear spectrum of discrete shorter products (Atkins et al. 1979), though these are rarely mentioned or studied. In a careful investigation using gel electrophoresis, a similar set of "paused" silk fibroin chains were isolated both from the intact silk gland in organ culture and when the message was translated in an in vitro system from rabbit reticulocytes (Lizardi et al. 1979). In this case, control experiments show that the paused chains are resident on active ribosomes and are intermediates in the synthesis of full-sized protein. The size distribution of RNAs in the fibroin message fraction shows no sign of the discontinuities of the nascent peptide chains. These latter two findings are particularly important, because they weigh strongly against an alternative explanation—paused chains occur because of specific, but spurious, endonuclease cleavage of the message. One would like more assurance that the ribosomes have been adequately halted during the isolation of nascent chains. Nevertheless, the intermediates characteristic of heterogeneous rates of translation have apparently been detected in several types of intact cells.

The distribution of transit times for *E. coli* β-galactosidase in vivo also has the qualities expected for a process that has a small number of slow steps that determine its variance, as discussed above. That is, the finishing times are very broadly distributed (Talkad et al. 1976) and are consistent with the idea that, during β-galactosidase synthesis, there are substantial pauses at average intervals of approximately 20 amino acids (as estimated by von Heijne et al. 1977), though this interval is not very precisely determined. Furthermore, the skewed Poisson shape of the β-galactosidase transit time distribution confirms that its synthesis is best approximated as a few slow steps. This finding is more remarkable because β-galactosidase is a large protein whose message transits actually consist of more than a thousand steps.

It is essential to realize that the currently accepted values for the step time of protein synthesis may be averages of a broad distribution. The mean step time or mean chain growth rate would not then be an accurate standard to which the kinetics of individual steps may be compared. The maximum velocity of translation may be much greater than mechanistic ideas have yet tried to accommodate, and total stops could be frequent events.

Transcriptional Pauses

Some translational pauses may be caused by a temporary halt of the transcriptional apparatus (Kassavetis and Chamberlin 1981), thereby halting the ribosomes following. All translational pauses are unlikely to be attributable to this cause, because messages, on the average, survive for several times their transit time (Kennell and Riezman 1977). This is also relevant only in prokaryotes, where transcription and translation are directly coupled. However, the potential influence of the transcriptional machinery in prokaryotes suggests an interesting possibility—that the product of the first transit of a message might be different from that of the latter ones.

Pauses and Errors

Now we wish to apply these ideas to questions of accuracy and error. It is likely that the codons over which the ribosome pauses play a more than proportionate role in erroneous translation. This should be true because the advance of the ribosome employs a repetitive reaction cycle. When the cycle stalls, it indicates that an unusual difficulty in translation has been encountered. Because the usual pathway of protein synthesis is temporarily blocked, there will be time for slow, possibly aberrant events, which are normally insignificant. The effects extend upstream from the first stalled ribosome because pauses both prolong transit of the message and redistribute the ribosomes so that they are usually at a pause. Because most ribosomes in transit can be paused ribosomes, it is more than likely that random events that produce errors will occur at these sites. Of the many alternative error-producing events, we provide a pair of arbitrarily chosen examples to help make this discussion concrete.

Consider a codon to be translated by a tRNA whose concentration is low. Low abundance slows the bimolecular reaction of that complexed tRNA with the ribosome. A low enough rate will slow the overall rate of translation at those codons and produce pauses. In actuality, the reaction of ternary complexes is so fast that slow ribosomal reaction rates may also be required to give a significant pause (see below). As the error equation implies, the error rate should be elevated at such a site because N/C is elevated.

Alternatively, consider a ribosome at a structural obstacle in the message. Although the energy available to translocate the ribosome is unknown (except as an upper limit), it may well be comparable to the energy required to denature the structures restraining the next codon. There will therefore be an unusually large activation requirement for translation of the next codon, and this implies a corresponding pause. In addition, translocation may not be complete. If only two coding nucleotides could be effectively drawn into the A site, the cognate tRNA would have no advantage over other tRNAs that read "two out of three" (Lagerkvist 1978) of the codon's nucleotides, or an out-of-phase codon might be read.

An important special case occurs when the step that is slowed to produce a pause is normally too fast for optimal accuracy. This may be the case for normal peptide bond formation (see discussion above). In this case, a pause might evolve to increase the accuracy of a potentially erroneous step.

This is not an argument that errors will happen only at pauses. There may be other classes of potential error-prone sites. However, the intermittent character of translation is evidence of a similarly distributed underlying tendency to error.

Only one system in which pauses have been characterized has been studied closely for translational error. Parker et al. (1980) have shown that there is a natural variant of MS2 coat protein that arises by mistranslation of AAU as lysine in infected cells. The site(s) of the substitution are incompletely characterized, but there apparently is a substantial contribution from an AAU codon that is also the first detected pause site in the protein (T. Johnston and J. Parker, pers. comm.; see also von Heijne 1978).

Permanently Halted Chains

The size distribution of total antibody-precipitable β-galactosidaselike material has also been determined in vivo and in vitro by gel electrophoresis (Manley

1978). The material reactive with antibody included a large number of discrete aminoterminal fragments of the complete protein. However, these experiments focused on the existence of *stable* products shorter than the complete protein. Longer labeling times, usually followed by chases, were employed. The discrete transients expected of briefly paused chains were not sought. The numerous shorter products that were found do not chase into full-length protein. They are instead the result of premature termination of chain extension by a ribosomal error or, less likely, on a shortened message. These experiments are therefore not directly relevant to hesitation during successful transits of a message. They do indicate that regions likely to prematurely curtail protein synthesis are also distributed nonuniformly, about once per 3000 codons. This agrees very well with the estimates of Menninger (1976) for the frequency of adventitious release of peptidyl tRNA from the ribosome. Premature halts, whatever their mechanism, are quantitatively as significant among translational errors as are misincorporated amino acids.

Pausing Implies Inefficient Catalysis

Translation with pauses is prone to error, as we have emphasized. But pauses also accumulate intermediates of the translation pathway and testify to steps with widely varied activation energies. Both conditions are diagnostic of nonoptimal, inefficient catalysis, in the terms of Albery and Knowles (1977). Because *E. coli* may devote material constituting 50% of its dry weight to the translational apparatus, we may wonder why selection has not rigorously optimized the catalytic efficiency of this particular pathway. One possibility is that *E. coli*'s evolutionary success depends mostly on a short generation time. The most rapid cell division may result from concentration on the optimization of the efficiency of a subset of codons. These optimized codons will be used selectively in the genes used to generate cell mass rapidly. The other class of slowly translated codons automatically created will be infrequent in major genes but concentrated in regulatory genes and other "finesse" functions, as is observed by Konigsberg and Godson (1983). The subject arises again in Optimization of Normal Coding, below.

The inhomogeneous distribution of codons returns us to the principal motif of this section. The relative rates of normal and aberrant processes determine the local tendency to error. These relative rates will vary widely, even within one message.

ACCURACY AND MESSAGE CONTEXT

The study of nucleotide context effects has a history in the study of suppressors. These studies have established that the same codon may be translated with different efficiencies when it occurs in different positions (i.e., in different "contexts"; Salzer 1969; Yahata et al. 1970). Put another way, the efficiency of a message sequence is determined in part by factors outside the codon being translated. The equation for error rate (above) defines the effects of significance to translational fidelity in a straightforward way: If message context can change the rate of normal translation of a codon relative to the rate of error-producing processes, then context has an effect on the error rate. We now treat evidence that suggests such a role.

Amino Acid Substitutions Should Be Context-dependent

The most studied error is that of amino acid substitution, which can occur because a noncognate tRNA can read a given codon (Yarus 1979a, but compare 1979b).

Context will affect the rate of amino acid substitution error if it can alter the relative translational efficiencies of two tRNAs that can act at the same codon. We must therefore decide whether context ever alters the action of tRNAs in this sense. This is not trivial because there are several possible modes of context action during termination suppression. Nearby sequences may act on

1. Release factors. The termination signal detected by release factors is assumed to extend outside the termination codon. Effects on the efficiency of suppressor tRNA are interpreted as effects on the ribosomal mechanism for chain termination, with which suppressor tRNAs compete.
2. Message structure or action. These effects are communicated through the message structure to the tRNA at the coding site of the ribosome—stacking of adjacent nucleotides on the codon in use or provision of a fourth base pair to the anticodon loop (Taniguchi and Weissman 1978) are mechanisms in this class.
3. tRNA-tRNA contacts. The close approach of tRNAs inside the ribosome (Fairclough and Cantor 1979; Johnson et al. 1982) suggests that certain pairs of tRNAs may fit together in a superior fashion when adjacent in the A and P sites. Sequence changes close to the test codon could then be acting by altering the identity of an adjacent tRNA in the translation sequence.

All three mechanisms are plausible, and some variation of all will probably occur. But only the latter two imply that context is significant to substitution error. Is there evidence that distinguishes the possibilities?

There are many measurements that compare the efficiency of two suppressors cognate to the same codon in two or more contexts. For example, Bossi and Roth (1980) found that *supB*, a lysine-inserting amber suppressor, was unchanged in efficiency when the HisD6404 message sequence was altered from UAGC to UAGA. The same context alteration increased the efficiency of *supE*, an independently derived suppressor tRNA, which inserts glutamine about tenfold. Because both amber suppressors compete with the same termination mechanism, this kind of observation cannot easily be explained as an effect on termination. A role for tRNA-tRNA interactions is easy to envision for context changes 5' to the test codon. These changes can alter the probability of effective occupation of the ribosomal A site. But for context changes 3' to the test codon (the termination codon), the implications are more extreme. In this case, the suppressor tRNA has already reached the P site. To alter the apparent efficiency of the test tRNA, the incoming tRNA must alter the probability of survival of the nascent peptide chain present or attached to the test tRNA. It is not implausible that tRNA-tRNA interactions could alter the likelihood of release of the growing chain from the ribosome. But the large magnitude of context effects (changes in efficiency of a factor of 2 are common) would imply that a single adverse tRNA-tRNA interaction will abort a nascent peptide. This seems unlikely because natural nascent chains are usually completed. Therefore, tRNA-tRNA interaction is most likely to mediate context effects for sequences 5' to the codon affected.

This is an essential point, because context changes of defined structure are all 3' (Bossi and Roth 1980) to the affected codon. Bossi (1983) has shown for UAG, and Miller and Albertini (1983) have shown for UGA that, in fact, more than one adjacent nucleotide affects the efficiency of suppression at an upstream terminator codon. We assume that these measurements model the selective effect of context on a mistranslating tRNA. Thus, the most likely possibility is that context alters the relative activities of tRNAs through effects on message activity. This implies a role for context in the distribution of substitution errors.

The context effect makes distinctions of great subtlety between different possible translations at a given message position. Feinstein and Altman (1978) have used the same ochre suppressor, reading either UAA or UAG at the same position in two message contexts. They observe a sixfold or greater difference in the *ratio* of activities on the two codons. In this comparison, the same tRNA is differently affected in the two message contexts even though tRNA-tRNA contacts are for the most part unaltered. The only change, in fact, is that due to the (presumably) small differences in the reading of two codons related by a frequent wobble, UAA/G. This suggests that context requires only small differences in the codon-anticodon region for substantial effects. This is easily envisioned in terms of mechanism 2 (above) but is incompatible with an explanation solely in the terms of mechanism 1—an effect on the termination mechanism—and explicable as an effect on tRNA-tRNA contact (mechanism 3) only with difficulty.

There are other observations in vitro and in vivo that support a context effect on relative translational rates. The Hirsh UGA suppressor (Hirsh 1971) normally translates UGA/G but is also active at UGU in vitro (Buckingham and Kurland 1977). The erroneous wobble to UGU appears fourfold less often in the context U/UGU/U than in G/UGU/G (Buckingham and Carrier 1983). There is presumably no involvement of release factors or termination in these results, despite the usual employment of this tRNA as a termination suppressor.

Even more to the point, M. Murgola (pers. comm.) has observed apparent context effects on missense suppressor tRNAs in vivo. A glycine-derived (reads GAA/G) and a lysine-derived (reads AAA/G) suppressor tRNA give different levels of *E. coli trpA* function, though acting at identical codons at two different positions. Because these suppressors compete with normal tRNAs for the same codon, the relative effect of context is unambiguous.

Premature Termination Is Expected to Be Both Frequent and Uneven

The extensive work on context effects and termination suppressors suggests the existence of another, little investigated type of translational error. Context effects have usually been measured as changes in the suppression of a termination codon (UAA, UGA, or UAG). In such comparisons, the competition between a suppressor tRNA and the polypeptide release factors (Caskey 1977) has been observed in different contexts. The outcome of this competition varies by orders of magnitude when sequences surrounding the termination codon change. It is only a small extrapolation to conclude that the frequency of mistaken action by the release factors (at sense codons) should be quite context-dependent and should vary over an order of magnitude or more, as do the efficiencies of nonsense suppressors. Both phenomena depend on the relative rate of translation of a sequence by a tRNA and release factors.

Termination May Be Frequent in Missense-prone Contexts

Manley (1978) has observed hotspots for the premature and permanent termination of β-galactosidase chains distributed over much of the message. One possible mechanism for the production of these aberrant products is the release of a growing peptide chain from the ribosome, still attached to its tRNA. This latter event is known to be frequent enough ($\sim 1/3000$ codons translated) to account for the β-galactosidase fragments (Menninger 1978). The discovery of a temperature-sensitive conditional lethal mutant of the *E. coli* enzyme that "scavenges" cytoplasmic peptidyl tRNA (Atherly and Menninger 1972) allows released peptidyl tRNAs to accumulate and be characterized in the mutant cell. Menninger has suggested that the release of peptidyl tRNA provides a route for the ejection of previous missense coding errors. This "ribosomal editor" therefore associates hotspots for miscoding with rapid release of nascent peptides (Caplan and Menninger 1979).

In summary, varied studies of suppressor tRNAs in different contexts suggest that the relative translational activities of tRNAs, and therefore substitution error rates, should be dependent on codon context. There is reason to believe that context can work through the structure and activity of the message, from the few cases where the evidence can be used to distinguish the mechanistic alternatives. Two other context mechanisms remain plausible and have not been critically tested.

Premature termination should also be context-dependent and yield discrete products. The relative frequencies of conceivable termination mechanisms are not known. In particular, the action of the release factors at amino acid codons, though likely, has not been demonstrated.

SUPPRESSION OF NATURAL TERMINATORS

Termination codons, in the usual course of events, are read by release factors, which trigger the hydrolysis of the aminoacyl bond connecting the completed polypeptide to the tRNA cognate to the carboxyterminal amino acid. The terminators are nevertheless codons, which should occasionally be translated as a result of the action of an aa-tRNA. The resulting extended protein is identical with a normal gene product except for its carboxyl extension. Evidence is accumulating that translation of terminators is used to make a variety of minor gene products, whose time of appearance and level of expression are automatically coupled to the major product. Perhaps the name "termination codon" will be too confining for these coding elements, once their full role is understood.

Normal Cellular tRNAs Act at Translation Terminator UGA

The introduction of termination suppressor tRNAs necessarily extends a certain fraction of cellular proteins, sometimes with observable consequences for cellular physiology. For example, amber suppressors alter stringent control in *E. coli* in proportion to their suppressor efficiency (Breeden and Yarus 1982). But the existence of normal cellular tRNAs that can potentially decode termination codons has also been established. The first known, and still the best-characterized case, is $tRNA^{Trp}$ of *E. coli*. Hirsh and Gold (1971) showed that this tRNA, which usually responds to UGG, can also respond to UGA in a few percent of cases in vitro. We have shown that this is also true in vivo, using a $tRNA^{Trp}$ gene under *lac* control (L. Raftery et al., unpubl.). The fully induced normal $tRNA^{Trp}$ gene gives easily measured levels of product from a *lacZ* UGA mutant gene.

The rabbit β-globin gene ends with UGA, followed 22 codons later by UAA (Czernilofsky et al. 1980). Geller and Rich (1980) detected a small amount of an extended globin among the products of reticulocytes labeled with methionine. Purification of a minor tryptophan acceptor from reticulocytes gave a fraction that was highly active in vitro, stimulating readthrough of the UGA terminator to give the extended β-globin. Evidently a normal, minor fraction of rabbit $tRNA^{Trp}$ possesses the same ability to misread the wobble position of UGA as does the *E. coli* tRNA.

An unusual tRNA discovered in the cytoplasmic compartment of bovine liver by Diamond et al. (1981) is probably also a member of this class of terminator readers. This molecule accepts serine and folds into a unique cloverleaf, requiring exceptional structures in the D-stem region and in the Tψ stem. Despite its acylation with serine, the tRNA has a CmCA anticodon. This is cognate to the UGG codon for tryptophan, though the tRNA is not bound to *E. coli* ribosomes in response to UGG or in response to any codon for serine. It does respond to UGA and will also strongly stimulate the formation of the UGA readthrough β-globin in an in vitro reticulocyte system. Diamond et al. suggest that the existence of a seryl tRNA exhibiting a tryptophan anticodon and acting at the stop codon UGA may be associated with the synthesis of phosphoseryl tRNA (Sharp and Stewart 1977). It appears, in any case, that tRNAs with CCA anticodons often can act at UGA. This tendency, though unrecognized by standard coding rules, is potentially available to cells to shape the distribution of translational products.

The *E. coli* RNA bacteriophage *Qβ* has apparently utilized this option to encode an essential protein. The virion contains not only 180 copies of the 14-kD coat protein but also requires at least one copy of the 38-kD A1 (also called IIb) protein in an infective particle (Hoffstetter et al. 1974). The A1 protein has the same amino terminus as the major coat subunit (Moore et al. 1971; Wiener and Weber 1971). It is a result of approximately 3% readthrough of the normal UGA terminator, ending at UAG 195 codons away (M. Billeter, cited in Kohli and Grosjean 1981). The molecular evidence is unusually complete and compelling in this case, lacking only the demonstration that the UGA terminator was translated as tryptophan. One apparent result of this mechanism is that streptomycin-resistant cells do not produce infective phage, in part because their more accurate ribosomes restrict the unusual translation of the coat terminator (Engleberg-Kulka et al. 1979) (for another potential explanation of this effect, see Shifts in Reading Frame, below).

A similar argument may be made for the coliphage λO cistron, whose product in vitro is 15% UGA readthrough. Though the requirement for the extended product of the *O* gene is not established, λ reproduction is also curtailed in streptomycin-resistant cells (Yates et al. 1977).

It has been suggested that the context of the UGA determines the likelihood of this type of readthrough (Engleberg-Kulka 1981), which is greatest at UGAA. It appears that UGAA is avoided in the average message of prokaryotes, though it is frequent in eukaryotes (Kohli and Grosjean 1981).

UAG Is Also Read As an Amino Acid

Although the UGA codon has been investigated most often, there are indications that UAG is also sometimes translated. There are two major peptide products of tobacco mosaic virus (TMV) RNA translation in vivo and in vitro. The 160-kD lesser product is an extension of the 110-kD major protein (Pelham 1978). Because the production of the larger material is stimulated by a yeast amber suppressor tRNA, and more weakly by a yeast ochre suppressor, the 160-kD material is probably attributable to readthrough of UAG termination of the 110-kD gene (see also Bienz and Kubli 1981). A tyrosine-acceptor tRNA that stimulates this readthrough in *Xenopus* oocytes has been purified from *Drosophila* and *Schizosaccharomyces*. These tRNAs have G-pseudo-U-A anticodons and are apparently translating UAG using a G/G wobble base pair (Bienz and Kubli 1981). If the anticodon G is modified to Q, the readthrough is considerably reduced. When readthrough of TMV RNA is studied in vitro, polyamines strongly modulate the level of the extended product (Morch and Benicourt 1980). Sensitivity to environment and to tRNA modification suggests that readthrough might be controlled in order to direct gene expression, though there is no specific evidence of this.

Quite analogous experiments in vitro using Moloney murine leukemia virus RNA indicate that a core-subunit reverse-transcriptase fusion protein of 180 kD is also a UAG extension product (Philipson et al. 1978). The large chain has the antigenic character of both component proteins. This kind of fusion in phase has the interesting property of linking two potentially functional domains to give a new oligofunctional protein. Such a molecule might often have novel and useful properties.

Though the data are much less complete in the case of UAG than UGA, this terminator can also be mistranslated as an amino acid. Again the active agent is a tRNA that can utilize an unusual wobble pairing (Bienz and Kubli 1981).

We have now considered the adventitious translation of amino acid codons as terminators, and, in this section, the translation of terminators as amino acids. There is yet a third way in which aberrant termination occurs; the next subject is frameshift translation.

SHIFTS IN READING FRAME

The production of a correct gene product usually requires that the reading frame set at initiation be retained. But translation depends on chemical discriminations that cannot be absolute. In fact, at equilibrium the ribosome presumably would be found nearly equally in all three frames, because an intrinsic sequence preference would itself be a potential source of frameshift errors. The translational apparatus therefore acquires its preference for one frame from its tRNAs. For this reason we expect frameshift errors among the other conceivable coding errors. These frameshift errors will likely be as context-dependent as is normal translation and therefore sporadic and prone to hotspots. Evidence tentatively identifying one such frameshift hotspot will be cited below.

In the average case, an out-of-frame termination codon is encountered 64/3 codons after such a frameshift, and the peptide comes to a novel end. In this section we treat potential uses of this event in control of gene expression.

A Unifying Idea About the Frameshift Pathway

The primary mechanism of frameshift error is not known, but one may base a guess on the results of studies of frameshift suppressor tRNAs (for review, see

Roth 1981). One type of frameshift mutation is an extra nucleotide in the message, e.g., GGGG (mutant) instead of the glycine codon GGG (Yourno 1972). The corresponding suppressor tRNA contains an extra nucleotide in its anticodon: CCCC (mutant) instead of CCC (Riddle and Carbon 1973). The straightforward conclusion is that the nature of the codon-anticodon complex determines the number of nucleotides translocated.

It is only a slight extension of this interpretation to suggest that reading an out-of-frame codon (using normal anticodon base-pairing) is a likely source of frameshift errors. This may have a particularly high probability when the cognate tRNA already in the A site can also read the message in an alternate phase, as happens at runs of the same nucleotide. Though one might have expected the ribosome to easily restrict reading to the three canonical coding nucleotides, the activity of the frameshift suppressor tRNA strongly suggests that adjacent nucleotides can be reached with unexpected ease.

We extend this idea further to subtler aspects of codon-anticodon interaction than the number and position of base pairs, as suggested by Kurland (1979). Thus, an abnormal tRNA-message complex gives rise to abnormal message translocation. The abnormality can be an extra nucleotide pair or pairing out of phase but also can be attempted pairing by a tRNA that is not cognate to the codon in phase (thus producing a relation between substitution error and frameshifts). Finally, it is likely that even normal tRNA-codon complexes will vary in their susceptibility to "interpretation" by the ribosome as signals for abnormal translocation.

There are therefore two types of information encoded in the tRNA/message, most plausibly in the area surrounding the anticodon/codon itself. The first is the specialized base-pairing information whose central aspect is the anticodon/codon. But the optimal action of the anticodon probably requires that the local tRNA sequence in the anticodon loop and stem be well suited to the anticodon sequence itself (Yarus 1982). The second type of information is that for the length of translocation, and this may also be encoded in the structure of the message and tRNA near the codon being translated. It may be useful to distinguish the coding and translocation instructions if, as seems likely, it turns out that they are encoded in somewhat distinguishable aspects of the RNAs.

Detection of Natural Frameshifts

Ambiguity in the translational step can be seen in the residual activity of frameshift mutants in the β-galactosidase gene (Atkins et al. 1972) and similarly in the 5% residual normal length protein made in vivo from frame-shifted yeast mitochondrial cytochrome-*c* oxidase genes (Fox and Weiss-Brummer 1980). Despite the potential for absolute disruption of the peptide chain by a frameshift, there is measurable enzyme activity in the cells carrying the frameshift mutation. The enzyme's temperature sensitivity is greater than normal. The small background of altered enzyme is made probably because an occasional natural frameshift error near the mutant site compensates the frameshift caused by the mutation. The tract of amino acid changes between the mutation and the frameshift that phenotypically suppresses it accounts for the temperature sensitivity of the enzyme. The background enzyme level is depressed in *rps*1 (*strA*) cells and is enhanced by *ram* mutations and streptomycin. This is consistent with the attribution of frameshift leakage to a tRNA-decoding error at a frequency determined by the accuracy of the ribosome (Gorini 1974). It is similarly consistent that the probability of frameshift may be increased or decreased during translation of MS2 bacteriophage RNA in vitro by the addition of specific purified tRNAs (Atkins et al. 1979). The active tRNAs differ for each frameshift studied. This is consistent with the error rate equation (see above, Accuracy and Kinetics). These are presumably "pool bias" experiments in which the use of a shift-prone codon-anticodon complex is forced by dramatically elevating the concentration and rate of reaction of the noncognate tRNA involved.

tRNA Selection Errors Produce Frameshifts In Vivo

Another experimental system that reveals a minority of frameshift translations has been extensively characterized by Weiss and Gallant (1983). They have used bacteriophage T4 *r*IIB frameshift mutants. Under the proper conditions, these mutants yield phage if the infected bacteria are subjected to a period of amino acid starvation, and the codons whose translation is depressed by starvation are near the frameshift alteration in the *r*IIB gene. Starvation presumably produces a pool bias (see above, Accuracy and Kinetics), which forces translation of the codons of the missing amino acid by noncognate tRNAs that are still acylated. Some of the resulting frameshifts suppress the nearby *r*IIB mutant, giving phage yields of up to 20% normal. This exemplifies the correlation between mistranslation and frameshift predicted by Kurland (1979). By starving for tryptophan, the unique UGG codon whose translation is affected by the starvation can be isolated for study. Starvation for leucine as well as tryptophan (but not other amino acids) strongly inhibits this phenotypic suppression. Weiss and Gallant (1983) propose that a leucine tRNA, possibly cognate to UUG, produces the frameshift when UGG cannot be translated in the usual way. The resulting pattern is entirely consistent with a context-dependent probability of frameshift that is strongly stimulated by forced miscoding. As might be predicted, the level of starvation-induced frameshift suppression is under the control of variables that affect the precision of ribosome/tRNA selections, such as the ppGpp level (Hall and Gallant 1972; O'Farrell 1978).

A Natural Frameshift Hotspot?

In several cases, genetic and biochemical data localize a frequent natural frameshift to a small region whose nucleotide sequence is known. In addition to the leaky yeast mitochondrial frameshift mutants mentioned above, a frameshift appears to be required to translate the MS2 coliphage lysis gene, and deletions have been used to identify the essential region of the message (Kastelein et al. 1982). The third case is a frameshifted and extended version of the T7 coliphage major capsid protein, which is also incorporated into the viral particle (J.J. Dunn and W. Studier, in prep.). This frameshift event occurs in some ribosomal transits, and the shift can be traced to a 23-bp sequence in the capsid gene. Remarkably, these three sequences have a shorter sequence in common, near or at the frameshift point: P/PUU/UUC (P=purine). As indicated by the division of this sequence into codons, it is also found in the same phase in all three cases, suggesting that it must be translated by a similar set of tRNAs to have its effect. This impression is reinforced by the size of the sequence (~2 codons), which suggests that the critical events involve two adjacent ribosome-bound tRNAs, tRNA$^{Ile/Val}$ and tRNA^{Phe}. Interpreting the sequence in this way reemphasizes that the translocational step size is encoded in the structure of these RNAs. Note that the essential tRNA can read the sequence in two alternate phases (as UUU), as well as in the normal phase (as UUC). It was suggested above that this situation may give rise to ambiguity in the reading frame.

A different type of evidence comes from studies of a group of weak, closely linked frameshift suppressors selected by J.F. Atkins et al. (unpubl.) in the *trpE* gene of *Salmonella*. These intragenic suppressors do not correct the shifted phase but instead seem to introduce a shift-prone sequence near the original mutant. Among the new sequences created are TTTC and AAAG, which can also be interpreted in alternative frames by the same tRNA.

Use of Frameshifting to Control Gene Expression

It is reasonable to look for a use for a translational pathway as frequent as frameshifting. Kastelein et al. (1982) have neatly dissected the expression of the MS2 lysis gene, using recombinant DNA techniques on a cloned viral genome under control of λ promoter p_L. The lysis gene overlaps the carboxyl end of the coat protein gene but is in a different reading frame. The entire lysis gene is not sufficient for its own expression. It is also required that the coat cistron be translated, though the coat protein itself is dispensable, because internally deleted coat cistrons also trigger the downstream lysis gene. However, a region of the coat gene just upstream of the initiation point of the lysis gene is essential; deletions cutting near or into the sequence A/AUU/UUC prevent expression *when the deletion is in phase*. When the deletion results in a frameshift, lysis function is always expressed. This is apparently because the ribosome pauses at one of two UAAs in the alternative reading frames. It then reinitiates at the nearby initiation codon of the lysis cistron, much as in the synthesis of a reinitiation fragment after termination at an amber or ochre mutant (Files et al. 1974). Kastelein et al. (1982) suggest that the essential sequence is a region of frequent frameshifting (effective in ~10% of transits [Beremand and Blumenthal 1979]) whose action causes ribosomes to pause at UAAs that are normally out of frame. Subsequently, the lysis gene product is initiated in a new reading frame.

The logical situation resembles that for suppression of natural terminators; frameshifts can be demonstrated in many systems but are of proven importance only to viruses, where they may conceivably be dictated by a compact mode of existence. On the other hand, great familiarity with the gene products of a sequence is needed to detect frameshift and suppression proteins, which are roughly an order of magnitude less abundant than products of the canonical reading frame. We may therefore expect more examples of essential proteins made by these mechanisms to turn up.

OPTIMIZATION OF NORMAL CODING

How may a system be constructed for accurate performance? In most of this review, we consider the kinetics and processes that comprise unusual or erroneous coding. But there is a complementary set of questions about normal codon choice, optimal tRNA concentrations, and other aspects of the translational milieu. These questions are germane, because an accurate system is one that *maximizes* the rate of correct reactions with respect to alternative ones. Enhancement of the cognate reaction is a strategy for accuracy equivalent to the inhibition of errors. Even in the current preliminary state of this subject, translation has been more deeply considered from this point of view than other biochemical systems. The concepts developed here might contribute something to any investigation of the accurate selection of other pathways—not only to a clearly parallel case such as transcription, but, less obviously, to all cases in which alternative biochemical possibilities are available.

Comparison of Natural Messages

What *is* the architecture of well-expressed mRNAs? What is the relation between codon choice and level of expression? Such questions have been approached by comparing the sequences of many translational units. But this comparison requires stringent assumptions. Promoters differ in structure, and message half-lives are different; most genes will probably be found to be under some sort of transcriptional control, however indirect. The structure of the translational initiation site is also critical and can be varied over wide limits to produce different protein chain initiation frequencies. Significant pauses at an internal codon will depress the level of expression of an otherwise ideally constructed message. Even terminators probably vary in their rates of action (see below), and if slow enough, can control the net rate of expression. These considerations emerge clearly in the explicit kinetic models of the translation process developed by Bergmann and Lodish (1979) and Harley et al. (1981).

Comparison of the messages for abundant and scarce gene products will therefore be confounded to an unknown degree by variation of other aspects of expression besides the rate of peptide chain extension. Nevertheless, such comparisons may be a useful first step, if we keep firmly in mind the implicit assumption that an abundant gene product possesses generally optimized expression.

The proposition that gene expression is regulated at this level is subject to direct experiment confirmation. The mean step time and/or pause times of weakly expressed genes should be greater than those of the mean gene. This measurement has not been made, to our knowledge, even though the natural strategy for various levels of expression raises profound questions of general interest.

Coding Specificity in Vivo Is Somewhat Uncertain

Below, we discuss the ways in which tRNAs are suited to the translation of particular messages. The matching of the tRNAs available to the codon distribution is complicated by current indications that we do not yet have an entirely reliable idea of how anticodons pair with codons in vivo. This subject entered an era of reevaluation with Lagerkvist's proposal (1978) that, when it will not produce a substitution error, tRNAs read "two out of three" of the codon's positions and are not very sensitive to the wobble nucleotide. This now seems to be a real, but slight, tendency that can be exaggerated in vitro by the absence of the tRNA that normally translates the codons reached by mismatching the wobble position (Goldman et al. 1979; but also see the discussion by Lustig et al. 1981). This is in accord with the prediction of the error rate equation, discussed above, in Accuracy and Kinetics.

In addition, the question of coding specificity has been examined in vivo, deleting tRNAs by selecting suppressor mutants. Murgola and Pagel (1980) have shown that *glyT* is the gene for the unique GGA-reading glycine tRNA in *E. coli* and cannot be altered unless the ability to read GGA is preserved. Sometimes the observed preservation of GGA reading, however, implied unorthodox coding at the first and second codon positions. This occurred in the absence of the normal GGA-reading tRNAs and therefore under conditions predicted by the error rate equation to maximize error.

Munz et al. (1981) inactivated the serine UCA-reading tRNAs (anticodons U*GA) of *Schizosaccharomyces* by mutation. This is a lethal event, despite the presence of a major tRNA that potentially reads these codons via standard wobble (anticodon IGA).

These results are incompatible with simple "two out of three" reading mechanism, but both also turned up coding specificities that were not predicted. Pending clarification of this subject, we proceed by making only conservative assumptions about the details of codon-anticodon pairing in vivo.

Evidence for an Optimal Translational Architecture

There is a variety of data most easily understood if the translation system of each organism has been individually optimized by evolution. Translation in vitro varies widely in efficiency for different messages, depending on the tRNA used (Sharma et al. 1976). The most efficient set of tRNAs is that from the same organism as the message, even when the rest of the translation system is from a different source (Le Meur et al. 1976). In special cases where a translation system is adapted to produce a limited set of products (reticulocyte globin or silk gland fibroin), the frequency of tRNAs expected to translate particular codons can be seen to correlate with the codon frequency (Chavancy et al. 1979). Reticulocyte tRNAs even seem to specialize progressively as globin production switches from the embryonic γ to the mature α and β globins (Hatfield et al. 1982). Finally, there is evidence that codon choice differs in the messages of rare and abundantly produced proteins. This deserves a separate discussion.

Selective Use of Codons

As soon as message sequences became known, it was clear that codons were not chosen randomly (Fiers et al. 1971). Organisms seem to have individual preferences that can be discerned in most of their genes (Grantham et al. 1981). The clearest tendency, which spans many organisms and on which there is general agreement, is the distinction between U and C in the wobble position of codons (Grosjean and Fiers 1982). Because NNU/C pairs of codons are translated by the same tRNA utilizing different wobble pairings, the NNU/C choice is not complicated by different tRNA concentrations or structures. Codon pairs NNA/G, in contrast, can be translated by different means, and, in fact, NNA and NNG are not as reliably distinguished as are NNU and NNC.

The preferences can be rationalized as a tendency to moderate interaction energy between codon and anticodon. If the first two codon nucleotides are A/U, highly expressed messages will selectively use C at the wobble position. If the first two are G/C, then codons used will selectively end with U. The pattern is weaker or fails to appear in codons with mixed composition. It is striking that the pattern is not simply less evident in genes more weakly expressed, as might perhaps be anticipated. Instead, the preferences *are* weaker, but distinctly reversed, with C preferred in G/C-rich codons (Grosjean and Fiers 1982) or strongly pairing codons (Grantham et al. 1981) and U in A/U-rich or more weakly pairing ones.

It is plausible that there is an intermediate, optimal codon composition for expression that is selected in strongly expressed genes. But it is less clear why genes whose products are made in small amounts would prefer codons at *both* extremes of composition. The idea that intermediate stability yields an optimized rate of translation implies that minor gene products will also be made less accurately. Their cognate translation will proceed slowly, and therefore their error rate will be elevated. Because the weakly expressed genes are repressors, primases, and so on, which might have effects on metabolism out of proportion to their numbers, it is unclear why slow translation of these genes is optimal. Perhaps only the maximal rate of growth matters in an evolutionary sense. The quality of these gene products, despite intuition, may not affect the growth rate (see above, Accuracy and Kinetics).

The striking thing about the use of the codons NNA/G lies not in the choice between them but in the fact that this pair is often avoided unless NNA/G are the only codons available (in *E. coli*; Grosjean and Fiers 1982). There are exceptions: CUG(Leu), GUA/G(Val), CCG(Pro), and GCA/G(Pro) are used at normal frequencies, but the other 7 comparable codon sets are all relatively infrequent in strongly expressed genes. Even among the pairs cited as an exception, CUA(Leu) is very infrequent. By comparison, there seems to be only one case in the entire coding table of strong prejudice against a pyrimidine-ending codon: CCC(Pro). This pattern, again, is altered in genes classified as weakly expressed, where these codons are used more often. However, the codon preference does not invert to prefer NNA/G among the low abundance set; it preserves the same direction. It is not evident why a purine in the message wobble position (pyrimidine in the anticodon wobble position) should be disfavored.

Quantitative Coadaptation of Codons and Messages

It is intuitively clear that frequent codons should require abundant tRNAs for accurate translation. This perception is borne out by the detailed conclusions of Post et al. (1979) and Ikemura (1981) on the use of codons in *E. coli*. But unbalancing the tRNA distribution will have the undesirable consequence of increasing the error rate at every codon that is *not* usually translated by the abundant tRNA (see the error rate equation). As a result, there is an optimal strategy. As a general assumption, the best compromise with regard to accuracy and rate of translation is to distribute tRNA abundances as the square root of the corresponding codon frequencies (von Heijne and Blomberg 1979). This conclusion should also hold for nucleotide abundances in replication and in all similarly constructed situations. The actual results seem more compatible with a linear relation, but the data are scattered (Chavancy and Garel 1981). One possible conclusion is that translational architecture has not always evolved to maximize speed and accuracy but that other idiosyncratic considerations can be dominant (the data scatter), as long as they are not grossly incompatible with rapid and precise translation (an increasing relation is observed). In fact, even very substantial departures from the optimum should not produce a large decrease in predicted accuracy.

There seems to be strong evidence from in vitro translation that the tRNA and mRNA of an organism are suited to each other. There is suggestive data to the effect that this tendency is also exploited in vivo. Measurement of the magnitude and mechanism of the resulting effects on the rate and accuracy of translation would be very interesting, especially if a new level of regulation of gene expression could be confirmed.

ACKNOWLEDGMENTS

We thank our colleagues who discussed results with us before publication or who supplied preprinted manuscripts: John Atkins, Lionello Bossi, Henri Grosjean, Jon Gallant, Emmanuel Goldman, Chuck Kurland, Manny Murgola, Tim Johnston, and Jack Parker. The preparation of this review was supported by U.S. Public Health Service research grants GM-30881 to M.Y. and GM-24983 and GM-32584 to R.C.T.

REFERENCES

Albery, W.J. and J.R. Knowles. 1977. Efficiency and evolution of enzyme catalysis. *Agnew. Chem. Int. Ed. Engl.* **16**: 285.

Andersson, D.I., K. Bohman, L.A. Isaksson, and C.G. Kurland. 1982. Translation rates and misreading characteristics of rpsD mutants in *E. coli. Mol. Gen. Genet.* **187**: 467.

Atherly, A.G. and J. Menninger. 1972. Mutant *E. coli* strain with temperature sensitive peptidyl-transfer RNA hydrolase. *Nat. New Biol.* **240**: 245.

Atkins, J.F., D. Elseviers, and L. Gorini. 1972. Low activity of β-galactosidase in frameshift mutants of *Escherichia coli. Proc. Natl. Acad. Sci.* **69**: 1192.

Atkins, J.F., R.F. Gesteland, B.R. Reid, and C.W. Anderson. 1979. Normal tRNAs promote ribosomal frameshifting. *Cell* **18**: 1119.

Beremand, M.N. and T. Blumenthal. 1979. Overlapping genes in RNA phage: A new protein implicated in lysis. *Cell* **18**: 257.

Bergmann, J.E. and H. Lodish. 1979. A kinetic model of protein synthesis application to hemoglobin synthesis and translation control. *J. Biol. Chem.* **254**: 11927.

Bienz, M. and E. Kubli. 1981. Wild-type tRNA^tyr reads the TMV RNA stop codon, but Q base-modified tRNA^tyr does not. *Nature* **294**: 188.

Bock, A., L.E. Faiman, and F.C. Neidhardt. 1966. Biochemical and genetic characterization of a mutant of *E. coli* with a temperature-sensitive valyl ribonucleic acid synthetase. *J. Bacteriol.* **92**: 1076.

Bollen, T., T. Cabezon, M. De Wilde, R. Villaroel, and A. Herzog. 1975. Alteration of ribosomal protein S17 by mutation linked to neamine resistance in *E. coli.* I. General properties of neaA mutants. *J. Mol. Biol.* **99**: 795.

Bossi, L. 1983. Context effects: Translation of UAG codon by suppressor tRNA is affected by the sequence following UAG in the message. *J. Mol. Biol.* **164**: 73.

Bossi, L. and J. Roth. 1980. The influence of codon context on genetic code translation. *Nature* **286**: 123.

Bradley, D., J. Park, and L. Soll. 1981. tRNA^Gln Su^{+2} mutants that increase amber suppression. *J. Bacteriol.* **145**: 704.

Breeden, L. and M. Yarus. 1982. Amber suppression relaxes stringent control by elongating stringent factor. *Mol. Gen. Genet.* **187**: 254.

Buckingham, R.H. and M.J. Carrier. 1983. The effect of codon context on mistranslation of messenger RNA. In *Interaction of translational and transcriptional controls in the regulation of gene expression* (ed. M. Grunberg-Manago and B. Safer). Elsevier Biomedical, New York. (In press.)

Buckingham, R. and H. Grosjean. 1983. The accuracy of messenger RNA:tRNA recognition. In *Accuracy in biology* (ed. D. Galas). Marcel Dekker, New York. (In press.)

Buckingham, R.H. and C.G. Kurland. 1977. Codon specificity of UGA suppressor tRNA^Trp from *Escherichia coli. Proc. Natl. Acad. Sci.* **74**: 5496.

Cabezon, T., A. Herzog, M. De Wilde, R. Villaroel, and A. Bollen. 1976. Cooperative control of translation fidelity by ribosomal proteins in *E. coli. Mol. Gen. Genet.* **144**: 59.

Campuzano, S., M.J. Cabanas, and J. Modolell. 1979. The binding of non-cognate Tyr-tRNA^Tyr to poly(uridylic acid)-programmed *Escherichia coli* ribosomes. *Eur. J. Biochem.* **100**: 133.

Caplan, A.B. and J.R. Menninger. 1979. Tests of the ribosomal editing hypothesis. Amino acid starvation differentially enhances the dissociation of peptidyl-tRNA from the ribosome. *J. Mol. Biol.* **134**: 621.

Caskey, C.T. 1977. Peptide chain termination. In *Molecular mechanisms of protein biosynthesis* (ed. H. Weissbach and S. Pestka), p. 443. Academic Press, New York.

Chaney, W.G. and A.J. Morris. 1979. Nonuniform size distribution of nascent peptides: The effect of messenger RNA structure upon the rate of translation. *Arch. Biochem. Biophys.* **194**: 283.

Chavancy, G. and J.P. Garel. 1981. Does quantitative tRNA adaptation to codon content in mRNA optimize the ribosomal translation efficiency? Proposal for a translation system model. *Biochimie* **63**: 187.

Chavancy, G., A. Chevallier, A. Fournier, and J.P. Garel. 1979. Adaptation of iso-tRNA concentration to mRNA codon frequency in the eukaryotic cell. *Biochimie* **61**: 71.

Crick, F.H.C. 1966. Codon-anticodon pairing: The wobble hypothesis. *J. Mol. Biol.* **19**: 548.

Cundliffe, E. 1981. Antibiotic inhibitors of ribosomal function. In *The molecular basis of antibiotic action,* 2nd ed. (ed. E.F. Gale et al.), p. 402. Wiley Interscience, New York.

Czernilofsky, A.P., A.D. Levinson, H.E. Varmus, J.M. Bishop, E. Tischer, and H.M. Goodman. 1980. Nucleotide sequence of an avian sarcoma virus oncogene (*src*) and proposed amino acid sequence for gene product. *Nature* **287**: 198.

Davies, J. and B.D. Davis. 1968. Misreading of ribonucleic acid code words induced by aminoglycoside antibiotics. *J. Biol. Chem.* **243**: 3312.

Davies, J., W. Gilbert, and L. Gorini. 1964. Streptomycin, suppression and the code. *Proc. Natl. Acad. Sci.* **51**: 883.

Davies, J., L. Gorini, and B.D. Davis. 1965. Misreading of RNA codewords induced by aminoglycoside antibiotics. *Mol. Pharmacol.* **1**: 93.

Davies, J., D.S. Jones, and H.G. Khorana. 1966. A further study of misreading of codons induced by streptomycin and neomycin using ribopolynucleotides containing two nucleotides in alternating sequence as templates. *J. Mol. Biol.* **18**: 48.

Diamond, A., B. Dudock, and D. Hatfield. 1981. Structure and properties of a bovine liver UGA suppressor serine tRNA with a tryptophan anticodon. *Cell* **25**: 497.

Dix, D.B., R.C. Thompson, E.R. Mackow, and F.N. Chang. 1983. Effect of ppGpp on the accuracy of protein biosynthesis. *Arch. Biochem. Biophys.* **223**: 319.

Edelmann, P. and J. Gallant. 1977. Mistranslation in *E. coli. Cell* **10**: 131.

Elias, P., F. Lustig, T. Axberg, B. Akesson, and U. Lagerkvist. 1979. Reading of the lysine codons in the MS2 coat protein cistron during protein synthesis in vitro. *FEBS Lett.* **98**: 145.

Engelberg-Kulka, H. 1981. UGA suppression by normal tRNA^Trp in *Escherichia coli:* Codon context effects. *Nucleic Acids Res.* **9**: 983.

Engelberg-Kulka, H., L. Dekel, M. Israel-Reches, and M. Belfort. 1979. The requirement of nonsense suppression for the development of several phages. *Mol. Gen. Genet.* **170**: 155.

Fairclough, R.H. and C.R. Cantor. 1979. The distance between the anticodon loop of two tRNAs bound to the 70S *E. coli* ribosome. *J. Mol. Biol.* **132**: 575.

Feinstein, S.I. and S. Altman. 1978. Context effects on nonsense codon suppression in *Escherichia coli. Genetics* **88**: 201.

Fersht, A. 1981. Enzymic editing mechanisms and the genetic code. *Proc. R. Soc. Lond. B* **212**: 351.

Fiers, W., R. Contreras, R. DeWachter, G. Haeeman, J. Merregaert, W. Minjou, and A. Vandenborghe. 1971. Recent progress in the sequence determination of bacteriophage MS2 RNA. *Biochimie* **53:** 495.

Files, J.G., U. Weber, and J.H. Miller. 1974. Translation reinitiation: Reinitiation of *lac* repressor fragments three internal sites early in the *lac* i gene of *Escherichia coli*. *Proc. Natl. Acad. Sci.* **71:** 667.

Fox, T.D. and B. Weiss-Brummer. 1980. Leaky +1 and −1 frameshift mutations at the same site in a yeast mitochondrial gene. *Nature* **288:** 60.

Galas, J.D. and W.B. Branscombe. 1976. Ribosome slowed by mutation to streptomycin resistance. *Nature* **262:** 617.

Gallant, J. and D. Foley. 1979. Stringent control of translational accuracy. In *Regulation of macromolecular synthesis by low molecular weight mediators* (ed. G. Koch and D. Richter), p. 5. Academic Press, New York.

Gallant, J. and I. Palmer. 1979. Error propagation in viable cells. *Mech. Ageing Dev.* **10:** 27.

Geller, A.I. and A. Rich. 1980. A UGA termination suppression tRNA[Trp] active in rabbit reticulocytes. *Nature* **283:** 41.

Gold, L., D. Pribnow, T. Schneider, S. Shinedling, B. Singer, and G. Stormo. 1981. Translation initiation in prokaryotes. *Annu. Rev. Microbiol.* **35:** 365.

Goldman, E. 1982. Effect of rate-limiting elongation on bacteriophage MS2 RNA-directed protein synthesis in extracts of *E. coli*. *J. Mol. Biol.* **158:** 619.

Goldman, E., W.M. Holmes, and G.W. Hatfield. 1979. Specificity of codon recognition by *E. coli* tRNA[Leu] isoaccepting species determined by protein synthesis in vitro directed by phage RNA. *J. Mol. Biol.* **129:** 567.

Gorini, L. 1970. The contrasting role of *strA* and *ram* gene products in ribosomal functioning. *Cold Spring Harbor Symp. Quant. Biol.* **34:** 101.

――――――. 1974. Streptomycin and misreading of the genetic code. In *Ribosomes* (ed. M. Nomura et al.), p. 791. Cold Spring Harbor Laboratory, Cold Spring Harbor, New York.

Grantham, R., C. Gautier, M. Gouy, M. Jacobzone, and R. Mercier. 1981. Codon catalog usage is a genome strategy modulated for gene expressivity. *Nucleic Acids Res.* **9:** r43.

Grosjean, H. and W. Fiers. 1982. Preferential codon usage in prokaryotic genes: The optimal codon-anticodon interaction energy and the selective codon usage in efficiently expressed genes. *Gene* **18:** 199.

Grosjean, D.H., S. DeHenau, and D.M. Crothers. 1978. On the physical basis for ambiguity in genetic coding interactions. *Proc. Natl. Acad. Sci.* **75:** 610.

Gupta, R.S. and D. Schlessinger. 1976. Coupling of rates of transcription, translation, and messenger ribonucleic acid degradation in streptomycin-dependent mutants of *E. coli*. *J. Bacteriol.* **125:** 84.

Hall, B. and J. Gallant. 1972. Defective translation in RC⁻ cells. *Nat. New Biol.* **237:** 131.

Harley, C.B., J.W. Pollard, C.P. Stanners, and S. Goldstein. 1981. Model for messenger RNA translation during amino acid starvation applied to the calculation of protein synthetic error rates. *J. Biol. Chem.* **256:** 10786.

Hatfield, D., F. Varricchio, M. Rich, and B.G. Forget. 1982. The aminoacyl-tRNA population of human reticulocytes. *J. Biol. Chem.* **257:** 3183.

Hirsh, D.I. 1971. Tryptophan transfer RNA as the UGA suppressor. *J. Mol. Biol.* **58:** 439.

Hirsh, D.I. and L. Gold. 1971. Translation of the UGA triplet in vitro by tryptophan transfer RNA's. *J. Mol. Biol.* **58:** 459.

Hoffstetter, H., H.-J. Monstein, and C. Weissman. 1974. The readthrough protein A₁ is essential for the formation of viable QB particles. *Biochim. Biophys. Acta* **374:** 238.

Hopfield, J.J. 1974. Kinetic proofreading: A new mechanism for reducing errors in biosynthetic processes requiring high specificity. *Proc. Natl. Acad. Sci.* **71:** 4135.

――――――. 1980. The energy relay: A proofreading scheme based on dynamic cooperativity and lacking all characteristic symptoms of kinetic proofreading in DNA replication and protein synthesis. *Proc. Natl. Acad. Sci.* **77:** 5248.

Ikemura, T. 1981. Correlation between the abundance of *E. coli* transfer RNAs and the occurrence of the respective codons in its protein genes. *J. Mol. Biol.* **146:** 1.

Jelenc, P.C. and C.G. Kurland. 1979. Nucleoside triphosphate regeneration decreases the frequency of translation errors. *Proc. Natl. Acad. Sci.* **76:** 3174.

Johnson, A.E., H.J. Adkins, E.A. Matthews, and C.R. Cantor. 1982. Distance moved by transfer RNA during translocation from the A site to the P site on the ribosome. *J. Mol. Biol.* **156:** 113.

Kassavetis, G. and M. Chamberlin. 1981. Pausing and termination of transcription within the early region of bacteriophage T7 DNA in vitro. *J. Biol. Chem.* **256:** 2777.

Kastelein, R.A., E. Remaut, W. Fiers, and J. van Duin. 1982. Lysis gene expression of RNA phage MS2 depends on frameshift during translation of the overlapping coat protein gene. *Nature* **295:** 35.

Kennell, D. and H. Riezman. 1977. Transcriptional and translational initiation frequencies of the *E. coli lac* operon. *J. Mol. Biol.* **114:** 1.

Kohli, J. and H. Grosjean. 1981. Usage of the three termination codons: Compilation and analysis of the known eukaryotic and prokaryotic termination sequences. *Mol. Gen. Genet.* **182:** 430.

Konigsberg, W. and G.N. Godson. 1983. Evidence for use of rare codons in the DNA G gene and other regulatory genes of *E. coli*. *Proc. Natl. Acad. Sci.* **80:** 687.

Kuhlberger, R., W. Piepersberg, A. Pretzet, P. Buckel, and A. Bock. 1979. Alteration of ribosomal protein L6 in gentamicin-resistant strains of *E. coli*. Effects on fidelity of protein synthesis. *Biochemistry* **18:** 187.

Kurland, C.G. 1979. Reading frame errors on ribosomes. In *Nonsense mutations and tRNA suppressors* (ed. J. Celis and J.D. Smith), p. 95. Academic Press, New York.

———. 1982. Translational accuracy in vitro. *Cell* **28:** 201.

Kurland, C.G. and J. Gallant. 1983. The secret life of the ribosome. In *Accuracy in biology* (ed. D. Galas). Marcel Dekker, New York. (In press.)

Ladner, J.E., A. Jack, J.D. Robertus, R.A. Brown, D. Rhodes, B.F.C. Clark, and A. Klug. 1975. Structure of yeast phenylalanine transfer RNA at 2.5 Å resolution. *Proc. Natl. Acad. Sci.* **72:** 4414.

Lagerkvist, U. 1978. "Two out of three": An alternative method for codon reading. *Proc. Natl. Acad. Sci.* **75:** 1759.

Le Meur, M-A., P. Gerlinger, and J.-P. Ebel. 1976. Messenger RNA translation in the presence of homologous and heterologous tRNA. *Eur. J. Biochem.* **67:** 519.

Lizardi, P., U. Mahdavi, D. Sheilds, and G. Candelas. 1979. Discontinuous translation of silk fibroin in a reticulocyte cell-free system and in intact silk gland cells. *Proc. Natl. Acad. Sci.* **76:** 6211.

Loftfield, R.B. and D. Vanderjagt. 1972. The frequency of errors in protein biosynthesis. *Biochem. J.* **128:** 1353.

Lustig, F., P. Elias, T. Axbery, T. Samuelsson, I. Tittawella, and U. Lagerkvist. 1981. Codon reading and translational error reading of the glutamine and lysine codons during protein synthesis in vitro. *J. Biol. Chem.* **256:** 2635.

Manley, J.L. 1978. Synthesis and degradation of termination and premature-termination of fragments of β-galactosidase in vitro and in vivo. *J. Mol. Biol.* **125:** 407.

Menninger, J. 1976. Peptidyl transfer RNA dissociates during protein synthesis from ribosomes of *Escherichia coli*. *J. Biol. Chem.* **251:** 3392.

———. 1978. The accumulation as peptidyl-transfer RNA of isoaccepting transfer RNA families in *E. coli* with temperature-sensitive peptidyl-transfer RNA hydrolase. *J. Biol. Chem.* **253:** 6808.

Miller, J.H. and A.M. Albertini. 1983. Effects of surrounding sequence on the suppression of nonsense codons. *J. Mol. Biol.* **164:** 59.

Mitra, S.K., F. Lustig, B. Akesson, T. Axberg, P. Elias, and U. Lagerkvist. 1979. Relative efficiency of anticodons in reading the valine codons during protein synthesis in vitro. *J. Biol. Chem.* **254:** 6397.

Moore, C.H., F. Farron, D. Bohnert, and C. Weissmann. 1971. Possible origin of a minor virus specific protein (A₁) in QB particles. *Nature* **234**: 204.

Morch, M.-D. and C. Benicourt. 1980. Polyamines stimulate suppression of amber termination codons in vitro by normal tRNAs. *Eur. J. Biochem.* **105**: 445.

Munz, P., U. Leupold, P. Agris, and J. Kohli. 1981. In vivo decoding rules in *Schizosaccharomyces pombe* are at variance with in vitro data. *Nature* **294**: 187.

Murgola, E.J. and F.T. Pagel. 1980. Codon recognition by glycine transfer RNAs of *E. coli* in vivo. *J. Mol. Biol.* **138**: 833.

Ninio, J. 1974. A semi-quantitative treatment of missense and nonsense suppression in the strA and ram ribosomal mutants of *E. coli*. *J. Mol. Biol.* **84**: 297.

—————. 1975. Kinetic amplification of enzyme discrimination. *Biochimie* **57**: 587.

O'Farrell, P.H. 1978. The suppression of defective translation by ppGpp and its role in the stringent response. *Cell* **14**: 545.

Orgel, L.E. 1963. The maintenance of the accuracy of protein synthesis and its relevance to ageing. *Proc. Natl. Acad. Sci.* **49**: 517.

Parker, J., T. Johnston, and P. Borgia. 1980. Mistranslation in cells infected with the bacteriophage MS2: Direct evidence of Lys for Asn substitution. *Mol. Gen. Genet.* **180**: 275.

Parker, J., J.W. Pollard, J.D. Friesen, and C.P. Stanners. 1978. Stuttering: High level mistranslation in animal and bacterial cells. *Proc. Natl. Acad. Sci.* **75**: 1091.

Pelham, H.R.B. 1978. Leaky UAG termination codon in tobacco mosaic virus RNA. *Nature* **272**: 469.

Philipson, L., P. Andersson, U. Olshevsky, R. Weinberg, D. Baltimore, and R. Gesteland. 1978. Translation of MuLV and MSV RNAs in nuclease-treated reticulocyte extracts: Enhancement of the *gag-pol* polypeptide with yeast suppressor tRNA. *Cell* **13**: 189.

Piepersberg, W., A. Bock, and H.G. Wittmann. 1975. Effects of different mutations in ribosomal protein S5 of *E. coli* on translation fidelity. *Mol. Gen. Genet.* **140**: 91.

Piepersberg, W., V. Noseda, and A. Bock 1979a. Bacterial ribosomes with two ambiguity mutations: Effects on translation fidelity, on the response to aminoglycosides and on the rate of protein synthesis. *Mol. Gen. Genet.* **171**: 23.

Piepersberg, W., D. Geyl, P. Buckel, and A. Bock. 1979b. In *Regulation of macromolecular synthesis by low molecular weight mediators* (ed. G. Koch and D. Richter), p. 39. Academic Press, New York.

Post, L.E., G.D. Strycharz, M. Nomura, H. Lewis, and P.P. Dennis. 1979. Nucleotide sequence of the ribosomal protein gene cluster adjacent to the gene for RNA polymerase subunit β in *Escherichia coli*. *Proc. Natl. Acad. Sci.* **76**: 1697.

Protzel, A. and A.J. Morris. 1974. Gel chromatographic analysis of nascent globin chains. *J. Biol. Chem.* **249**: 4594.

Riddle, D. and J. Carbon. 1973. Frameshift suppression: A nucleotide addition in the anticodon of a glycine transfer RNA. *Nat. New Biol.* **242**: 230.

Rosenberger, R.F. and G. Foskett. 1981. An estimate of the frequency of in vivo transcription errors at a nonsense codon in *Escherichia coli*. *Mol. Gen. Genet.* **183**: 561.

Rosset, R. and L. Gorini. 1969. A ribosomal ambiguity mutation. *J. Mol. Biol.* **39**: 95.

Roth, J.R. 1981. Frameshift suppression. *Cell* **24**: 601.

Ruusala, T., M. Ehrenberg, and C.G. Kurland. 1982. Is there proofreading during polypeptide synthesis? *EMBO J.* **1**: 741.

Salzer, W. 1969. The influence of the reading context upon the suppression of nonsense codons. *Mol. Gen. Genet.* **105**: 125.

Samuelsson, T., P. Elias, F. Lustig, T. Axberg, G. Folsch, B. Akesson, and U. Lagerkvist. 1980. Aberrations of the classic codon reading scheme during protein synthesis in vitro. *J. Biol. Chem.* **255**: 4583.

Savageau, M.A. 1981. Accuracy of proofreading with zero energy cost. *J. Theor. Biol.* **93**: 179.

Sharma, O.K., D.N. Beezley, and M.K. Roberts. 1976. Limitation of reticulocyte tRNA in the translation of heterologous mRNA's. *Biochemistry* **15:** 4313.

Sharp, S.J. and T.S. Stewart. 1977. The characterization of phosphoseryl tRNA from lactating bovine mammary gland. *Nucleic Acids Res.* **4:** 2123.

Strigini, P. and L. Gorini. 1970. Ribosomal mutations affecting efficiency of amber suppression. *J. Mol. Biol.* **47:** 517.

Tabor, H., C.W. Tabor, M.S. Cohn, and E.W. Hafner. 1981. Streptomycin resistance (rpsL) produces an absolute requirement for polyamines for growth of an *E. coli* strain unable to synthesize putrescine and spermidine. *J. Bacteriol.* **147:** 702.

Tai, P.C., B.J. Wallace, and B.D. Davis. 1978. Streptomycin causes misreading of natural messenger by interacting with ribosomes after initiation. *Proc. Natl. Acad. Sci.* **75:** 275.

Talkad, V., E. Schneider, and D. Kennel. 1976. Evidence for variable rates of ribosome movement in *E. coli. J. Mol. Biol.* **104:** 299.

Tanaka, N., H. Masukawa, and H. Umezawa. 1967. Structural basis of kanamycin for miscoding activity. *Biochem. Biophys. Res. Commun.* **26:** 544.

Taniguchi, T. and C. Weissman. 1978. Site-directed mutations in the initiator region of the bacteriophage Qβ cistron and the effect on ribosome binding. *J. Mol. Biol.* **118:** 533.

Thompson, R.C. and D.B. Dix. 1982. Accuracy of protein biosynthesis. A kinetic study of the reactions of poly(U)-programmed ribosomes with a leucyl-tRNA$_2$ elongation factor Tu GTP complex. *J. Biol. Chem.* **257:** 6677.

Thompson, R.C. and A.M. Karim. 1982. The accuracy of protein biosynthesis is limited by its speed: High fidelity selection by ribosomes of aminoacyl-tRNA ternary complexes containing GTP [γ^{-s}]. *Proc. Natl. Acad. Sci.* **79:** 4922.

Thompson, R.C. and P.J. Stone. 1977. Proofreading of the codon-anticodon interaction of ribosomes. *Proc. Natl. Acad. Sci.* **74:** 198.

Thompson, R.C., D.B. Dix, and J.F. Eccleston. 1980. Single turnover kinetic studies of guanosine triphosphate hydrolysis and peptide formation in the elongation factor Tu-dependent binding of aminoacyl-tRNA to *E. coli* ribosomes. *J. Biol. Chem.* **255:** 11088.

Thompson, R.C., D.B. Dix, R.D. Gerson, and A.M. Karim. 1981a. A GTPase reaction accompanying the rejection of Leu-tRNA$_2$ by UUU-programmed ribosomes. *J. Biol. Chem.* **256:** 81.

―――――― . 1981b. Effect of Mg^{2+} concentration, polyamines, streptomycin, and mutations in ribosomal proteins on the accuracy of the two-step selection of aminoacyl-tRNAs in protein biosynthesis. *J. Biol. Chem.* **256:** 6676.

Uhlenbeck, O.C., F.H. Martin, and P. Doty. 1971. Self-complementary oligoribonucleotides: Effects of helix defects and guanylic acid−cytidylic acid base pairs. *J. Mol. Biol.* **57:** 215.

von Heijne, G. and C. Blomberg. 1979. The concentration dependence of the error frequencies and some related quantities in protein synthesis. *J. Theor. Biol.* **78:** 113.

von Heijne, G., L. Nilsson, and C. Blomberg. 1977. Translation and messenger RNA secondary structure. *J. Theor. Biol.* **68:** 321.

―――――― . 1978. Models for mRNA translation: Theory versus experiment. *Eur. J. Biochem.* **92:** 397.

Wagner, E.G.H. and C.G. Kurland. 1980. Translation accuracy enhanced in vitro by (p)ppGpp. *Mol. Gen. Genet.* **180:** 139.

Wagner, E.G.H., M. Ehrenberg, and C.G. Kurland. 1982. Kinetic suppression of translation errors by (p)ppGpp. *Mol. Gen. Genet.* **185:** 269.

Wagner, E.G., P. Jelenc, M. Ehrenberg, and C.G. Kurland. 1980. Rate of elongation of polyphenylalanine in vitro. *Eur. J. Biochem.* **122:** 193.

Weiss, R. and J. Gallant. 1983. Mechanism of ribosome frameshifting during translation of the genetic code. *Nature* **302:** 389.

Wiener, A.M. and K. Weber. 1971. Natural read-through at the UGA termination signal of Qβ coat protein cistron. *Nature* **234:** 206.

Yahata, H., Y. Ocada, and A. Tsugita. 1970. Adjacent effect on suppression efficiency. II. Study on ochre and amber mutants of T4 phage lysozyme. *Mol. Gen. Genet.* **106:** 208.

Yarus, M. 1972. Intrinsic precision of aminoacyl-tRNA synthesis through parallel systems of ligands. *Nat. New Biol.* **239:** 106.

————. 1979a. The relationship of the accuracy of aminoacyl-tRNA synthesis to that of translation. In *tRNA: Structure, properties, and recognition* (P. Shimmel et al.), p. 501. Cold Spring Harbor Laboratory, Cold Spring Harbor, New York.

————. 1979b. The accuracy of translation. *Prog. Nucleic Acid Res. Mol. Biol.* **23:** 195.

————. 1982. Translation efficiency of transfer RNA's: Uses of an extended anticodon. *Science* **218:** 646.

Yarus, M., C. McMillan III, S. Cline, D. Bradley, and M. Snyder. 1980. Construction of a composite tRNA gene by anticodon loop transplant. *Proc. Natl. Acad. Sci.* **77:** 5092.

Yates, J.L. 1979. Role of ribosomal protein S12 in discrimination of aminoacyl-tRNA. *J. Biol. Chem.* **254:** 11550.

Yates, G.L., W.R. Gette, M.E. Furth, and M. Nomura. 1977. Effects of ribosomal mutations on the read-through of a chain termination signal: Studies on the synthesis of bacteriophage -O gene protein in vitro. *Proc. Natl. Acad. Sci.* **74:** 689.

Yegian, C.D. and G.S. Stent. 1969. An unusual condition of leucine transfer RNA appearing during leucine starvation of *E. coli. J. Mol. Biol.* **39:** 45.

Yourno, J. 1972. Externally suppressible +1 "glycine" frameshift possible quadruplet isomers for glycine and proline. *Nat. New Biol.* **239:** 219.

Zengel, J.M., R. Young, P.P. Dennis, and M. Nomura. 1977. Role of ribosomal protein S12 in peptide chain elongation: Analysis of pleiotropic, streptomycin-resistant mutants of *Escherichia coli. J. Bacteriol.* **123:** 1320.

Attenuation in Bacterial Operons

Carl E. Bauer, Jannette Carey, Lawrence M. Kasper,
Steven P. Lynn, Daryle A. Waechter, and Jeffrey F. Gardner
Department of Microbiology
University of Illinois
Urbana, Illinois 61801

INTRODUCTION

Expression of bacterial genes and operons is often controlled by molecular events that regulate the frequencies of transcription initiation and termination at specific promoter and terminator sites. Transcriptional repressors and activators are well-known regulatory molecules that determine the frequency of transcription initiation. Repressors reduce initiation by preventing RNA polymerase binding to the promoter, whereas activators stimulate transcription by interacting with DNA, RNA polymerase, or both. Regulation of gene expression by molecular events that act on transcription termination has become evident only more recently. Antitermination involves interactions between RNA polymerase and accessory factors that lead to a modified transcription complex that no longer recognizes transcription termination signals. In contrast, attenuation, which is the subject of this paper, is the regulation of transcription termination by molecular interactions that are believed to govern the formation of the transcription termination signal itself. The term "attenuator" is generally used to describe a site where regulation of transcription termination occurs within an operon.

Attenuation was discovered during investigations on the regulation of the histidine (*his*) operon of *Salmonella typhimurium* and the tryptophan (*trp*) operon of *Escherichia coli*. Kasai (1974) showed that the regulatory region of the *his* operon contains a transcriptional barrier that he described as an attenuator site. He found that a small deletion (*hisO* 1242) between the promoter and the first gene allowed greatly enhanced transcription of the structural genes in vitro compared with wild type. In the *trp* system, Yanofsky and his collaborators (Jackson and Yanofsky 1973; Bertrand et al. 1976) isolated a set of deletions that extended from the *trp* regulatory region into the structural genes but did not affect repression at the *trp* operator. Some of these deletions showed a significant increase in expression of the intact *trp* structural genes. This result implied that a regulatory site distinct from the operator was deleted in the mutants. They also observed that RNA fragments derived from the 5′ end of the in vivo *trp* mRNA were more abundant than fragments from the coding region (Bronson et al. 1973; Squires et al. 1976). In vivo and in vitro transcription studies showed that an attenuator site is located between the

65

promoter-operator region and the first structural gene of the operon (Bertrand et al. 1976, 1977; Lee et al. 1976, 1978; Squires et al. 1976; Lee and Yanofsky 1977). In subsequent years, attenuator sites have been discovered in several operons.

In this paper we review genetic systems that are regulated by attenuation. Our emphasis is directed toward operons that encode enzymes necessary for the biosynthesis of amino acids, since most of the recent information on attenuation has come from study of these systems. In addition, we describe other interesting systems in which gene expression appears to be regulated by similar attenuation mechanisms.

ATTENUATION CONTROL IN AMINO ACID BIOSYNTHESIS

Features of Attenuated Operons

The regulatory regions of amino acid biosynthetic operons controlled by attenuation have the following characteristic structural features (Lee and Yanofsky 1977; Barnes 1978; DiNocera et al. 1978; Zurawski et al. 1978a; Gardner 1979; Gemmill et al. 1979; Lawther and Hatfield 1980; Nargang et al. 1980; Wessler and Calvo 1981; Friden et al. 1982; Hauser and Hatfield 1983). The promoter used for transcription initiation is typically located 150 or more nucleotides upstream of the first structural gene (Fig. 1A). This arrangement defines the so-called "leader region" in which attenuation takes place. The junction between the leader region and the first structural gene is marked by a site of transcription termination that has the features usually associated with ρ-independent terminators (Rosenberg and Court 1979; Fig. 1A): a G+C-rich region containing dyad symmetry followed by an A+T-rich region of 4−9 bp in length. When transcribed into RNA, this region has the potential to form a hairpin structure (3:4 in Fig.1) immediately followed by a stretch of uridine residues. This RNA configuration has been described as the "termination structure" (Yanofsky 1981), because of its role in transcription termination. The current model (Platt 1981; Platt and Bear, this volume) for the function of this RNA structure in transcription termination is that the hairpin forms in the transcript as soon as the polymerase has moved far enough downstream, a distance estimated by Ryan and Chamberlin (1983) to be within 10 nucleotides of the growing point. Polymerase recognizes hairpin formation (Ryan and Chamberlin 1983) and may pause in response to it. The nucleotides incorporated into the transcript immediately after the hairpin are uridines, which form uniquely unstable base pairs with the template deoxyriboadenosines (Martin and Tinoco 1980). This unstable arrangement following a hairpin is thought to signal termination either by enhancing dissociation of the transcript or by inducing a conformational change in the polymerase, or both (Platt 1981).

When transcription termination occurs at the attenuator, it gives rise to a short "leader transcript," ranging in length from 140 nucleotides to 200 nucleotides, depending upon the operon (See Fig. 2). The leader region

A)

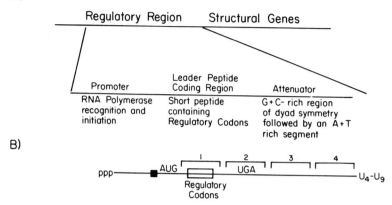

C) Sufficient Amino Acid Supply   Amino Acid Limitation

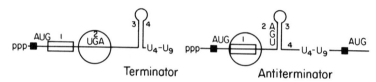

No Translation
(Superattenuation)

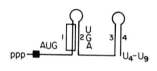

Terminator

Figure 1   Schematic representation of the regulatory regions of amino acid biosynthetic operons and the model for attenuation. (*A*) The regulatory region contains a
promoter, leader-peptide-coding region, and attenuator site. (*B*) The leader transcript
contains a ribosome recognition site (■) followed by a methionine codon (AUG), the
regulatory codons (□), and a translation termination codon (UGA). Regions of potential
secondary structure (1, 2, 3, and 4) are shown by brackets. The 3' terminus of the
transcript contains a run of uridine residues of variable length. (*C*) The model for
attenuation regulation under various physiological conditions is shown. If a sufficient
supply of amino acids is present, complete translation of the leader peptide results in
formation of hairpin 3:4 and transcription termination. If the cell is limited for the
appropriate amino acid(s), ribosome stalling in the stretch of regulatory codons results
in formation of hairpin 2:3 and transcription into the structural genes. Initiation of
translation of the structural genes occurs at a ribosome recognition sequence immediately preceding the first gene. Under conditions where translation of the leader peptide
does not occur, 1:2 and 3:4 base-pairing leads to transcription termination. See the text
for further details.

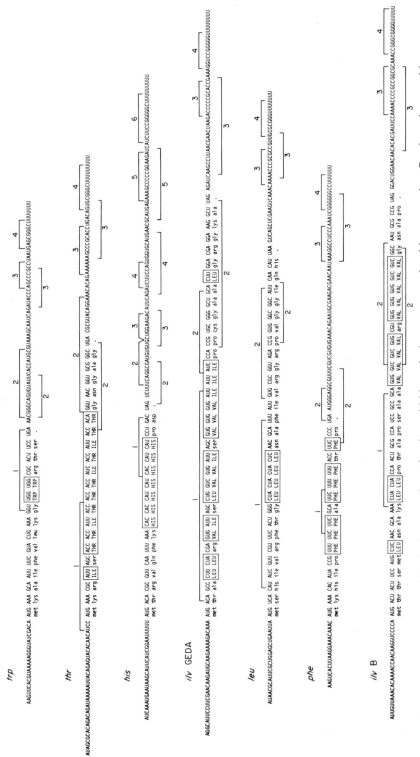

Figure 2 Nucleotide sequences of the leader transcripts of amino acid biosynthetic operons regulated by attenuation. Brackets above and below the sequences show the regions that can base-pair in the termination and antitermination configurations, respectively. The regulatory codons (capital letters) are boxed.

68

encodes several other self-complementary sequences; the leader transcript can therefore potentially form several intramolecular RNA secondary structures in addition to the termination structure. Figure 1B shows an arrangement typical of amino acid biosynthetic operons regulated by attenuation, in which segments 1 and 2 or 2 and 3 can base pair. Formation of hairpin 1:2 is compatible with termination structure (3:4) formation, whereas 2:3 formation is mutually exclusive with both 1:2 and 3:4. Figure 2 shows the RNA sequences of the leader transcripts encoded by amino acid biosynthetic operons known to be regulated by attenuation. Regions of representative transcripts that can base pair are shown in Figure 3 in configurations leading to either transcriptional readthrough or termination. The *S. typhimurium his* operon is more complex, since more secondary structures are possible, but similar alternatives can be drawn (Fig. 3).

The leader transcript contains a short open reading frame punctuated by start and stop codons. The initiation codon is preceded by a signal thought to be important for ribosome initiation of peptide synthesis: a purine-rich tract of 4–7 nucleotides centered 5–9 nucleotides 5' to the AUG (Shine and Dalgarno 1974). The leader peptide coding region includes an unusually large fraction of codons for the amino acid that is the end product of the operon's biosynthetic enzymes. These "regulatory codons" are clustered within the leader peptide in a position at least partially overlapping RNA segment 1 (Fig. 2). For example, the *E. coli trp* leader peptide contains 2 adjacent tryptophan codons, and the *S. typhimurium his* leader contains 7 consecutive histidine codons. The leader peptides of the *E. coli thr* and *ilvGEDA* operons, which are regulated by more than one amino acid (multivalent control), contain regulatory codons for two and three different amino acids, respectively (see Fig. 2).

Attenuation Model

The structural features mentioned above are incorporated into the model for the mechanism of attenuation (Lee and Yanofsky 1977; Keller and Calvo 1979; Oxender et al. 1979; Johnston and Roth 1981b; Kolter and Yanofsky 1982). When the supply of an amino acid is abundant, its biosynthetic enzymes are not required. Transcription initiated at the promoter proceeds until the translation initiation signal of the leader mRNA is synthesized and ribosomes bind to this region. When segment-2 transcription is complete, RNA polymerase may pause in response to formation of hairpin 1:2. The exact role of pausing, if it occurs in vivo, is not yet understood, and the order of the pausing and ribosome loading steps is uncertain, but pausing may allow coupling of transcription with translation (Winkler and Yanofsky 1981; Kolter and Yanofsky 1982). When transcription resumes, the ribosome likewise proceeds through the regulatory codons to the stop codon and may either dissociate quickly, allowing hairpin 1:2 to reform, or remain at the stop codon, masking segment 2. In either case, when segment-4 transcription is complete, hairpin

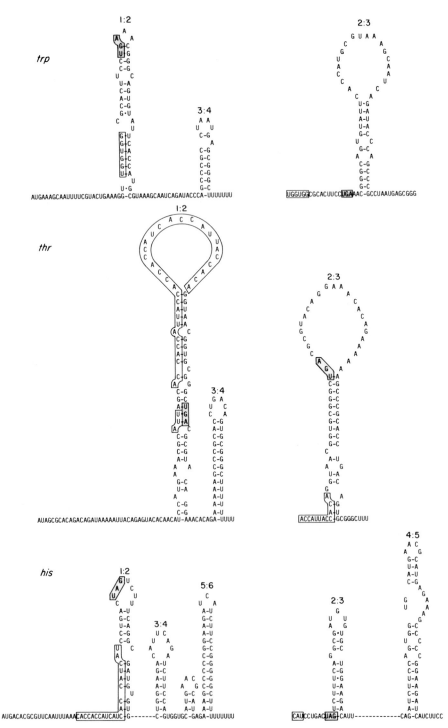

Figure 3 (See facing page for legend.)

3:4 forms, generating a transcription termination signal prior to the structural genes, and RNA polymerase terminates in response to this signal.

When the amino acid(s) in question is limiting and operon expression is required, how is the attenuation signal circumvented? When ribosomes translating the leader peptide reach the regulatory codons, they find the cognate aminoacyl tRNAs in short supply due to the amino acid limitation and will therefore stall in this region of the transcript. The regulatory codons are situated so that a stalled ribosome masks segment 1 of the leader transcript (Fig. 1), preventing formation of hairpin 1:2. Consequently, when segment-3 transcription is complete, hairpin 2:3 forms. Thus, when segment 4 is synthesized it cannot pair with 3 to form the termination structure, and transcription proceeds into the structural genes. Because 2:3 formation precludes formation of the terminator, hairpin 2:3 has been called the ''antitermination structure.'' The downstream genes contain their own translation initiation signals and, therefore, direct protein synthesis in the usual fashion.

The attenuation model implies that, in attenuated operons at least, the rate of transcription elongation must be delicately balanced with the kinetics of RNA hairpin formation and dissociation. Thus, when operon expression is not needed and ribosomes have arrived at the leader peptide stop codon, they must either remain bound to it, masking segment 2, or dissociate rapidly. In the latter case, hairpin 1:2 must reform immediately, and it must be stable long enough so that no significant dissociation of 1:2 occurs prior to formation of 3:4. If hairpin 1:2 forms too slowly or dissociates too rapidly, segments 1, 2, and 3 might simultaneously become available to base-pair before segment-4 synthesis is complete. Because of their comparable stabilities, hairpin 2:3 is probably as likely to form as hairpin 1:2, allowing wasteful readthrough when operon expression is unnecessary. When operon expression is required and ribosomes are stalled on segment 1, a similar kinetic argument can be made. Although the antiterminator structure is predicted to have a lower total free energy than the termination structure, hairpin 2:3 must form rapidly enough to prevent 3:4 pairing, and it must have a lifetime long enough to preclude terminator formation at least until transcription has proceeded beyond the influence of the termination signal.

Although we can calculate the stability of an RNA secondary structure from its sequence, we do not know the relationship between equilibrium stability and the rates of helix formation and melting. We could, in principle, derive this relationship empirically from observations of RNA secondary structure rearrangements during RNA synthesis, but the data are too sparse to permit generalizations. Some hairpin structures seem to dissociate in short times, allowing rearrangement, as is found for $Q\beta$ midivariant RNA during replica-

Figure 3 Alternative RNA secondary structures in the *trp*, *thr*, and *his* leader transcripts. The termination (*left*) and antitermination (*right*) structures (defined by the brackets in Fig. 2) are shown.

tion (Kramer and Mills 1981), whereas other hairpins such as those in nascent tRNA transcripts do not appear to rearrange once formed (Boyle et al. 1980). The low but significant readthrough seen under conditions of amino acid excess in several attenuated operons suggests that some dissociation of 1:2 may occur before synthesis of segment 4 is complete, allowing 2:3 formation. However, other explanations have also been proposed (see below) to account for at least some of this readthrough.

EVIDENCE FOR THE ATTENUATION MODEL DERIVED FROM ANALYSIS OF MUTANTS

The major features of the attenuation model are supported by analyses of a variety of mutations. In the following discussion, representative mutations are grouped according to the aspects of the model they support. Consequently, this discussion is intended not as a comprehensive description of all known mutations but, rather, to highlight those that most persuasively demonstrate the features of the model.

Mutations Affecting Termination

Mutations in sequences encoding the terminator stem (Figs. 1 and 4) permit increased transcriptional readthrough into the structural genes. Many such mutations have been identified in several of the bacterial amino acid biosynthetic operons regulated by attenuation (Stauffer et al. 1978; Gardner 1979; Gemmill et al. 1979; Johnston and Roth 1981a,b; S. Lynn et al., in prep.). In most cases, the efficiency of readthrough correlates with lower predicted stability for the terminator hairpin in the transcript from the mutant. Not all terminator stem mutations support this interpretation, however. In the *E. coli thr* operon, several mutations that reduce the predicted stability of the terminator hairpin do not increase operon expression (S. Lynn et al., in prep.). It is still possible, therefore, that specific sequence or conformational determinants play a role in transcription termination.

Mutations in the A+T-rich tract following the region of dyad symmetry in the attenuator of the *E. coli trp* operon also reduce transcription termination. Deleting 4 of the 8 A-T bp of the sequence reduces transcription termination from 85% to about 55% in vivo (Bertrand et al. 1977). Changing the second of the 8 A-T bp to a C-G, resulting in a U→G change in the transcript (*trpL*135; see Fig. 4), results in a threefold to fourfold increase in operon expression in vivo and a decrease in termination in vitro (Zurawski and Yanofsky 1980).

Mutations Altering RNA Secondary Structures

Because the decision to terminate transcription depends on RNA secondary structure alternatives, termination is expected to be influenced by mutations in sequences outside the terminator region itself. One *E. coli* mutant (*trpL*75; see

Fig. 4) has a single base-pair change that is predicted to destabilize only the antiterminator stem (Zurawski et al. 1978b). This strain is unable to respond to tryptophan starvation by increasing operon expression, as if antitermination is ineffective even when ribosomes are stalled on segment 1. Apparently the mutation alters the kinetics of the antiterminator helix sufficiently that segment 3 is available to pair with segment 4, forming the terminator, before polymerase has proceeded beyond the termination signal. Several similar mutations, all of which disrupt or delete structures that compete with terminator stem formation, have been identified in the analogous stems of the *S. typhimurium his* operon leader (stems 2, 4, and 5; see Fig. 4). Interestingly, these mutations are suppressible by amber suppressors even though they neither map in the leader peptide nor create nonsense codons. Ribosomes translating the leader peptide in suppressor-bearing strains presumably read through the UAG termination codon, translate up to a second in-frame UGA codon immediately preceding the 5:6 hairpin, and disrupt terminator formation (Johnston and Roth 1981b). Furthermore, a double mutant containing the *hisO*9712 and *hisO*9713 mutations has been constructed. Strains carrying each of these mutations alone show reduced *his* operon expression, whereas the double mutant shows wild-type expression (H.M. Johnston and J.R. Roth, pers. comm.). The double mutant substantially restores the predicted stability of the antiterminator stem by creating a new RNA base pair in which GC is replaced by AU. This double mutant therefore supports the notion that RNA base-pairing rather than base sequence governs the regulatory decision.

Analysis of deletions in the *trp* leader region of *Serratia marcescens* is consistent with the proposed roles of segments 1, 2, and 3 in the regulation of transcription termination (Stroynowski and Yanofsky 1982). Mutants with deletions originating from positions $+10$ to $+20$ in the 5' end of the leader and extending various lengths into segments 1, 2, or 3 were analyzed for their effects on *trp* operon expression. Deletions that extend into region 1 increase operon expression, presumably by increasing the frequency of antiterminator formation. Deletions that remove both segments 1 and 2 show reduced operon expression because antiterminator formation is not possible. Similarly, transcription termination at the *E. coli thr* attenuator is efficient when regions 1 and 2 are removed (Chapman and Gardner 1981). These observations indicate that sequences upstream are not required for efficient termination and that the RNA secondary structure encoded by segments 3 and 4 is sufficient for termination. As expected, deletions that remove part or all of segment 3 or the entire attenuator show increased expression of the structural genes (Chapman and Gardner 1981; Johnston and Roth 1981a,b; Parsot et al. 1982; Stroynowski and Yanofsky 1982).

Evidence for Translation of the Leader Peptide Coding Region

The ability of cells to utilize the leader peptide ribosome-binding site to initiate protein synthesis was demonstrated using an *E. coli* strain containing a

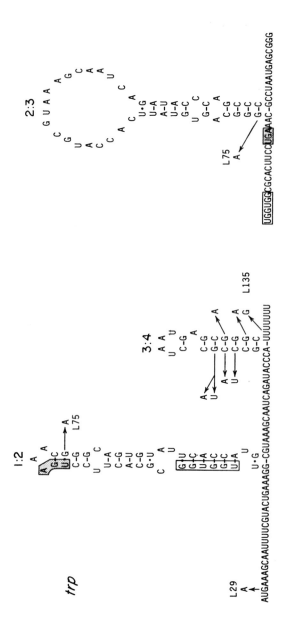

trp

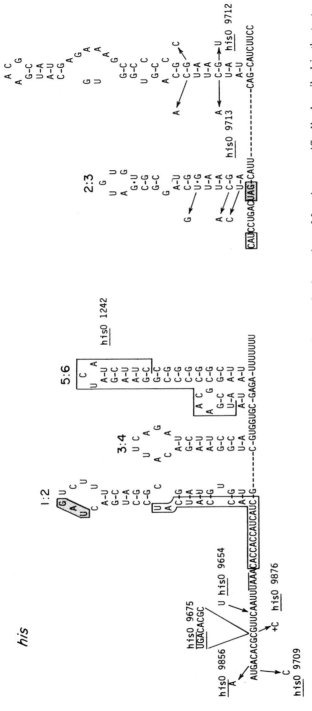

Figure 4 Representative regulatory mutations in the *trp* (*top*) and *his* (*bottom*) leader regions. Mutations specifically described in the text are indicated by their allele numbers. The *hisO9675* mutation is an 8-bp insertion that introduces a UGA termination codon (underlined). The *hisO9876* mutation is a +1 frameshift mutation that introduces a UAA codon downstream in the new reading frame (overlined). The *hisO1242* deletion (indicated by the large bracket) deletes most of the 5:6 structure.

75

deletion extending from within the leader peptide to the 5' end of the *trpE* gene (Miozzari and Yanofsky 1978), as well as with a strain in which the 5' end of the *lacI* gene was fused to the leader peptide in-frame (Schmeissner et al. 1977). In both cases, the hybrid proteins were expressed. Although these results suggest that the leader peptide could be made in vivo, they do not establish that leader peptide synthesis is necessary for the attenuation decision.

The requirement for translation of the leader peptide is implied by several mutations that alter ribosome binding or initiation determinants. Mutations in amino acid operons of several bacteria have been identified that delete or substitute a new codon for the AUG start codon of the leader peptides, thus reducing or preventing translation initiation. In the *E. coli trpL*29 mutant (Fig. 4), expression of the *trp* enzymes is reduced to 30% of wild type (Zurawski et al. 1978b). In the *S. typhimurium his* operon, two different mutations (*hisO*9709 and *hisO*9856 in Fig. 4) alter the leader peptide initiation codon but display intermediate operon expression (Johnston and Roth 1981b). Levels of *his* enzymes in these strains are high enough to confer a His$^+$ phenotype but too low to confer resistance to the histidine analog aminotriazole. The two mutant start codons AUA and ACG may therefore be functional but inefficient initiators for protein synthesis.

The effects of mutations that alter the ribosome recognition sequence upstream of the initiation codon also support the idea that translation is crucial in determining the level of expression of the structural genes. In the *S. marcescens trp* operon, both the complementarity of the Shine and Dalgarno region to the 16S ribosomal RNA (rRNA) and its distance from the AUG were altered by deletion (Stroynowski et al. 1982). The fused deletion endpoints in these mutants in some cases created new polypurine tracts at new distances from the initiation codon. The regulatory phenotype of these mutants appears to be correlated with how well the fusion sequence restores the consensus translation initiation determinants. As the ribosome-binding consensus sequence deteriorates, the mutants exhibit increased termination at the attenuator compared with wild type. By inference then, there must be some mechanism in the wild-type cell by which translation of the leader peptide permits occasional transcription into the structural genes even when excess tryptophan is present.

An explanation for this phenotype, which has been called superattenuation, has been proposed (Stroynowski et al. 1982). If translation did not occur, the nascent transcript would form structure 1:2 as soon as segment-2 synthesis was complete. Although 1:2 and 2:3 are of comparable stability, 2:3 formation is presumed to be kinetically blocked by prior formation of 1:2, so in the absence of translation, structure 3:4 forms in virtually every transcript. During translation, however, ribosomes stopped at the wild-type stop codon would mask segment 2. If the ribosome remains at the stop codon until segment-3 synthesis is complete, the first three segments would be free to base-pair when the ribosome dissociates. Due to their similar stabilities, 1:2 and 2:3 are equally likely to be present in the transcript while segment 4 is being synthesized. The low level of readthrough that does occur in wild-type cells in the presence of tryptophan has been attributed to transcripts bearing structure 2:3 due to rapid dissociation of ribosomes from the stop codon. When translation is reduced or abolished by mutation, the opportunity to form 2:3 is similarly reduced, and these mutants display the superattenuation phenotype. It has been suggested (Stroynowski et al. 1982) that superattenuation may restrict transcription of amino acid biosynthetic operons in response to any limitation in the cell's translation machinery since, under such a limitation, transcription would be wasteful.

Support for the Role of Ribosome Movement

Analysis of nonsense mutations at various locations in the *S. typhimurium his* leader region indicates that the level of *his* operon expression is correlated with the position at which translation terminates (Johnston and Roth 1981b). Thus, when ribosomes stop at nonsense codons preceding the tandem histidine codons, all segments of the transcript are available for base-pairing in the order they are synthesized, and formation of the terminator is allowed. These early nonsense mutations (*hisO* 9675 and *hisO* 9654; see Fig. 4) confer a His$^-$ phenotype. A nonsense mutation situated farther along the transcript (*hisO* 9876) confers a His$^+$ aminotriazole-sensitive phenotype, indicating that perhaps ribosomes stopped at this position occasionally preclude terminator stem formation by interfering with 1:2 formation. These mutations support the idea that ribosome movement along the transcript during peptide synthesis, rather than the peptide itself, mediates the attenuation response. This conclusion is buttressed by the finding that one of the His$^-$ nonsense mutations (*hisO* 9876) is the result of an 8-bp insertion. When the resulting frameshift mutation is suppressed, a His$^+$ phenotype results despite a dramatically different leader-peptide sequence.

Effects of Mutations in Unlinked Genes

The general model for attenuation predicts that mutations that alter the transcriptional or translational machinery may affect the expression of operons regulated by attenuation. Several such classes have been identified. Rifampicin-resistant strains of *E. coli* have mutations in the β subunit of RNA polymerase, some of which result in either increased or decreased levels of *trp* operon enzymes compared with wild type (Yanofsky and Horn 1981). One mutant polymerase that shows a decreased termination frequency in vivo and in vitro also has a decreased length of pausing at hairpin 1:2, whereas another mutant enzyme with increased termination frequency has an increased pause length (R. Fisher and C. Yanofsky, pers. comm.). Each kind of mutant shows the expected increase or decrease in termination frequency upon in vitro transcription of template lacking the 1:2 hairpin coding sequence, but deletion of the sequence encoding the termination structure abolishes the unique termination properties of the mutants.

It is possible that other transcription factors are involved in the attenuation response. The NusA protein appears to be a component of elongating RNA polymerase (Greenblatt and Li 1981). Ward and Gottesman (1981) demonstrated that the *nusA*1 mutant has a slightly decreased termination frequency at the *E. coli trp* terminator in vivo. Yet in vitro transcription studies showed that NusA protein affects pausing and not termination at the *trp* leader region (Farnham et al. 1982). Thus, it has been proposed that increased readthrough at the *trp* terminator in the *nusA*1 mutant in vivo is the result of decreased pausing at hairpin 1:2 (Farnham et al. 1982). This interpretation, if correct, strengthens the suggestion that pausing functions in vivo to allow coupling of translation with transcription.

Mutations altering generalized translational components of the cell affect the expression of several amino acid biosynthetic operons as well. Examples include mutations in tRNAs, aminoacyl-tRNA synthetases, and tRNA modification enzymes (Cortese et al. 1974; Wyche et al. 1974; Morse and Morse 1976; Johnson et al. 1977; Lawther and Hatfield 1977, 1980; Yanofsky and Soll 1977; Davidson and Williams 1979; Turnbough et al. 1979). These observations are not surprising, since attenuation, as described above, is influenced by the extent of tRNA charging and the rate of protein synthesis.

IN VITRO TRANSCRIPTION STUDIES SUPPORT THE ATTENUATION MODEL
Length of Transcripts

Transcription studies on several amino acid biosynthetic operons (*his*, *ilvB*, *ilvGEDA*, *leu*, *phe*, *thr*, and *trp*) from *E. coli* or other enteric species have supported the model of regulation by an attenuation mechanism. In vitro transcription from these operons shows that the leader RNA transcripts terminate upstream of the structural genes at high efficiencies (Lee et al. 1976; Squires et al. 1976; Lee and Yanofsky 1977; Zurawski et al. 1978a; Gemmill et al. 1979; Lawther and Hatfield 1980; Nargang et al. 1980; Freedman and Schimmel 1981; Wessler and Calvo 1981; Friden et al. 1982; Gardner 1982; Lynn et al. 1982; Searles et al. 1982; Hauser and Hatfield 1983). In each case, a sequence showing homology with the consensus promoter sequence (Rosenberg and Court 1979) is located at the appropriate distance from the attenuator. Sequencing of the 5' and 3' ends of the terminated transcripts isolated in vitro using DNA templates from the *his*, *thr*, and *trp* operons has established that the transcript is initiated upstream of the leader polypeptide and terminates within the stretch of uridine nucleotides immediately following the 3:4 hairpin (Lee et al. 1976; Squires et al. 1976; Freedman and Schimmel 1981; Frunzio et al. 1981; Gardner 1982). In addition, the leader RNAs synthesized in vitro from these operons are approximately the same size as the in vivo transcripts (Bertrand et al. 1977; Freedman and Schimmel 1981; Frunzio et al. 1981; Lynn et al. 1982). The 3' termini of the in vivo *trp* leader transcript are within

1 or 2 nucleotides of those found on the in vitro transcript (Bertrand et al. 1977), whereas the 5' termini are identical (Squires et al. 1976).

Secondary Structure of Transcripts

Several lines of evidence support the role for RNA secondary structure in transcription termination suggested by mutational analysis. Partial RNase T1 digestion of the terminated leader RNA isolated from in vitro transcription of the *trp* and *thr* operons reveals nuclease-resistant regions consistent with the presence of hairpins 1:2 and 3:4 (Lee and Yanofsky 1977; Oxender et al. 1979; Gardner 1982). Stability of RNA secondary structures is implicated in termination as well. Farnham and Platt (1980, 1982) have shown that termination is affected when RNA base analogs that either strengthen or weaken RNA-RNA interactions are incorporated into the transcript during in vitro transcription. This result points to a direct role for RNA secondary structure in transcription termination and implies that the previous observation of decreased termination at the *E. coli trp* attenuator in vitro in the presence of the ribonucleotide analog inosine triphosphate (Lee and Yanofsky 1977) was mediated by reduced stability of RNA secondary structure.

Evidence for formation of the 1:2 hairpin during in vitro transcription was obtained by transcribing the *E. coli trp* leader region in the presence of an oligodeoxynucleotide complementary to segment 1 (Winkler et al. 1982). This addition resulted in an increased frequency of readthrough past the attenuator. Competition for binding to segment 1 by the oligonucleotide apparently frees segment 2 to participate in the antiterminator hairpin (2:3) and, thus, indirectly inhibits formation of the 3:4 terminator.

RNA Polymerase Pausing

Measurement of the rates of in vitro transcription elongation in the leader regions of the *thr* and *trp* operons reveals a site of RNA polymerase pausing just beyond the 1:2 hairpin (Farnham and Platt 1981; Winkler and Yanofsky 1981; Gardner 1982). Pausing in the *trp* leader region is relieved by addition of an oligodeoxynucleotide complementary to segment 1 (R. Fisher and C. Yanofsky, pers. comm.). Since the oligonucleotide appears to compete with 1:2 hairpin formation, this result implies that pausing is due to RNA secondary structure. RNA polymerase pausing is detectable at many hairpin structures (Rosenberg et al. 1978; Farnham and Platt 1980, 1981; Kingston and Chamberlin 1981; Winkler and Yanofsky 1981) and may thus be a natural response of the transcription complex. Direct evidence for pausing at hairpin 1:2 in vivo has not been obtained, however, so its exact role in attenuation is uncertain.

IN VITRO EVIDENCE FOR TRANSLATION OF THE LEADER PEPTIDE

Early work by Artz and Broach (1975) established that translation is important in regulation of the *S. typhimurium his* operon in a cell-free system. Recently, *trp* operon attenuation in a DNA-dependent extract was shown to depend on the amount of Trp-tRNA[Trp] present (Das et al. 1982). Ribosome binding to the leader peptide Shine and Dalgarno region had also been demonstrated (Platt et al. 1976). Despite this substantial evidence that attenuation control requires translation of the leader peptide, isolation of the peptide itself has only very recently been accomplished.

Das et al. (1983) have shown that the leader peptide is synthesized in a DNA-dependent cell-free extract by using a DNA template specially constructed to circumvent an anticipated difficulty. A sequence at the 3' end of the leader transcript can base-pair with the leader peptide ribosome-binding region and could thereby prevent translation after the first ribosome has traversed the coding region (Yanofsky 1981). The template used to demonstrate leader peptide synthesis included the *S. typhimurium trp* promoter, operator, and leader peptide coding region, but all sequences downstream from the coding region were deleted. This deletion presumably allowed multiple rounds of translation and allowed detection of the peptide by SDS-polyacrylamide gel electrophoresis. When a separate template containing the distal sequence complementary to the Shine and Dalgarno region was tested, leader peptide synthesis was reduced by about 90%. This result confirms that the ribosome-binding domain could be masked by a downstream sequence and suggests that once the decision to terminate transcription has been made, no further translation occurs.

Several criteria indicate that the leader peptides of *E. coli*, *S. typhimurium*, and *S. marcescens* were correctly synthesized in vitro (Das et al. 1983). Sequence analysis of the initiation dipeptides and tripeptides produced in a defined protein-synthesizing system revealed only the appropriate amino termini. Each peptide could be labeled in a crude protein synthesis extract only by radioactive amino acids predicted to be present. The half-life of the *E. coli* peptide was shown to be only 3−4 minutes, suggesting another explanation for the previous difficulty in isolating the peptide and providing additional evidence for ribosome movement rather than peptide function as the mediator of the attenuation response.

ATTENUATION CONTROLS A VARIETY OF SYSTEMS

If one considers attenuation to be regulated transcription termination within an operon, then many other systems appear to use this mode of control. Some of these systems seem to use mechanisms completely analogous to the mechanism we have described for amino acid biosynthetic operons, whereas for other systems quite different models have been developed. In discussing them here, we will try to indicate in each case how well the evidence supports the proposed mechanism. Although there are now additional operons in which an internal termination site has been found (e.g., Barry et al. 1979, 1980; Nath and Guha 1982), we have not discussed those in which the evidence for regulation at that site is preliminary.

Phenylalanyl-tRNA Synthetase

The first example of the use of regulatory codons to control termination decision-making in an operon for gene products other than amino acid biosynthetic enzymes has recently come to light. The *E. coli pheS,T* operon encodes the small and large subunits, respectively, of phenylalanyl-tRNA synthetase. Mutational analysis had indicated that upstream sequences influenced *pheS,T* expression (Plumbridge and Springer 1982). The sequence of the region preceding *pheS* has now been completed (M. Grunberg-Manago, pers. comm.) and reveals features characteristic of attenuated operons. Two regions of G+C-rich dyad symmetry are present, and the second of these is followed by an A+T-rich stretch. When transcribed, this region could form the usual pattern of RNA structures: either hairpins 1:2 plus 3:4 followed by uridines, or the alternative hairpin 2:3. An open reading frame containing five phenylalanine codons, three of which are consecutive, is preceded by a translation initiation signal. The phenylalanine codons are appropriately placed to permit ribosome masking of segment 1 by a stalled ribosome. Involvement of these structural features in attenuation control of *pheS,T* expression is further supported by mapping the 5′ end of the terminated in vitro transcript to the upstream promoter start site and localizing the 3′ end to a site near the attenuator (M. Grunberg-Manago, pers. comm.). Fusion of the control region to *lacZ* showed that β-galactosidase expression in vivo is derepressed in the presence of the *trpX* (*miaA*; Eisenberg et al. 1979) mutation, which is defective in modification of *trp* and *phe* tRNAs, providing the strongest support for attenuation mediated by translation of the regulatory codons.

Ampicillin Resistance

The *E. coli* chromosomal *ampC* gene encodes β-lactamase, a periplasmic protein capable of hydrolyzing penicillin and related antibiotics (Normark and Burman 1977). β-lactamase production is under growth-rate-dependent regulation (Jaurin and Normark 1979), as are the cell's transcriptional and translational components. Ampicillin-resistant mutations mapping to the region immediately upstream of *ampC* are *cis*-dominant and abolish growth rate control (Jaurin et al. 1981). DNA sequencing of the control region reveals a promoterlike sequence separated from the structural gene by an attenuatorlike sequence. A short transcript terminated within the presumed attenuator has been detected in vitro. The 5′-end sequence of this transcript and readthrough (runoff) RNAs showed that both initiate at the presumptive promoter. One mutation that causes a mismatch in the presumed terminator stem reduces production of the terminated leader transcript and increases the amount of readthrough transcript. This mutation increases *ampC* expression in vivo.

Although the RNA secondary structure corresponding to the *ampC* terminator was detectable with RNases, the 41-nucleotide leader transcript contains no sequences that could form RNA secondary structure alternatives that might preclude terminator formation. Furthermore, no open reading frame is present in the leader region. A novel variation of the attenuation model has been proposed (Jaurin et al. 1981) to account for the growth-rate-dependent regulation apparently mediated by the leader region. The 5' end of the terminated transcript contains an initiation codon for protein synthesis followed immediately by an ochre codon. These codons are presumed to support formation of a translation initiation complex with assistance from a weak, but appropriately placed, Shine and Dalgarno sequence. Ribosomes nonproductively bound here would mask enough of the terminator hairpin to permit readthrough into the β-lactamase gene. The investigators propose that ribosome binding in the *ampC* leader region is proportional to the intracellular concentration of ribosomes, which is under growth-rate control.

Erythromycin Resistance

The mechansim for inducible resistance to erythromycin, reviewed elsewhere in this volume (Weisblum), has been called translational attenuation. Erythromycin binds sensitive 50S ribosomal subunits (Lai and Weisblum 1971) and is proposed (Horinouchi and Weisblum 1980; Hahn et al. 1982) to thereby cause stalling during translation of a leader peptide preceding the erythromycin methylase structural gene. The stalled ribosome is postulated to elicit unmasking of downstream ribosome-binding determinants via RNA secondary structure rearrangements, permitting translation of the methylase. This protein confers resistance to erythromycin by specific methylation of the 23S rRNA. Because induction of resistance occurs at subinhibitory concentrations of the drug, sensitive ribosomes are still available to translate the methylase mRNA once it is unmasked. In the absence of erythromycin, unimpeded translation of the leader region promotes formation of an RNA secondary structure that masks the ribosome-binding site of the methylase gene.

Pyrimidine Biosynthesis

The *pyrBI* operon of *E. coli* encodes the two subunits for aspartate transcarbamoylase, an enzyme that catalyzes the conversion of carbamoylphosphate to carbamoylaspartate, the first committed step in de novo synthesis of UTP (Pauza et al. 1982). When UTP levels are high, in vitro transcription of the operon terminates within a leader region preceding the structural genes (Turnbough et al. 1983). DNA sequence analysis of the leader region revealed a terminatorlike sequence 23 bp 5' to the structural genes with a G+C-rich region of dyad symmetry followed by a run of 8 thymidines (Roof et al. 1982; Navre and Schachman 1983; Turnbough et al. 1983). A second region of weaker symmetry followed by an interrupted stretch of 7 thymidines is also present approximately 20 bp upstream of the attenuatorlike sequence. The size of in vitro transcripts from the *pyrBI* region indicates that transcription terminates at the attenuator site and that, in the presence of ITP instead of GTP, a longer transcript is made. Furthermore, at low UTP levels RNA polymerase appears to pause in the interrupted run of Ts preceding the attenuator (Turnbough et al. 1983). Limiting nucleotide triphosphate concentrations have been observed to enhance pausing in vitro in other systems as well (Kassavetis and Chamberlin 1981).

The *pyrBI* leader transcript contains an open reading frame of 44 codons preceded by a ribosome initiation signal. This coding sequence extends to within 6 bp of the *pyrBI* structural genes. The putative peptide contains runs of 3 and 2 consecutive phenylalanine codons encoded by the runs of Ts corresponding to the pause and termination sites, respectively. The regulatory effect, however, is presumed to be due not to ribosome stalling at these phenylalanine codons but, rather, to polymerase either pausing at the first or terminating at the second run of Ts that comprise the phenalanine codons. A model for control of *pyrBI* expression proposed by Turnbough et al. (1983) suggests that when UTP levels are low, polymerase pausing within the interrupted run of thymidines allows ribosomes to initiate on the nascent transcript and translate up to the paused polymerase. The polymerase eventually proceeds beyond the pause site and transcribes the attenuator region. However, the ribosome closely follows the transcription complex, preventing formation of the termination structure and allowing readthrough into the structural genes. When UTP levels are adequate, pausing does not occur, coupling to ribosomes is prevented, and attenuator formation blocks transcription prior to the structural genes.

Ribosomal Proteins

The S10 operon of *E. coli* encodes 11 ribosomal proteins whose expression is regulated by one of the encoded proteins, L4, at the level of both transcription (Zengel et al. 1980) and translation (Yates and Nomura 1980). Lindahl et al. (1983) have recently shown that in the presence of excess L4, only short leader transcripts are produced in vivo from a promoter site located about 200 bp before the coding sequence for S10, the first structural gene. The termination site for this leader transcript was inferred from its length to be immediately 5' to the S10 ribosome-binding domain. The leader transcript contains most of a 32-codon open reading frame, but this is preceded by no discernible Shine and Dalgarno sequence. However, several regions of the leader transcript share homology with the L4-binding site on 23S rRNA. The longest contiguous region of homology in the leader RNA occurs at the termination site and can be drawn in a hairpin followed by a run of 4 uridines. The investigators postulate that this hairpin may be intrinsically unstable, allowing readthrough, unless L4 is bound to it. In the presence of excess L4 the stabilized termination structure blocks transcription into the structural genes.

Yeast *LEU2*

The DNA sequence of the control region upstream of the yeast *LEU2* gene (Andreadis et al. 1982) reveals striking similarities to control regions of prokaryotic amino acid biosynthetic operons regulated by attenuation. This finding is surprising because, in eukaryotes, translation and transcription occur in physically separate cellular compartments. The *LEU2* reading frame was established by aligning the DNA sequence with short peptides recovered from hydrolysates of the *LEU2*-gene product, isopropylmalate dehydrogenase, allowing the presumptive start site for translation to be inferred. The DNA sequence 5' to this start site encodes a potential 23-amino-acid peptide containing 6 leucine codons, two of which are in tandem. The sequence near the 3' end of the peptide coding region can be drawn in a hairpin loop followed by 4 thymidines, and when transcribed would resemble several other eukaryotic (and prokaryotic) RNA polymerase terminators. Recent S1 mapping indicates that the 5' end of one of the several RNAs transcribed from the *LEU2* region originates just upstream of the peptide coding sequence (A. Andreadis and P.R. Schimmel, pers. comm.). Furthermore, Northern blot analysis of *LEU2* transcripts isolated from repressed and derepressed (leucine-starved) cells suggests that perhaps half the variation in *LEU2* enzyme levels in response to leucine can be accounted for by regulation at the level of transcription (A. Andreadis and P.R. Schimmel, pers. comm.). These observations can be accommodated by any of several models (Andreadis et al. 1982; Yanofsky 1983).

SUMMARY

Expression of many amino acid biosynthetic operons appears to be controlled by attenuation. In these systems the attenuation decision is based on the level of charging of tRNAs with the regulatory amino acids. Because the extent of tRNA aminoacylation is closely dependent on the concentration of the cognate amino acid, attenuation provides a rapid, sensitive, and efficient means of regulating gene expression that responds to the cell's competence for carrying out protein synthesis. Coupling the ability of ribosomes to translate the regulatory codons with the extent of transcription termination at the attenuator allows for the possibility of global control by limitation in any component of the cell's translational machinery.

The occurrence of attenuation in diverse systems indicates that it is a more general control device than might be imagined from its specific mechanism in amino acid biosynthetic operons. The systems we discussed illustrate several variations on the general theme and suggest that many more variations may yet be discovered. For example, when our understanding of factors involved in ribosome stalling is extended, we may find that translational attenuation mechanisms are in widespread use in polycistronic prokaryotic operons. Translational attenuation could occur in eukaryotes as well, since it can proceed entirely in the cytoplasm, but the specific details of the prokaryotic mechanism probably will not apply, since our current model for 80S ribosome initiation suggests that, with few exceptions, only the 5'-proximal AUG is efficiently recognized (Kozak 1981).

In the case of *pyrB*I, like the amino acid operons, ribosomes can be thought of as antitermination or antipausing factors when their action is coupled to the transcription elongation complex. Similarly, ribosomal protein L4 appears to act like a termination factor. Surely other cellular components with analogous roles, as well as some with as yet undefined roles, will be identified in the coming years.

ACKNOWLEDGMENTS

We thank our colleagues for valuable comments on the manuscript and are grateful to C. Yanofsky, J. Roth, P. Schimmel, and M. Grunberg-Manago for providing unpublished information. Work carried out in our laboratory is supported by the National Science Foundation.

REFERENCES

Andreadis, A., Y.-P. Hsu, G.B. Kohlhaw, and P. Schimmel. 1982. Nucleotide sequence of yeast LEU2 shows 5'-noncoding region has sequences cognate to leucine. *Cell* **31**: 319.

Artz, S.W. and J.R. Broach. 1975. Histidine regulation in *Salmonella typhimurium*: An activation-attenuation model of gene regulation. *Proc. Natl. Acad. Sci.* **72**: 3453.

Barnes, W.M. 1978. DNA sequence from the histidine operon control region: Seven histidine codons in a row. *Proc. Natl. Acad. Sci.* **75**: 4281.

Barry, G., C.L. Squires, and C. Squires. 1979. Control features within the *rplJL-rpoBC* transcription unit of *Escherichia coli*. *Proc. Natl. Acad. Sci.* **76**: 4922.

Barry, G., C. Squires, and C.L. Squires. 1980. Attenuation and processing of RNA from the *rplJL-rpoBC* transcription unit of *Escherichia coli*. *Proc. Natl. Acad. Sci.* **77**: 3331.

Bertrand, K., C. Squires, and C. Yanofsky. 1976. Transcription termination *in vivo* in the leader region of the tryptophan operon of *Escherichia coli*. *J. Mol. Biol.* **103**: 319.

Bertrand, K., L.J. Korn, F. Lee, and C. Yanofsky. 1977. The attenuator of the tryptophan operon of *Escherichia coli*. Heterogeneous 3'-OH termini *in vivo* and deletion mapping of functions. *J. Mol. Biol.* **117**: 227.

Boyle, J., G.T. Robillard, and S.-H. Kim. 1980. Sequential folding of transfer RNA: A nuclear magnetic resonance study of successively longer tRNA fragments with a common 5' end. *J. Mol. Biol.* **139**: 601.

Bronson, M.J., C. Squires, and C. Yanofsky. 1973. Nucleotide sequences from tryptophan messenger RNA of *Escherichia coli*: The sequence corresponding to the amino-terminal region of the first polypeptide specified by the operon. *Proc. Natl. Acad. Sci.* **70**: 2335.

Chapman, J. and J.F. Gardner. 1981. Secondary lambda attachment site in the threonine operon attenuator of *Escherichia coli*. *J. Bacteriol.* **146**: 1046.

Cortese, R., R. Landsberg, R.A. Von der Haar, H.E. Umbarger, and B.N. Ames. 1974. Pleiotropy of *his*T mutants blocked in pseudouridine synthesis in tRNA: Leucine and isoleucine-valine operons. *Proc. Natl. Acad. Sci.* **71**: 1857.

Das, A., I.P. Crawford, and C. Yanofsky. 1982. Regulation of tryptophan operon expression by attenuation in cell-free extracts of *Escherichia coli*. *J. Biol. Chem.* **257**: 8795.

Das, A., J. Urbanowski, H. Weissbach, J. Nestor, and C. Yanofsky. 1983. *In vitro* synthesis of the tryptophan operon leader peptides of *Escherichia coli*, *Serratia marcescens* and *Salmonella typhimurium*. *Proc. Natl. Acad. Sci.* **80**: 2879.

Davidson, J.P. and L.S. Williams. 1979. Regulation of isoleucine and valine biosynthesis in *Salmonella typhimurium*: The effect of *his*U on repression control. *J. Mol. Biol.* **127**: 229.

DiNocera, P.P., F. Blasi, R. DiLauro, R. Frunzio, and C.B. Bruni. 1978. Nucleotide sequence of the attenuator region of the histidine operon of *Escherichia coli* K-12. *Proc. Natl. Acad. Sci.* **75**: 4276.

Eisenberg, S.P., L. Soll, and M. Yarus. 1979. The effect of an *E. coli* regulatory mutation on tRNA structure. *J. Mol. Biol.* **135**: 111.

Farnham, P.J. and T. Platt. 1980. A model for transcription termination suggested by studies on the *trp* attenuator *in vitro* using base analogs. *Cell* **20**: 739.

————— . 1981. Rho-dependent termination: Dyad symmetry in DNA causes RNA polymerase to pause during transcription *in vitro*. *Nucleic Acids Res.* **9**: 563.

————— . 1982. Effects of DNA base analogs on transcription termination at the tryptophan operon attenuator of *Escherichia coli*. *Proc. Natl. Acad. Sci.* **79**: 998.

Farnham, P.J., J. Greenblatt, and T. Platt. 1982. Effects of NusA protein on transcription termination in the tryptophan operon of *Escherichia coli*. *Cell* **29**: 945.

Freedman, R. and P. Schimmel. 1981. *In vitro* transcription of the histidine operon. *J. Biol. Chem.* **256**: 10747.

Friden, P., T. Newman, and M. Freundlich. 1982. Nucleotide sequence of the *ilv*B promoter-regulatory region: A biosynthetic operon controlled by attenuation and cyclic AMP. *Proc. Natl. Acad. Sci.* **79**: 6156.

Frunzio, R., C.B. Bruni, and F. Blasi. 1981. *In vivo* and *in vitro* detection of the leader RNA of the histidine operon of *Escherichia coli* K-12. *Proc. Natl. Acad. Sci.* **78**: 2767.

Gardner, J.F. 1979. Regulation of the threonine operon: Tandem threonine and isoleucine codons in the control region and translational control of transcription termination. *Proc. Natl. Acad. Sci.* **76**: 1706.

————— . 1982. Initiation, pausing, and termination of transcription in the threonine operon regulatory region of *Escherichia coli*. *J. Biol. Chem.* **257**: 3896.

Gemmill, R.M., S.R. Wessler, E.B. Keller, and J.M Calvo. 1979. *leu* operon of *Salmonella typhimurium* is controlled by an attenuation mechanism. *Proc. Natl. Acad. Sci.* **76**: 4941.

Greenblatt, J. and J. Li. 1981. Interaction of the sigma factor and the *nus*A gene protein of *E. coli* with RNA polymerase in the initiation-termination cycle of transcription. *Cell* **24**: 421.

Hahn, J., G. Grandi, T.J. Gryczan, and D. Dubnau. 1982. Translational attenuation of *erm*C: A deletion analysis. *Mol. Gen. Genet.* **186**: 204.

Hauser, C.A. and G.W. Hatfield. 1983. Nucleotide sequence of the *ilv*B multivalent attenuator region of *Escherichia coli* K-12. *Nucleic Acids Res.* **11**: 127

Horinouchi, S. and B. Weisblum. 1980. Posttranscriptional modification of mRNA conformation: Mechanism that regulates erythromycin-induced resistance. *Proc. Natl. Acad. Sci.* **77**: 7079.

Jackson, E. and C. Yanofsky. 1973. The region between the operator and first structural gene of the tryptophan operon of *Escherichia coli* may have a regulatory function. *J. Mol. Biol.* **76**: 89.

Jaurin, B. and S.J. Normark. 1979. *In vivo* regulation of chromosomal β-lactamase in *Escherichia coli*. *J. Bacteriol.* **138**: 896.

Jaurin, B., T. Grundstrom, T. Edlund, and S. Normark. 1981. The *E. coli* β-lactamase attenuator mediates growth rate-dependent regulation. *Nature* **290**: 221.

Johnson, E.J., G.N. Cohen, and I. Saint-Girons. 1977. Threonyl-transfer ribonucleic acid synthetase and the regulation of the threonine operon of *Escherichia coli*. *J. Bacteriol.* **129**: 66.

Johnston, H.M. and J.R. Roth. 1981a. Genetic analysis of the histidine operon control region of *Salmonella typhimurium*. *J. Mol. Biol.* **145:** 713.

_____ . 1981b. DNA sequence changes of mutations altering attenuation control of the histidine operon of *Salmonella typhimurium*. *J. Mol. Biol.* **145:** 735.

Kasai, T. 1974. Regulation of the expression of the histidine operon in *Salmonella typhimurium*. *Nature* **249:** 523.

Kassavetis, G.A. and M.J. Chamberlin. 1981. Pausing and termination of transcription within the early region of bacteriophage T7 DNA *in vitro*. *J. Biol. Chem.* **256:** 2777.

Keller, E.B. and J.M. Calvo. 1979. Alternative secondary structures of leader RNAs and the regulation of the *trp*, *phe*, *his*, *thr*, and *leu* operons. *Proc. Natl. Acad. Sci.* **76:** 6186.

Kingston, R.E. and M.J. Chamberlin. 1981. Pausing and attenuation of *in vitro* transcription in the *rrn*B operon of *E. coli*. *Cell* **27:** 523.

Kolter, R. and C. Yanofsky. 1982. Attenuation in amino acid biosynthetic operons. *Annu. Rev. Genet.* **16:** 113.

Kozak, M. 1981. Mechanism of mRNA recognition by eukaryotic ribosomes during initiation of protein synthesis. *Curr. Top. Microbiol. Immunol.* **93:** 81.

Kramer, F.R. and D.R. Mills. 1981. Secondary structure formation during RNA synthesis. *Nucleic Acids Res.* **9:** 5109.

Lai, C.J. and B. Weisblum. 1971. Altered methylation of ribosomal RNA in an erythromycin-resistant strain of *Staphylococcus aureus*. *Proc. Natl. Acad. Sci.* **68:** 856.

Lawther, R.P. and G.W. Hatfield. 1977. Biochemical characterization of an *Escherichia coli his*T strain. *J. Bacteriol.* **130:** 552.

_____ . 1980. Multivalent translational control of transcription termination at attenuator of *ilv*GEDA operon of *Escherichia coli* K-12. *Proc. Natl. Acad. Sci.* **77:** 1862.

Lee, F. and C. Yanofsky. 1977. Transcription termination at the *trp* operon attenuators of *Escherichia coli* and *Salmonella typhimurium*: RNA secondary structure and regulation of termination. *Proc. Natl. Acad. Sci.* **74:** 4365.

Lee, F., K. Bertrand, G. Bennett, and C. Yanofsky. 1978. Comparison of the nucleotide sequences of the initial transcribed regions of the tryptophan operons of *Escherichia coli* and *Salmonella typhimurium*. *J. Mol. Biol.* **121:** 193.

Lee, F., C.L. Squires, C. Squires, and C. Yanofsky. 1976. Termination of transcription *in vitro* in the *Escherichia coli* tryptophan operon leader region. *J. Mol. Biol.* **103:** 383.

Lindahl, L, R. Archer, and J.M. Zengel. 1983. Transcription of the S10 ribosomal protein operon is regulated by an attenuator in the leader. *Cell* **33:** 241.

Lynn, S.P., J.F. Gardner, and W.S. Reznikoff. 1982. Attenuation regulation in the *thr* operon of *Escherichia coli* K-12: Molecular cloning and transcription of the controlling region. *J. Bacteriol.* **152:** 363.

Martin, F. and I. Tinoco. 1980. DNA-RNA hybrid duplexes containing oligo(dA:rU) sequences are exceptionally unstable and may facilitate termination of transcription. *Nucleic Acids Res.* **8:** 2295.

Miozzari, G.F. and C. Yanofsky. 1978. Translation of the leader region of the *Escherichia coli* tryptophan operon. *J. Bacteriol.* **133:** 1457.

Morse, D.E. and A.N.C. Morse. 1976. Dual-control of the tryptophan operon is mediated by both tryptophanyl-tRNA synthetase and the repressor. *J. Mol. Biol.* **103:** 209.

Nargang, F.E., C.S. Subrahmanyam, and H.E. Umbarger. 1980. Nucleotide sequence of *ilv*GEDA operon attenuator region of *Escherichia coli*. *Proc. Natl. Acad. Sci.* **77:** 1823.

Nath, S.K. and A. Guha. 1982. Abortive termination of *bio*BFCD RNA synthesized *in vitro* from the *bio*ABFCD operon of *Escherichia coli* K-12. *Proc. Natl. Acad. Sci.* **79:** 1786.

Navre, M. and H.K. Schachman. 1983. Synthesis of aspartate transcarbamoylase in *Escherichia coli*: Transcriptional regulation of the *pyr*B-*pyr*I operon. *Proc. Natl. Acad. Sci.* **80:** 1207.

Normark, S. and L.G. Burman. 1977. Resistance of *Escherichia coli* to penicillins: Fine-structure mapping and dominance of chromosomal beta-lactamase mutations. *J. Bacteriol.* **132:** 1.

Oxender, D.L., G. Zurawski, and C. Yanofsky. 1979. Attenuation in the *Escherichia coli* tryptophan operon: Role of RNA secondary structure involving the tryptophan codon region. *Proc. Natl. Acad. Sci.* **76:** 5524.

Parsot, C., I. Saint-Girons, and P. Cossart. 1982. DNA sequence change of a deletion mutation abolishing attenuation control of the threonine operon of *E. coli* K-12. *Mol. Gen. Genet.* **188:** 455.

Pauza, C.D., M.J. Karels, M. Navre, and H.K. Schachman. 1982. Genes encoding *Escherichia coli* aspartate transcarbamoylase: The *pyr*B-*pyr*I operon. *Proc. Natl. Acad. Sci.* **79:** 4020.

Platt, T. 1981. Termination of transcription and its regulation in the tryptophan operon of *E. coli*. *Cell* **24:** 10.

Platt, T., C. Squires, and C. Yanofsky. 1976. Ribosome-protected regions in the leader-*trp*E sequence of *Escherichia coli* tryptophan operon messenger RNA. *J. Mol. Biol.* **103:** 411.

Plumbridge, J.A. and M. Springer. 1982. *Escherichia coli* phenylalanyl-tRNA synthetase operon: Characterization of mutations isolated on multicopy plasmids. *J. Bacteriol.* **152:** 650.

Roof, W.D., K.F. Foltermann, and J.R. Wild. 1982. The organization and regulation of the *pyr*BI operon in *E. coli* includes a *rho*-independent attenuator sequence. *Mol. Gen. Genet.* **187:** 391.

Rosenberg, M. and D. Court. 1979. Regulatory sequences involved in the promotion and termination of RNA transcription. *Annu. Rev. Genet.* **13:** 319.

Rosenberg, M., D. Court, H. Shimatake, C. Brady, and D.L. Wulff. 1978. The relationship between function and DNA sequence in an intercistronic regulatory region of phage λ. *Nature* **272:** 414.

Ryan, T. and M.J. Chamberlin. 1983. Transcription analysis with heteroduplex *trp* attenuator templates indicates that the transcript stem and loop structure serves as the termination signal. *J. Biol. Chem.* **258:** 4690.

Schmeissner, U., D. Ganem, and J.H. Miller. 1977. Genetic studies of the *lac* repressor. II. Fine structure deletion map of the *lac*I gene, and its correlation with the physical map. *J. Mol. Biol.* **109:** 303.

Searles, L.L., S.R. Wessler, and J.M. Calvo. 1982. Transcription attenuation is the major mechanism by which the *leu* operon of *Salmonella typhimurium* is controlled. *J. Mol. Biol.* **163:** 377.

Shine, J. and L. Dalgarno. 1974. The 3'-terminal sequence of *Escherichia coli* 16S ribosomal RNA: Complementarity to nonsense triplets and ribosome binding sites. *Proc. Natl. Acad. Sci.* **71:** 1342.

Squires, C., F. Lee, K. Bertrand, C.L. Squires, M.J. Bronson, and C. Yanofsky. 1976. Nucleotide sequence of the 5' end of tryptophan messenger RNA of *Escherichia coli*. *J. Mol. Biol.* **103:** 351.

Stauffer, G.V., G. Zurawski, and C. Yanofsky. 1978. Single base-pair alterations in the *Escherichia coli trp* operon leader region that relieve transcription termination at the *trp* attenuator. *Proc. Natl. Acad. Sci.* **75:** 4833.

Stroynowski, I. and C. Yanofsky. 1982. Transcript secondary structures regulate transcription termination at the attenuator of the *Serratia marcescens* tryptophan operon. *Nature* **298:** 34.

Stroynowski. I., M. van Cleemput, and C. Yanofsky. 1982. Superattenuation in the tryptophan operon of *Serratia marcescens*. *Nature* **298:** 38.

Turnbough, C.L., Jr., K.L. Hicks, and J.P. Donahue. 1983. Attenuation control of *pyr*BI operon expression in *Escherichia coli* K-12. *Proc. Natl. Acad. Sci.* **80:** 368.

Turnbough, C.L., Jr., R.J. Neill, R. Landsberg, and B.N. Ames. 1979. Pseudouridylation of tRNAs and its role in regulation in *Salmonella typhimurium*. *J. Biol. Chem.* **254:** 5111.

Ward, D.F. and M.E. Gottesman. 1981. The *nus* mutations affect transcription termination in *Escherichia coli*. *Nature* **292:** 212.

Wessler, S.R. and J.M. Calvo. 1981. Control of *leu* operon expression in *Escherichia coli* by a transcription attenuation mechanism. *J. Mol. Biol.* **149:** 579.

Winkler, M.E. and C. Yanofsky. 1981. Pausing of RNA polymerase during *in vitro* transcription of the tryptophan operon leader region. *Biochemistry* **20:** 3738.

Winkler, M.E., K. Mullis, J. Barnett, I. Stroynowski, and C. Yanofsky. 1982. Transcription termination at the tryptophan operon attenuator is decreased *in vitro* by an oligomer complementary to a segment of the leader transcript. *Proc. Natl. Acad. Sci.* **79:** 2181.

Wyche, J.H., B. Ely, T.A. Cebula, M.C. Snead, and P.E. Hartman. 1974. Histidyl-transfer ribonucleic acid synthetase in positive control of the histidine operon in *Salmonella typhimurium. J. Bacteriol.* **117:** 708.

Yanofsky. C. 1981. Attenuation in the control of expression of bacterial operons. *Nature* **289:** 751.

_____ . 1983. Prokaryotic mechanisms in eukaryotes? *Nature* **302:** 751.

Yanofsky, C. and V. Horn. 1981. Rifampicin resistance mutations that alter the efficiency of transcription termination at the tryptophan operon attenuator. *J. Bacteriol.* **145:** 1334.

Yanofsky, C. and L. Soll. 1977. Mutations affecting tRNA^Trp and its charging and their effect on regulation of transcription termination at the attenuator of the tryptophan operon. *J. Mol. Biol.* **113:** 663.

Yates, J.L. and M. Nomura. 1980. *E. coli* ribosomal protein L4 is a feedback protein. *Cell* **21:** 517.

Zengel, J.M., D. Mueckl, and L. Lindahl. 1980. Protein L4 of the *E. coli* ribosome regulates an eleven gene r protein operon. *Cell* **21:** 523.

Zurawski, G. and C. Yanofsky. 1980. *Escherichia coli* tryptophan operon leader mutations, which relieve transcription termination, are *cis*-dominant to *trp* leader mutations, which increase transcription termination. *J. Mol. Biol.* **142:** 123.

Zurawski, G., K. Brown, C. Killingly, and C. Yanofsky. 1978a. Nucleotide sequence of the leader region of the phenylalanine operon of *Escherichia coli. Proc. Natl. Acad. Sci.* **75:** 4271.

Zurawski, G., D. Elseviers, G.V. Stauffer, and C. Yanofsky. 1978b. Translational control of transcription termination at the attenuator of the *Escherichia coli* tryptophan operon. *Proc. Natl. Acad. Sci.* **75:** 5988.

Inducible Resistance to Macrolides, Lincosamides, and Streptogramin Type-B Antibiotics: The Resistance Phenotype, Its Biological Diversity, and Structural Elements That Regulate Expression

Bernard Weisblum
Pharmacology Department
University of Wisconsin Medical School
Madison, Wisconsin 53706

INTRODUCTION

Reports of erythromycin-resistant strains of *Staphylococcus aureus* began to appear in the clinical literature in the mid-1950s, shortly after the introduction of this drug into clinical practice (Chabbert 1956; Jones et al. 1956; Garrod 1957). The mechanism of resistance in these clinical isolates can be ascribed to reduced affinity between erythromycin and its binding site on the 50S subunit of the ribosome (Saito et al. 1969; Weisblum et al. 1971), which in turn results from a specific structural modification of the ribosome (Lai and Weisblum 1971; Lai et al. 1973a,b; Saito and Mitsuhashi 1980). The structural modification, methylation of adenine in 23S ribosomal RNA (rRNA), confers resistance to three chemically distinct groups of antibiotics that inhibit 50S ribosome subunit function: the macrolides, lincosamides, and streptogramin type-B (MLS) antibiotics. The material presented in this paper has been organized under the headings of three genera (namely *Stapphylococcus, Streptococcus*, and *Streptomyces*), which illustrate the biological diversity of MLS resistance, the more general form of clinical erythromycin resistance, with emphasis on the variations of inducer and induced phenotype that occur in these organisms, followed by a discussion of the structure and function of genetic control elements, studied in model systems, which regulate expression of this interesting resistance mechanism at the molecular level.

MLS RESISTANCE IN *Staphylococcus* SPECIES

Historical Background

Early reports of erythromycin-resistant *S. aureus* noted that such strains were either coresistant to other macrolides or that they could rapidly become so

(Chabbert 1956; Jones et al. 1956; Garrod 1957). It was also noted that such strains were already resistant, or could rapidly become resistant, to the streptogramin type-B antibiotics and to the lincosamides (Jones et al. 1956; Barber and Waterworth 1964; Desmyter and Reybrouck 1964; Griffith et al. 1965; Bourse and Monier 1968; Goldmann and Heiss 1971). In light of subsequent findings based on a survey involving antibiotics with different modes of action (Weisblum and Demohn 1969), these results can be summarized as follows: *S. aureus* cells become resistant to erythromycin by a single biochemical alteration of the ribosome, which reduces the affinity between drug and ribosome. By this single alteration, N^6-methylation of adenine in 23S rRNA, cells become more generally resistant to all macrolides, as well as to lincosamides and streptogramin type-B antibiotics. These groups of antibiotics comprise three of at least ten chemically distinct classes of antibiotics, which inhibit protein synthesis by their action on the 50S ribosomal subunit. In view of the apparent relationship defined by observed coresistance patterns, these three groups of inhibitors were designated collectively as the MLS antibiotics (Weisblum 1975).

MLS resistance in *S. aureus* can be expressed either constitutively or inducibly. In the latter case, erythromycin, to the exclusion of most other MLS antibiotics, appears to have a high level of activity as inducer, and cells that become induced become coresistant to all of the MLS antibiotics. Starting with an initially inducible population, constitutively resistant mutants can be obtained by using MLS antibiotics with low inducing activity for selection. The distribution of MLS resistance is not limited to *S. aureus* alone and occurs in other staphylococci, including *S. epidermidis* (Parisi 1981) and *S. hycis* (DeVriese 1976).

The descriptive terminology used in early reports of MLS resistance in *S. aureus* included references to "dissociated" and "generalized" or "double" resistance, as well as to an apparent "antagonism" between erythromycin and other MLS antibiotics (Chabbert 1956; Garrod 1957; Bourse and Monier 1968). The terms dissociated and generalized or double resistance were used to describe strains that we have come to recognize as inducibly and constitutively resistant, respectively. The term dissociated reflected the observation that disks containing spiramycin, a macrolide, produced a large clear inhibition zone, whereas disks containing an equivalent amount of erythromycin produced only a small inhibition zone indicative of resistance. The term antagonism referred to the distorted "D"-shaped (rather than circular) inhibition zone surrounding a spiramycin disk on a lawn of inducible *S. aureus* facing an erythromycin disk placed nearby. We now recognize that this phenomenon reflects induction of resistance by erythromycin to spiramycin in situ and that it can be produced more generally by use of a pair of MLS antibiotics, one that induces efficiently and one that does not. Although these designations provided formally accurate descriptions of various aspects of induced MLS resistance at the cellular level, terminology descriptive of the mechanism of induction at the molecular level would require further biochemical analysis.

Inducibility of MLS Resistance

Weaver and Pattee (1964), using a turbidimetric assay of growth, described time and concentration requirements for appearance of resistance in broth cultures and drew the important conclusion that expression of erythromycin resistance in the dissociated strains resembled expression of inducible enzymes. The studies of Kono et al. (1966) indicated that ribosomes from MLS-resistant staphylococci had a lower affinity for erythromycin, and the studies of Nakajima et al. (1968) noted that erythromycin remained unchanged after incubation with extracts of resistant cells. Together, these findings indicated that unlike clinical resistance to penicillin, streptomycin, or chloramphenicol, in which conversion of the drug to an inactive derivative was responsible for the resistance phenotype, MLS resistance involved a modification of the target of antibiotic action, the ribosome, instead. This provided a useful framework for further biochemical studies of MLS resistance. Weisblum et al. (1971) extended these studies by measuring efficiency of plating (eop) as a function of inducing conditions and correlated reduced eop with appearance of ribosomes incapable of binding erythromycin. Studies of the requirements for induction of MLS resistance, whether tested by measurement of turbidity or by increased eop, indicated that the process occurred optimally ($\sim$100% eop) if the culture was incubated in erythromycin-containing medium (0.01−0.1 μg/ml) for a period of about 1 hour. What kind of structural modification of the ribosome occurs during 1 hour of incubation that results in 100% eop? Moreover, how does erythromycin at subinhibitory concentrations switch on expression of the gene(s) that confer resistance?

23S rRNA Methylation and MLS Resistance

Studies of rRNA methylation provided some answers to the above questions. 23S rRNA from induced or constitutively resistant cells was found to contain the modified base N^6,N^6-dimethylamino purine (N^6,N^6-dimethyladenine), absent from the 23S rRNA of sensitive or uninduced cells (Lai and Weisblum 1971; Lai et al. 1973a). Use of inducible cells provided a handle for switching on the resistance phenotype in connection with rRNA labeling for analytical studies and the preparation of substrates for ribosome reconstitution experiments.

That modification of 23S rRNA is both necessary and sufficient for the resistance phenotype was demonstrated by ribosome reconstitution experiments (Lai et al. 1973b) in which 23S rRNA from three sources, namely, uninduced cells, induced cells, and constitutively resistant cells, was reconstituted in vitro with the total ribosomal protein fraction plus 5S rRNA from sensitive cells of *Bacillus stearothermophilus*. The resultant reconstituted 50S ribosome subunits, when tested for antibiotic-resistant protein synthesis, were found to be resistant or sensitive according to the phenotype of the cells from which the 23S rRNA used for reconstitution was obtained.

To formulate an explicit model for the mechanism of induction at the molecular level required the nucleotide sequence of the regulatory region(s) responsible for the inducible phenotype. It was therefore desirable to obtain a minimal DNA sequence that could specify inducible MLS resistance. One candidate, plasmid pE194 (molecular mass, 2.4 MD), discovered by Iordanescu (1976) and described in further detail by Iordanescu and Surdeanu (1980), was found to specify inducible MLS resistance in *S. aureus* in a form indistinguishable from that reported in previous studies. We therefore posed the question, Does plasmid pE194 alone confer inducible MLS resistance, or does it work in concert with specific functions determined by chromosomal genes or with genes associated with other plasmids?

In an important technical advance, Ehrlich (1977) reported that plasmids from *S. aureus* could be introduced into *Bacillus subtilis* by transformation where they were capable of replication and expression of antibiotic resistance genes as in the original host. To test whether pE194 contained all necessary genetic information to confer inducible MLS resistance, we examined a strain of *B. subtilis* into which Gryczan and Dubnau (1978) introduced plasmid pE194 by transformation followed by induction with erythromycin (0.05 μg/ ml) and selection for erythromycin resistance (10 μg/ml). When the *B. subtilis* transformant carrying pE194 was tested (Weisblum et al. 1979b), inducible MLS resistance associated with dimethyladenine (m_2^6A) in a form indistinguishable from that of the original *S. aureus* was found. These observations provided evidence that all necessary requirements for the inducible MLS phenotype were specified by pE194 and that additional factors, if any, were universally present in bacteria.

An upper limit to the DNA sequence length required for inducible MLS resistance was reduced from the full length of pE194 (3728 nucleotides) to the length of the largest of three pE194 subfragments, fragment A (1442 nucleotides), obtained by digestion with *Taq*I restriction endonuclease (Horinouchi and Weisblum 1980, 1982a). Transformant cells of *B. subtilis*, into which pE194 *Taq*I fragment A was introduced by subcloning in a *Cla*I site of plasmid pC194 (Horinouchi and Weisblum 1982b), were found to express inducible MLS resistance. The sequence of pE194 *Taq*I fragment A revealed one open reading frame capable of specifying a protein with a molecular mass of 29,000 daltons, suggesting that this protein was the putative methylase whose existence was predicted by our earlier studies.

In parallel studies of pE194 in the *B. subtilis* background, we isolated a series of high-copy-number mutants of this plasmid (Weisblum et al. 1979b), one of which, *cop-6*, following introduction into *B. subtilis* CU403, made it technically possible to obtain minicells containing a correspondingly high intracellular plasmid concentration and to visualize plasmid-determined proteins with a higher sensitivity than in the original lower-copy parent. Using the minicell-producing *cop-6* transformant, Shivakumar et al. (1979) demonstrated that a 29K protein was synthesized in response to inducing concentrations of erythromycin and that deletion mutations (produced by excision with restriction endonuclease *Hinc*II at a unique site in the 29K determinant followed by excision and religation), which led to loss of erythromycin resistance, also resulted in reduced size (higher mobility) of the 29K band. Moreover, Shivakumar et al. (1980) made the interesting observation that synthesis of such defective 29K proteins was hyperinducible. By use of *B. subtilis* cells carrying the *cop-6* mutant as enriched starting material, Shivakumar and Dubnau (1981) have also purified the MLS methylase predicted by our studies.

Inducers and Noninducers of MLS Resistance

The distinction between inducing and noninducing MLS antibiotics has direct application in studies of the MLS system. Noninducing MLS antibiotics have been useful in selecting constitutively resistant mutants, since cultivation of an inducible strain in the presence of a noninducing MLS antibiotic selects cells in which methylation is expressed continuously. On the other hand, growth of inducible strains of *S. aureus* in medium containing high concentrations of erythromycin (e.g., 50 μg/ml) does not select constitutively resistant mutants as one might expect. The cells that finally grow in the presence of high erythromycin concentrations appear to have become induced in situ and, when retested, appear indistinguishable from the inducible parent strain.

Other MLS antibiotics shown to have inducing activity in *S. aureus* include the macrolide antibiotic oleandomycin (Weaver and Pattee 1964) and the lincosamide antibiotic celesticetin (Allen 1977). Notably lacking inducing activity in *S. aureus* are numerous other macrolides, including tylosin and carbomycin, and lincosamides, lincomycin and its currently used semisynthetic derivative clindamycin, and virginiamycin S, a representative of the streptogramin type-B family of antibiotics. In addition, mutants of *S. aureus* have been obtained in which the macrolide leucomycin (Saito et al. 1970), or carbomycin and lincomycin (Tanaka and Weisblum 1974), induce.

Different classes of mutations in pE194 can be isolated, depending on the cell background in which selection is performed. For example, selection using *S. aureus* carrying pE194 in medium containing either tylosin or clindamycin yielded only constitutively resistant mutants that mapped in the control region for methylase synthesis (Gryczan et al. 1980; Horinouchi and Weisblum 1980, 1981; Hahn et al. 1982). On the other hand, growth of *B. subtilis* carrying pE194 in tylosin-containing medium yielded *cop* mutants with approximately fivefold elevated pE194 copy number (Weisblum et al. 1979b). The Cop phenotype was maintained if plasmid DNA obtained from the resistant mutant strain was reintroduced into *B. subtilis* by transformation, indicating involvement of plasmid, rather than host cell functions, in the *cop* mutation.

Induction of MLS Resistance
and Inhibition of Protein Synthesis Are Inseparable

Pestka et al. (1976) compared activities for induction and inhibition in a series of 57 erythromycin congeners and concluded that the two activities were closely correlated. This provocative observation posed a paradox: How does one switch on protein synthesis using an inhibitor of protein synthesis? One simple (but incorrect) answer would be to postulate the existence of a regulator protein, e.g., repressor, which becomes fully saturated with inducer at concentrations that only partially saturate the ribosomes. In such an interaction the inducer would bind to the repressor with the same specificity with which it binds to the ribosome. Such a model would be consistent with the observed requirement that erythromycin induces resistance optimally at subinhibitory concentrations but would require that a "simple" regulatory molecule distinguish between inhibitors and noninhibitors of protein synthesis with a degree of accuracy comparable to that of the ribosome. Moreover, if inhibition of protein synthesis is functionally correlated with induction, how can one account for an observed induction optimum $10^{-8}-10^{-7}$ M, at least one order of magnitude lower than the threshold for inhibition of protein synthesis ($10^{-7}-10^{-6}$ M)?

Pestka's (1976) data pointed to a correlation between inhibitory activity and inducing activity in the erythromycin series; however, they did not explain why other MLS antibiotics whose inhibitory potency was comparable to that of erythromycin lacked demonstrable inducing activity in *S. aureus*. Thus, there appear to be at least two components capable of contributing to induction specificity: one that allows quantitative comparison within the erythromycin series, and another that distinguishes one subset of MLS antibiotics (members of which induce in a given system) from another subset (members of which do not). The regulatory macromolecule with which erythromycin interacts during induction turns out to be the ribosome itself, as is discussed in further detail below.

MLS RESISTANCE IN Streptococcus SPECIES

Initial reports of MLS resistance in *Streptococcus pyogenes* called attention to the observation that these strains were resistant to both erythromycin and lincomycin (Sanders et al. 1968; Hyder and Streitfeld 1973; Dixon and Lipinski 1974; Malke 1974). Streptococci resistant to erythromycin and lincomycin, when tested with a streptogramin type-B antibiotic, are invariably found resistant to this group of antibiotics as well (Malke 1979). Although combined resistance to MLS antibiotics could be attributed to a coincidental occurrence of unrelated individual determinants, we tested the hypothesis that streptococci became MLS resistant by a mechanism similar to that found in the staphylococci. We demonstrated the presence of m_2^6A in 23S rRNA obtained from *S. faecalis* D-5 constitutively resistant, owing to the presence of a plasmid, pAMβ1, reported by Clewell et al. (1974), and the absence of methylated adenine in a strain, DS-5, from which pAMβ1 was eliminated (Graham and Weisblum 1979b). Clewell (1981) has reviewed the area of streptococcal plasmids including numerous examples that specify MLS resistance.

It was noted by Dixon and Lipinski (1974) and by Malke et al. (1981) that certain resistant strains of *S. pyogenes* were inducible by both erythromycin and lincomycin. Since test disks containing each of the MLS antibiotics produce turbid inhibition zones, we infer that all three classes of MLS antibiotics can induce in situ; this interpretation is supported by studies from our laboratory in which an inducible MLS determinant specified by *S. sanguis* plasmid pAM77 (Yagi et al. 1978) was used as a model test system, and induction was measured both by turbidimetric assay and by demonstrating appearance of a 29K protein in minicells grown under inducing conditions (Horinouchi et al. 1983).

Relatedness between MLS Determinants from Streptococci

At this time we can recognize inducible MLS resistance in *S. pyogenes*, specified by plasmid pAC-1 (17 MD), and in *S. sanguis*, specified by plasmid pAM77 (4.5 MD), as well as constitutive MLS resistance in *S. faecalis*, specified by plasmid pAMβ1. Studies of these and other streptococcal MLS determinants have demonstrated a high degree of sequence homology. Thus, Yagi et al. (1975) have shown at least 95% sequence homology by electron microscope analysis of heteroduplex molecules formed in vitro between pAMβ1 and pAC-1. Sequence homology by the DNA blot method has been shown for the MLS determinants of resistant *S. pyogenes*, *S. faecalis*, *S. sanguis*, and *S. pneumoniae* (*Streptococcus* group), all of which cross-hybridize with the MLS determinant carried by *S. aureus* plasmid pI258 (Weisblum et al. 1979a). Sequence homology between the streptococcal MLS determinants and the determinant specified by staphylococcal plasmid pI258 suggests that extensive genetic exchange has occurred in nature involving both streptococci and staphylococci as well. These retrospective analytical studies are supported by reports of direct transfer of MLS determinants in the laboratory (LeBlanc et al. 1978; Gibson et al. 1979; Engel et al. 1980; Horodniceanu et al. 1981). For reasons discussed below, we were unable to demonstrate hybridization between the *Streptococcus* group determinants (which include the staphylococcal determinant of pI258) and the determinant specified by *S. aureus* plasmid pE194.

Inducible MLS Determinant from Streptococcal Plasmid pAM77

Y. Yagi and D.B. Clewell (unpubl.) introduced plasmid pAM77 into *B. subtilis* by transformation, and Horinouchi et al. (1983) showed that the induced MLS phenotype in *B. subtilis* resembled the phenotype seen in the original host *S. sanguis* A1 by Yagi et al. (1978). Following introduction of plasmid pAM77 into a minicell-producing strain of *B. subtilis*, enhanced synthesis of only one of about five proteins was seen in response to erythromycin, lincomycin, or a streptogramin type-B antibiotic used for induction. The induced protein migrates on polyacrylamide gel electrophoresis with a mobility indistinguishable from that of the 29K pE194 methylase from *S. aureus*.

We have determined the DNA sequence that encodes the pAM77 methylase and compared both the DNA and amino acid sequences with the corresponding methylase sequences of pE194. The two methylase amino acid and DNA structural gene sequences are shown in Figure 1. From the DNA sequences we deduce that the two methylases differ in number by one amino acid (244 for the pE194 methylase and 245 for the pAM77 methylase). Moreover, half of the amino acid residues in both proteins are identical in sequence. Comparison of the respective DNA sequences reveals a 50% mismatch in the nucleotide sequences, sufficient to reduce hybrid stability to a level below that demonstrable by the usual conditions of stringency. The longest cluster of identical amino acids contains 9 residues; the corresponding structural gene sequence contains 14 sequential identical nucleotides in this region. The two methylase structural genes are sufficiently similar that we must conclude that they share a common ancestral sequence.

Further comparison of the two methylase structural gene sequences shows that the pE194 methylase sequence contains codons with an average of 29% G+C, whereas the pAM77 methylase utilizes codons with an average of 35% G+C. The patterns of codon use vary in parallel with the DNA base compositions of *S. aureus* and *S. pyogenes* (32% and 43%, respectively). These findings suggest that the two methylases have adapted to optimal expression in cells with significantly different percentages of G+C by means of codon usage appropriate to the base composition of the host organism.

Plasmid Sequence Determines Specificity of Induction

The observation that erythromycin, lincomycin, and streptogramin type-B antibiotics induce methylase synthesis in *S. sanguis* prompts us to inquire into the basis for the wide discrepancy between inducing activities of different MLS antibiotics whose potencies as inhibitors of protein synthesis are nearly equivalent. We can readily identify three factors that may play a role in the observed differences in relative inducing activity of different MLS antibiotics: (1) differences in the amino acid composition of the control peptide and the relative efficiency with which a particular MLS antibiotic can inhibit incorporation; (2) differences in ribosomes of various species reflected in characteristic spectra of inhibition; and (3) other cellular differences, such as unequal permeability to different MLS antibiotics.

The host cell background contribution to induction of the pE194 and pAM77 MLS determinants can be controlled by studying induction specificity in the same host cell background. pE194 and pAM77 introduced into *B. subtilis* by

```
        1 - 20
pE194   Met Asn Glu  Lys Asn Ilu Lys  His  Ser Gln Asn Phe  Ilu Thr  Ser Lys His Asn Ilu Asp
        ATG AAC GAG  AAA AAT ATA AAA  CAC  AGT CAA AAC TTT  ATT ACT  TCA AAA CAT AAT ATA GAT
pAM77   Met Lys      Lys Asn Ilu Lys  Tyr  Ser Gln Asn Phe  Leu Thr  Asn Glu Lys Val Leu Asn
        ATG AAA      AAA AAT ATA AAA  TAT  TCT CAA AAC TTT  TTA ACG  AAT GAA AAG GTA CTC AAC

        21 - 40
        Lys Ilu Met Thr Asn Ilu Arg  Leu Asn Glu His Asp Asn Ilu Phe Glu Ilu Gly Ser Gly
        AAA ATA ATG ACA AAT ATA AGA  TTA AAT GAA CAT GAT AAT ATC TTT GAA ATC GGC TCA GGA
        Gln Ilu Ilu Lys Gln Leu Asn  Leu Lys Glu Thr Asp Thr Val Tyr Glu Ilu Gly Thr Gly
        CAA ATA ATA AAA CAA TTG AAT  TTA AAA GAA ACC GAT ACC GTT TAC GAA ATT GGA ACA GGT

        41 - 60
        Lys Gly His Phe Thr Leu Glu Leu Val Lys Arg Cys Asn Phe Val Thr Ala Ilu Glu Ilu
        AAA GGC CAT TTT ACC CTT GAA TTA GTA AAG AGG TGT AAT TTC GTA ACT GCC ATT GAA ATA
        Lys Gly His Leu Thr Thr Lys Leu Ala Lys Ilu Ser Lys Gln Val Thr Ser Ilu Glu Leu
        AAA GGG CAT TTA ACG ACG AAA CTG GCT AAA ATA AGT AAA CAG GTA ACG TCT ATT GAA TTA

        61 - 80
        Asp His Lys Leu Cys Lys Thr Thr Ala Glu Asn Lys Leu Val Asp His Asp Asn Phe Gln Val
        GAC CAT AAA TTA TGC AAA ACT ACA GAA AAT AAA CTT GTT GAT CAC GAT AAT TTC CAA GTT
        Asp Ser His Leu Phe Asn Leu Ser Ser Glu Lys Leu Lys Leu Asn Ilu Arg Val Thr Leu
        GAC AGT CAT CTA TTC AAC TTA TCT TCA GAA AAA TTA AAA CTG AAC ATT CGT GTC ACT TTA

        81 - 100
        Leu Asn Lys Asp Ilu Leu Gln Phe Lys Phe Pro Lys Asn Gln Ser Tyr Lys Ilu Tyr Gly
        TTA AAC AAG GAT ATA TTG CAG TTT AAA TTT CCT AAA AAC CAA TCC TAT AAA ATA TAT GGT
        Ilu His Gln Asp Ilu Leu Gln Phe Gln Phe Pro Asn Lys Gln Arg Tyr Lys Ilu Val Gly
        ATT CAC CAA GAT ATT CTA CAG TTT CAA TTC CCT AAC AAA CAG AGG TAT AAA ATT GTT GGG

        101- 120
        Asn Ilu Pro Tyr Asn Ilu Ser Thr Asp Ilu Ilu Arg Lys Ilu Val Phe Asp Ser Ilu Ala
        AAT ATA CCT TAT AAC ATA AGT ACG GAT ATA ATA CGC AAA ATT GTT TTT GAT AGT ATA GCT
        Ser Ilu Pro Tyr His Leu Ser Thr Gln Ilu Ilu Lys Lys Val Val Phe Glu Ser His Ala
        AGT ATT CCT TAC CAT TTA AGC ACA CAA ATT ATT AAA AAA GTG GTT TTT GAA AGC CAT GCG

        121- 140
        Asn Glu Ilu Tyr Leu Ilu Val Glu Tyr Gly Phe Ala Lys Arg Leu Leu Asn Thr Lys Arg
        AAT GAG ATT TAT TTA ATC GTG GAA TAC GGG TTT GCT AAA AGA TTA TTA AAT ACA AAA CGC
        Ser Asp Ilu Tyr Leu Ilu Val Glu Glu Gly Phe Tyr Lys Arg Thr Leu Asp Ilu His Arg
        TCT GAC ATC TAT CTG ATT GTT GAA GGA GAA GGA TTC TAC AAG CGT ACC TTG GAT ATT CAC CGT

        141- 160
        Ser Leu Ala Leu Leu Leu Met Ala Glu Val Asp Ilu Ser Ilu Leu Ser Met Val Pro Arg
        TCA TTG GCA TTA CTT TTA ATG GCA GAA GTT GAT ATT TCT ATA TTA AGT ATG GTT CCA AGA
        Ser Leu Gly Leu Leu Leu His Thr Gln Val Ser Ilu Gln Gln Leu Leu Lys Leu Pro Ala
        TCA CTA GGG TTG CTC TTG CAC ACT CAA GTC TCG ATT CAG CAA TTG CTT AAG CTG CCA GCG

        161- 180
        Glu Tyr Phe His Pro Lys Pro Lys Val Asn Ser Ser Leu Ilu Arg Leu Ser Arg Lys Lys
        GAA TAT TTT CAT CCT AAA CCT AAA GTG AAT AGC TCA CTT ATC AGA TTA AGT AGA AAA AAA
        Glu Cys Phe His Pro Lys Pro Lys Val Ser Val Leu Ilu Lys Leu Thr Arg His Thr
        GAA TGC TTT CAT CCT AAA CCA AAA GTA AAC AGT GTC TTA ATA AAA CTT ACC CGC CAT ACC

        181- 200
        Ser Arg Ilu Ser His Lys Asp Lys Gln Lys Tyr Asn Tyr Phe Val Met Lys Trp Val Asn
        TCA AGA ATA TCA CAC AAA GAT AAA CAA AAG TAT AAT TAT TTC GTT ATG AAA TGG GTT AAC
        Thr Asp Val Pro Asp Lys Tyr Trp Lys Leu Tyr Thr Tyr Phe Val Ser Lys Trp Val Asn
        ACA GAT GTT CCA GAT AAA TAT TGG AAG CTA TAT ACG TAC TTT GTT TCA AAA TGG GTC AAT

        201- 220
        Lys Glu Tyr Lys Lys Ilu Phe Thr Lys Asn Gln Phe Asn Asn Ser Leu Lys His Ala Gly
        AAA GAA TAC AAG AAA ATA TTT ACA AAA AAT CAA TTT AAC AAT TCC TTA AAA CAT GCA GGA
        Arg Glu Tyr Arg Gln Leu Phe Thr Lys Asn Gln Phe His Gln Ala Met Lys His Ala Lys
        CGA GAA TAT CGT CAA CTG TTT ACT AAA AAT CAG TTT CAT CAA GCA ATG AAA CAC GCC AAA

        221- 240
        Ilu Asp Asp Leu Asn Asn Ilu Ser Phe Glu Gln Phe Leu Ser Leu Phe Asn Ser Tyr Lys
        ATT GAC GAT TTA AAC AAT ATT AGC TTT GAA CAA TTC TTA TCT CTT TTC AAT AGC TAT AAA
        Val Asn Asn Leu Ser Thr Val Thr Tyr Glu Gln Val Leu Ser Ilu Phe Asn Ser Tyr Leu
        GTA AAC AAT TTA AGT ACC GTT ACT TAT GAG CAA GTA TTG TCT ATT TTT AAT AGT TAT CTA

        241- 244
        Leu Phe Asn          Lys END
        TTA TTT AAT          AAG TAA
        Leu Phe Asn Gly Arg  Lys END
        TTA TTT AAC GGG AGG   AAA TAA
```

Figure 1 Comparison of the pE194 and pAM77 methylase structural gene and predicted amino acid sequences. The MLS methylases of pE194 and pAM77 are displayed in parallel, with boxed regions indicating regions of identity of the two polypeptide sequences.

99

transformation both express inducible MLS resistance. As noted previously, similar patterns of induction were observed for pE194 irrespective of whether the plasmid was in *S. aureus* or in *B. subtilis*; erythromycin appeared to have maximum inducing activity in both cases. In the case of *B. subtilis* minicells containing pAM77, we found that erythromycin, lincomycin, and streptogramin type B also induced resistance and synthesis of a 29K protein with specificity characteristic of the original *S. sanguis* background (Horinouchi et al. 1983). Thus, it appears that both the MLS resistance phenotype per se and the characteristic pattern of induction are determined by the plasmid and are presumably associated with the resistance control elements of the system. The structure and organization of the pE194 and pAM77 MLS control elements are described in further detail below in the context of their respective nucleotide sequences.

Transposable Elements Associated with Streptococcal MLS Resistance

A noteworthy feature of the inducible MLS determinant associated with Tn*917* in *S. faecalis* is the observation of Tomich and Clewell (1980) that erythromycin induces both MLS resistance and transposition. This raises the possibility (discussed below) of the coordinated regulation of expression of several genes, i.e., operons, for which MLS antibiotics can provide the inductive stimulus mediated by an attenuator-type mechanism.

Docherty et al. (1981) have reported the presence of inducible MLS resistance in several natural isolates of *Bacillus licheniformis*. One such determinant, from *B. licheniformis* 749, was subcloned, using pBD9 as cloning vector, and compared by DNA blot analysis with both pE194 and pAM77 with respect to sequence homology. Unlike the pE194 and pAM77 determinants, the *B. licheniformis* determinant migrates with an apparent molecular mass of 35,000 daltons. Since hybridization could not be demonstrated between the *B. licheniformis* MLS determinant and pE194 and pAM77 under the standard stringency conditions used, three distinct classes of MLS determinants were therefore defined. In view of the observed lack of hybridization between the pAM77 and pE194 MLS determinants, which appear to be clearly related as shown in Figure 1, caution should be exercised in defining distinct classes that may form part of a continuum of clearly interrelated structural gene sequences presumably derived from a common ancestral gene.

At this point, clinical MLS-resistant isolates of *Clostridium welchii* (Brefort et al. 1977), *Bacteroides fragilis* (Privitera et al. 1979), and *Corynebacterium diphtheriae* (Coyle et al. 1979) have also been reported. In studies of *Mycoplasma pneumoniae*, Niitu et al. (1974) have reported strains coresistant to erythromycin and lincomycin. Since streptogramin B-type antibiotics were not tested, MLS resistance in this medically important group of organisms remains presumptive.

MLS RESISTANCE IN *Streptomyces* SPECIES

Inducible and Constitutive MLS Resistance in *Streptomyces*

Inducible MLS resistance is widely distributed in the genus *Streptomyces*. The first suggestion of the presence of MLS resistance in a streptomycete was based on the report of Teraoka and Tanaka (1974) that ribosomes from *Streptomyces erythreus* (an organism used for industrial production of erythromycin) were resistant to both erythromycin and lincomycin. We therefore tested the resistance of intact cell suspensions of *S. erythreus* to MLS antibiotics (including vernamycin B) and found apparent constitutive resistance associated with the presence of m_2^6A (Graham and Weisblum 1979a). The adenine methylase predicted by these studies has been purified by Skinner and Cundliffe (1982), who used the enzyme preparation to methylate *Bacillus stearothermophilus* 23S rRNA. This 23S rRNA was reconstituted into active ribosomes, which were shown to be resistant to MLS antibiotics when tested using an in vitro synthesizing system.

In examining producers of other MLS antibiotics—notably *S. fradiae* and *S. cirratus*, which produce the macrolides tylosin and cirramycin, respectively, *S. lincolnensis*, which produces lincomycin, and *S. loidensis*, which produces vernamycin B—we noted that resistance was not expressed constitutively, as in the case of *S. erythreus*, and that m_2^6A could not be found in our 23S rRNA preparations except at low levels. Instead, we found monomethyladenine (m^6A) in *S. fradiae* and *S. cirratus*, whereas *S. lincolnensis* and *S. loidensis* contained neither m^6A nor m_2^6A. Since we had only seen m_2^6A in association with MLS resistance previously, these observations posed the question of the connection between m^6A and MLS resistance in the two macrolide producers and the significance of the absence of methylated adenine in the *S. lincolnensis* and *S. loidensis* samples. Before undertaking biochemical tests of adenine methylation in 23S rRNA, we examined several different *Streptomyces* species to test further for inducibility of MLS resistance.

The study of MLS resistance in *Streptomyces* presents us with an interesting challenge. Since *Streptomyces* synthesize antibiotics, and in particular MLS antibiotics, a study of induced MLS resistance in these organisms must take into account the possibility of self-induction by endogenously synthesized antibiotics. We were therefore faced with two problems: (1) to study induction of MLS resistance in a background with minimal possibility of endogenous induction; and (2) to study induction in *Streptomyces* that synthesize MLS antibiotics and, if possible, to correlate induction with the commitment to enter antibiotic production, a major redirection in the metabolic machinery of the producing organism.

Patterns of Inducibility

Since it would be difficult to prove that a given strain of *Streptomyces* does not produce any MLS antibiotics, we chose to work with a common strain not known to produce MLS antibiotics for our induction studies. One such strain, *S. viridochromogenes* NRRL 2860, was found in preliminary disk sensitivity tests to show tylosin-inducible resistance to carbomycin and erythromycin. Having established that tylosin had inducing activity in this strain, we determined eop (tested at a tylosin concentration of 25 μg/ml) as a function of tylosin concentration used for induction and as a function of induction time. From these studies we learned that exposure of a test culture to tylosin at a concentration optimum between 0.1 μg/ml and 0.5 μg/ml over a period of 1 hour sufficed to increase eop from less than 1% to approximately 100% (Fujisawa and Weisblum 1981). Concomitant with induction of MLS resistance in this organism, we noted the appearance of m^6A rather than m_2^6A. Moreover, we noted that the macrolide antibiotic carbomycin had low inducing activity in *S. viridochromogenes* manifested by a large clear inhibition zone and therefore used this antibiotic in an attempt to select constitutively resistant mutants. One such mutant that we analyzed for methylation showed the presence of m^6A in 23S rRNA from cells grown in the absence of added antibiotic. *S. viridochromogenes* thus presented us with two surprises: first, that tylosin rather than erythromycin could function effectively as an inducer; and second, that the induced modification of the ribosome resulted mainly in formation of m^6A rather than m_2^6A. A summary of some of the different methylation phenotypes found in *Streptomyces* is presented below.

S. *lincolnensis* NRRL 2936, a strain that synthesizes lincomycin, was also found to methylate adenine inducibly. In this case 23S rRNA lacks methylated adenine, and when lincomycin was tested for inducing activitiy it was found that addition of lincomycin to the growth medium (final concentration, 1 μg/ml) resulted in the appearance of m^6A in 23S rRNA. Likewise, selection of *S. lincolnensis* with maridomycin yielded a strain that, when tested, had m_2^6A in 23S rRNA without a requirement for induction by added antibiotic. Since *S. lincolnensis* is destined to make lincomycin late during its growth cycle, we have the task of resolving an apparent paradox in which *S. lincolnensis* requires lincomycin to become induced and requires induction before going into production. It is conceivable that *S. lincolnensis* initially makes subinhibitory inducing levels of antibiotic. Alternatively, synthesis of an endogeneous inducer could serve to initiate expression of MLS resistance, lincomycin production, and other late functions, in parallel rather than in series. Irrespective of the details of the process by which cells both become resistant and enter the productive phase, we speculated whether constitutively resistant mutants might have an advantage over their inducible parents that would permit them to start production earlier and possibly make higher levels of antibiotic. This was found to be the case. For example, a maridomycin-resistant mutant of *S. lincolnensis* was found to produce about fivefold as much lincomycin as its wild-type parent (M. Mayford and B. Weisblum, unpubl.).

The case of the macrolide producers presents additional complexity. Examination of 23S rRNA reveals the presence of methylated adenine suggestive of constitutive expression, whereas the same strains appear inducible when tested

Table 1   Patterns of adenine methylation in 23S rRNA of *Streptomyces*

Streptomyces	Strain	Methylation	Control	MLS antibiotic produced
erythreus	NRRL 2338	m_2^6A	constitutive	erythromycin
lincolnensis	NRRL 2936	m^6A	inducible	lincomycin
fradiae	NRRL 2702	$m^6A - m_2^6A$	(self?) inducible	tylosin
cirratus	ATCC 21731	m^6A	(self?) inducible	cirramycin
hygroscopicus	IFO 12995	$m^6A + m_2^6A$	(self?) inducible	maridomycin
diastaticus	NRRL 2560	m_2^6A	inducible	virginiamycin
viridochromogenes	NRRL 2860	m^6A	inducible	none known

by the disk method (Fujisawa and Weisblum 1981). In addition, *S. fradiae* and *S. hygroscopicus* are capable of modifying adenine in 23S rRNA to form both m^6A and m_2^6A. A possible interpretation of these observations would be that the macrolide producers synthesize endogenous inducers and that both mono-methylating and dimethylating enzymes are present in association with MLS resistance.

In *Streptomyces* spp. inducible MLS resistance can be seen in its most diverse and general form, summarized in Table 1. We have noted variations in modes of expression, variety of inducers, and patterns of rRNA methylation. Our findings in this group of organisms therefore prompt a redefinition of the MLS resistance phenotype to account for the experimental observations that (1) distinct subsets of MLS antibiotics, not found previously to induce, effectively induce resistance, and (2) that the induced ribosomal alteration can involve monomethylation or dimethylation of adenine. This broader description of the MLS phenotype subsumes all the variant phenotypes found to date.

A MODEL FOR TRANSLATIONAL ATTENUATION CONTROL OF MLS RESISTANCE

Structure and Function of the pE194 MLS Attenuator

Mechanisms of gene regulation at the level of mRNA can be subdivided into four general classes, according to whether they affect (1) the start of synthesis of the specific mRNA, (2) the completion of synthesis of the specific mRNA, (3) the utilization of specific mRNA, or (4) the stability of the mRNA. The term attenuation applied to gene regulation has largely been used to describe phenomena related to termination of transcription and its reversal under inducing conditions associated with mechanism 2 (Kasai 1974; Yanofsky 1981; Kolter and Yanofsky 1982). In view of structural and functional similarities between the regions that control amino acid biosynthesis (by arrested transcription) and MLS resistance (by arrested translation), the control mechanism for MLS resistance is referred to as translational attenuation.

Plasmid pE194 *Taq*I fragment A (1442 bp), capable of specifying inducible resistance, provided us with experimental material that could be sequenced conveniently (Horinouchi and Weisblum 1980). By subcloning this fragment into a ClaI site of pC194, both the MLS determinant and its control elements were functionally mapped. The control sequence itself was further localized in a region comprising about 10% of the length of fragment A by a second set of experimental constructions in which pE194 was cut at two sites using restriction endonucleases and religated using complementary fragments derived, respectively, from the inducible parent strain and from a constitutively resistant mutant. On the basis of the DNA sequence that we determined for pE194 *Taq*I fragment A and for 11 constitutively resistant mutants, it was possible to propose tentatively an explicit model for regulation of MLS resistance falling into category 3 above. This postulates that the methylase mRNA is synthesized in an inactive conformation and that the process of induction converts it to an active form (Figs. 2, 3, and 4). A similar model has also been proposed by Gryczan et al. (1980) and by Hahn et al. (1982), and a comparison of similarities and differences between the two models is presented in Figure 3.

According to the model we proposed, mRNA encoding the MLS methylase is synthesized constitutively in a completed but inactive form. The 5' end of this mRNA contains a set of four inverted complementary repeat sequences capable of assuming alternative conformations. The four sequences, designated 1, 2, 3, and 4, can associate by virtue of their complementarity as 1+2 and 3+4 or as 2+3. An additional set of inverted complementary repeat sequences, A and A', bearing no sequence similarity to 1, 2, 3, or 4, is also present. The numbering scheme used for our model was chosen to facilitate comparison between inducible MLS resistance and the tryptophan operon attenuator, discussed in further detail below.

We postulate further that the 1+2, 3+4 association pattern corresponds to the nascent inactive conformation of the control region. This nascent state of the messenger is inactive because the ribosome loading site for synthesis of the MLS methylase labeled SD-2 (Shine and Dalgarno 1974) is sequestered in the loop formed by the association of 3+4. Activation of the methylase mRNA is achieved by a conformational rearrangement of the inverted complementary repeat sequences, which results in association of 2+3, thereby unmasking the sequestered ribosome loading site. The conformational rearrangement of the control region results from hindered translation of a 19-amino-acid "leader" or "control" peptide encoded by sequences A and 1; if the ribosomes stall during translation of these sequences, sequence 2 is freed from its pairing with sequence 1. Stalling of the ribosome is brought about by erythromycin.

The model proposed by Gryczan et al. (1980) and by Hahn et al. (1982) postulates an essentially similar mechanism of SD-2 unmasking; the intermediate states of the regulatory region that they propose are summarized in Figure 3b. They postulate three conformations, labeled A, B, and C, of which B and C correspond to our conformations I and V, respectively. Conformation A, however, differs from the conformations that we propose and has the interesting property that it could form during transcription, leaving SD-2 transiently unmasked. In the absence of induction, A could equilibrate with inactive

```
Mbol         -35          SstI      -10    +1      SD-1              MetGlyIluPheSerIluPheValIluSe
GATCACTCATCATGTTCATATTTATCAGAGCTCGTGCTATAATTATACTAATTTTATAAGGAGGAAAAAATATGGGCATTTTTAGTATTTTTGTAATCAG
---------+---------+---------+---------+---------+---------+---------+---------+---------+---------+  100
CTAGTGAGTAGTACAAGTATAAATAGTCTCGAGCACGATATTAATATGATTAAAATATTCCTCCTTTTTTATACCCGTAAAAATCATAAAAACATTAGTC
                                                                                            (A)
rThrValHisTyrGlnProAsnLysLysEND          Hinf I                      SD-2         MetAsnGluL
CACAGTTCATTATCAACCAAACAAAAAATAAGTGGTTATAATGAATCGTTAATAAGCAAAATTCATATAACCAAATTAAAGAGGGTTATAATGAACGAGA
---------+---------+---------+--●-----+---------+--●●-----+---------+---------+--●+---------+---------+  200
GTGTCAAGTAATAGTTGGTTTGTTTTTATTCACCAATATTACTTAGCAATTATTCGTTTTAAGTATATTGGTTTAATTTCTCCCAATATTACTTGCTCT
     (1)                     (2)                   (3)                   (4)

ysAsnIluLysHisSerGlnAsnPheIluThrSerLysHisAsnIluAspIluIluMetThrAsnIluArgLeuAsnGluHisAspAsnIluPheGluIl
AAAATATAAAACACAGTCAAAACTTTATTACTTCAAAACATAATATAGATAAAATAATGACAAATATAAGATTAAATGAACATGATAATATCTTTGAAAT
---------+---------+---------+---------+---------+---------+---------+---------+---------+---------+  300
TTTTATATTTGTGTCAGTTTTGAAATAATGAAGTTTTGTATTATATCTATTTTATTACTGTTTATATTCTAATTTACTTGTACTATTATAGAAACTTTA
     (A')

uGlySerGlyLysGlyHisPheThrLeuGluLeuValLysArgCysAsnPheValThrAlaIluGluIluIluAspHisLysLeuCysLysThrThrGluAsn
         HaeIII
CGGCTCAGGAAAAGGCCATTTTACCCTTGAATTAGTAAAGAGGTGTAATTTCGTAACTGCCATTGAAATAGACCATAAATTATGCAAAACTACAGAAAAT
---------+---------+---------+---------+---------+---------+---------+---------+---------+---------+  400
GCCGAGTCCTTTTCCGGTAAAATGGGAACTTAATCATTTCTCCACATTAAAGCATTGACGGTAACTTTATCTGGTATTTAATACGTTTTGATGTCTTTTA

LysLeuValAspHisAspAsnPheGlnValLeuLysAspIluLeuGlnPheLysPheProLysAsnGlnSerTyrLysIluTyrGlyAsnIluProT
     BclI
AAACTTGTTGATCACGATAATTTCCAAGTTTTAAACAAGGATATATTGCAGTTTAAATTTCCTAAAAACCAATCCTATAAAATATATGGTAATATACCTT
---------+---------+---------+---------+---------+---------+---------+---------+---------+---------+  500
TTTGAACAACTAGTGCTATTAAAGGTTCAAAATTTGTTCCTATATAACGTCAAATTTAAAGGATTTTTGGTTAGGATATTTTATATACCATTATATGGAA

yrAsnIluSerThrAspIluIluIluArgLysIluValPheAspSerIluAlaAsnGluIluIluTyrLeuIluIluValGluTyrGlyPheAlaLysArgLeuLeuAs
ATAACATAAGTACGGATATAATACGCAAAATTGTTTTTGATAGTATAGCTAATGAGATTTATTTAATCGTGGAATACGGGTTTGCTAAAAGATTATTAAA
---------+---------+---------+---------+---------+---------+---------+---------+---------+---------+  600
TATTGTATTCATGCCTATATTATGCGTTTTAACAAAAACTATCATATCGATTACTCTAAATAAATTAGCACCTTATGCCCAAACGATTTTCTAATAATTT

nThrLysArgSerLeuAlaLeuLeuLeuMetAlaGluValAspIluSerIluLeuSerMetValProArgGluTyrPheHisProLysProLysValAsn
TACAAAACGCTCATTGGCATTACTTTTAATGGCAGAAGTTGATATTTCTATATTAAGTATGGTTCCAAGAGAATATTTTCATCCTAAACCTAAAGTGAAT
---------+---------+---------+---------+---------+---------+---------+---------+---------+---------+  700
ATGTTTTGCGAGTAACCGTAATGAAAATTACCGTCTTCAACTATAAAGATATAATTCATACCAAGGTTCTCTTATAAAAGTAGGATTTGGATTTCACTTA

SerSerLeuIluArgLeuSerArgLysLysSerArgIluSerHisLysAspLysGlnLysTyrAsnTyrPheValMetLysTrpValAsnLysGluTyrL
                                                                           HincII
AGCTCACTTATCAGATTAAGTAGAAAAAAATCAAGAATATCACACAAAGATAAACAAAAGTATAATTATTTCGTTATGAAATGGGTTAACAAAGAATACA
---------+---------+---------+---------+---------+---------+---------+---------+---------+---------+  800
TCGAGTGAATAGTCTAATTCATCTTTTTTTAGTTCTTATAGTGTGTTTCTATTTGTTTTCATATTAATAAAGCAATACTTTACCCAATTGTTTCTTATGT

ysLysIluPheThrLysAsnGlnPheAsnAsnSerLeuLysHisAlaGlyIluIluAspAspLeuAsnAsnIluSerPheGluGlnPheLeuSerLeuPheAs
AGAAAATATTTACAAAAAATCAATTTAACAATTCCTTAAAACATGCAGGAATTGACGATTTAAACAATATTAGCTTTGAACAATTCTTATCTCTTTTCAA
---------+---------+---------+---------+---------+---------+---------+---------+---------+---------+  900
TCTTTTATAAATGTTTTTTAGTTAAATTGTTAAGGAATTTTGTACGTCCTTAACTGCTAAATTTGTTATAATCGAAACTTGTTAAGAATAGAGAAAAGTT

                                                                      HinfI
nSerTyrLysLeuPheAsnLysEND                                          ClaITaqI
TAGCTATAAATTATTTAATAAGTAAGTTAAGGGATGCATAAACTGCATCCCTTAACTTGTTTTTCGTGTGCCTATTTTTTGTGAATCGAT
---------+---------+---------+----●+--------+---------+---------+---------+---------+---------+  990
ATCGATATTTAATAAATTATTCATTCAATTCCCTACGTATTTGACGTAGGGAATTGAACAAAAAGCACACGGATAAAAAACACTTAGCTA
```

Figure 2 The pE194 methylase promoter, attenuator, and structural gene sequence. The promoter-attenuator region of the pE194 MLS methylase is shown, indicating the −35, −10, and +1 positions associated with the promoter and the two Shine and Dalgarno sequences, SD-1 and SD-2, associated with the attenuator. The location of +1 has been tested and verified by the S1 mapping technique (M. Mayford, unpubl.). The inverted complementary repeat sequences associated with the attenuator are designated A, 1, 2, 3, 4, and A′, corresponding to the numbering scheme used in the text and in Figs. 3 and 4. (●) The nucleotides on which the inverted repeats center. An inverted complementary repeat sequence centering on nucleotide 941 is postulated to serve as transcription terminator for the methylase mRNA. Unique sites for restriction endonucleases SstI, HaeIII, BclI, and HincII are also shown. The nucleotides are numbered arbitrarily starting at the MboI site nearest the methylase promoter.

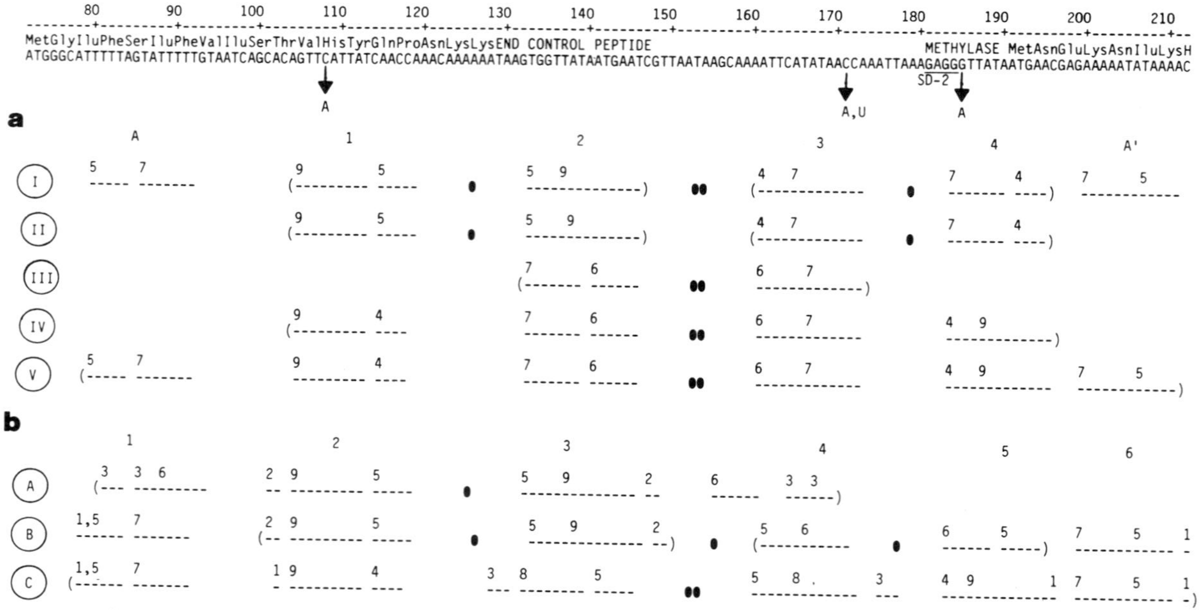

Figure 3 The pE194 methylase attenuator region. Alternative conformations at the 5′ end of the pE194 methylase mRNA are shown in relation to the control peptide and methylase structural gene sequences as proposed (*a*) by Horinouchi and Weisblum (1980, 1981) and (*b*) by Gryczan et al. (1980) and Hahn et al. (1982). Inverted complementary repeat sequences are postulated to associate around three centers of symmetry (●) located around positions 55, 82-83, and 108. Inverted complementary repeat sequences in the model of Horinouchi and Weisblum (1980, 1981) are labeled A, 1, 2, 3, 4, and A′, whereas the model proposed by Gryczan et al. (1980) and Hahn et al. (1982) uses the numbers 1–6 to designate these regions. (---) Nucleotides involved in pairing; numerals above the dashed lines indicate the number of clustered nucleotides, each in relation to their counterparts, similarly numbered on the opposite side of the center of symmetry. Conformations B and C (*b*) (Hahn et al. 1982) correspond to conformations I and V (*a*), respectively (Horinouchi and Weisblum 1980, 1981). Conformation A (*b*) is postulated to constitute an active nascent conformation, possibly responsible for basal level (uninduced level) of methylase synthesis by transiently stabilizing the 5′ end of the mRNA in a way that leaves SD-2, the Shine and Dalgarno sequence of the methylase, accessible to ribosomes. Locations of three point mutations to constitutive expression are also shown.

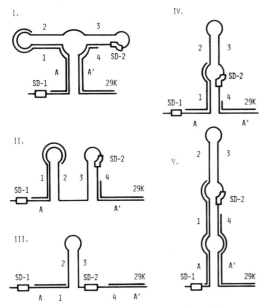

Figure 4   Potential conformations of pE194 attenuator region. Five possible conformations of the pE194 attenuator region are shown in schematic form. According to the
model, activation of the nascent methylase mRNA involves the transition I to II to III,
whereas deactivation following removal of the inductive stimulus involves the transition
III to IV to V, rather than III to II to I. We feel that the former is favored, since the reaction can go energetically downhill, whereas the latter is unlikely since it would
require an energy of activation to separate 2+3. Double lines over sequences A, 1, 4,
and A′ indicate ability to code for polypeptide. The model proposed by Gryczan et al.
(1980) and Hahn et al. (1982) postulates two (inactive) conformations, B and C, similar
to conformations I and V shown above. An active conformation, A (see Fig. 3b), is
similar to and could represent a precursor of II. If so, under noninducing conditions, we
might find the progression of conformations A to II to I; whereas under inducing
conditions, the progression of conformations A to III.

conformation II and ultimately with I. Prior to equilibration with conformation
II, an undetermined amount of transient basal level translation would be
expected. Alternatively, in the presence of inducer, we would expect that the
transition to (active) conformation III would occur preferentially.

   The unmasking of SD-2 can occur either indirectly or directly, i.e., either by
preemptive unmasking in which dissociation of 1+2 allows 2 to preempt 3,
which frees 4, or by direct dissociation of 3+4, which can result from
mutation in the sequence of 3+4, weakening their association and spontaneously unmasking SD-2 with a higher degree of probability than in the wild-
type inducible form. Basal level expression could occur by spontaneous
dissociation of 3+4 or without any dissociation but as an inherent feature of
conformation A (shown in Fig. 3b).

The translational attenuation model for regulation of MLS resistance resembles the transcriptional attenuation models for control of amino acid biosynthesis (for reviews, see Yanofsky 1981; Kolter and Yanofsky 1982), in that both involve hindered ribosome function in the course of synthesis of a control peptide, which in turn leads to a modification of the secondary structure of mRNA with predictable biological consequences.

We postulate that hindered translation of the MLS control peptide mediates the induction process for MLS resistance but with two important features distinguishing this process from induction of the tryptophan operon: (1) In the case of induced MLS resistance, ribosome function is hindered by the inducer, erythromycin, whereas in the case of the tryptophan operon, the ribosome is hindered by a deficiency of tryptophanyl tRNA (or other aminoacyl tRNAs in their respective operons); and (2) reorientation of the MLS control region results in unmasking the sequestered ribosome loading site for synthesis of the MLS methylase, whereas reorientation of the attenuator regions associated with biosynthesis of tryptophan, histidine, phenylalanine, threonine, and iso-leucine-valine results in completion of the respective mRNAs otherwise prematurely terminated by the association of 3+4 (or its functional equivalent), which in turn supplies the signal for termination of mRNA synthesis.

Critical Test of the Model

The proposed model predicts that mutation to constitutive expression of MLS resistance would be most effective for nucleotide changes in sequences 1, 3, and 4, but not in 2. Thus, nucleotide changes in 1 would weaken its association with 2 and favor preemptive association of 2 with 3. However, base alterations in 2, which weaken association with 1, would likewise weaken the association of 2 with 3, thereby reducing the efficiency of the 2+3 preemptive pairing reaction. Mutations that directly affect either 3 or 4 would reduce the energy of association of 3 with 4, bypassing the preemptive reaction.

In a test of this model, 11 independent constitutively resistant mutants were selected randomly, and the DNA sequence of the MLS control region was determined. In agreement with predictions, none of these 11 mutants involved sequence 2. Alterations in sequences 3 and 4 were found in nine of the mutants analyzed (five in sequence 2 and four in sequence 4), as shown in Figure 2.

Analysis of the remaining two mutants was particularly informative. One of these, 3 bases from the 5' end of sequence 1 (UUCAUUAUCAACCA), had a change from C to A. Insofar as this mutation would have a destabilizing effect on the pairing of the first 2 residues of sequence 1 (UU), as well as the third, we infer that a significant level of induction occurs when the control region is disrupted as far as the third of the 14 bases in sequence 1. We are still uncertain whether ribosome stall at some preferred codon in the control peptide is most effective in inducing MLS resistance. However, Steitz (1969) has reported that bound ribosomes were capable of protecting between 10 and 20 nucleotides at the 5' end of R17 RNA phage gene sequences from ribonuclease digestion. Thus, a ribosome stalled between the initiator codon of the control peptide and the first codon in sequence 1 (GUU, valine) should, in principle, be able to cover and, therefore, presumably disrupt the pairing of 1 + 2. We do not yet know the extent to which constitutive MLS resistance resulting from this point mutation is mediated by a change in the amino acid sequence of the putative control peptide, histidine (CAU) to asparagine (AAU), as distinct from the more direct impact on destabilization of 1+2. In the other constitutive mutant, 62 nucleotides that specify the control peptide (19 amino acids), including all of sequence 1, were excised leaving a control region containing intact sequences 2, 3, and 4. In the nascent methylase mRNA specified by this mutant, 2+3 would form as soon as sequence 3 was synthesized, resulting in a nascent fully active conformation of the message.

Deletion Analysis of the Control Region

A deletion of sequence 1, including the control peptide, was obtained as a spontaneous mutation to constitutive expression selected in *S. aureus* with clindamycin (Horinouchi and Weisblum 1981). Extending the deletion to include both sequences 1 and 2 should yield a mutant in which SD-2 remains trapped by the association of 3+4. With no apparent means for preemptive attack by 2 on 3+4, the model would predict that resistance would be partially repressed but inducible either by "spontaneous" dissociation of 3+4 or by a process in which utilization of SD-2 directly facilitated its further utilization. In such a mutant, basal expression of methylase synthesis owing to spontaneous dissociation of 3+4 would yield some low level of methylated ribosomes. Thus, sensitive ribosomes complexed with erythromycin and in the process of translating at the start of the methylase gene might contribute directly to dissociation of 3+4 in a way that would permit methylated ribosomes resulting from basal level expression to synthesize additional levels of methylase ("direct" or "translational" activation). Since the amino acids encoded at the amino end of the methylase differ from the amino acid sequence in the control region, we might expect this process of direct activation to have a different antibiotic concentration dependence. A possible manifestation of this phenomenon may be implicated in studies of inducible MLS resistance in *S. pyogenes* by Dixon and Lipinski (1974), who noted concentric rings of apparent alternating resistant growth and inhibition surrounding the lincomycin test disk. They termed this phenomenon "zonal resistance." Malke (1978) extended these studies with eop assays in which he noted relative maxima of growth in two lincomycin concentration ranges, around 0.06 μg/ml and greater than 100 μg/ml.

By cutting the pE194 control region with *Sst*I (*Sac*I) restriction endonuclease (see Fig. 2) and resecting with BAL 31 exonuclease, Hahn et al. (1982) obtained a series of mutants from which sequences 2 and 3 were successively removed. Deletion of sequence 1 yielded the constitutive phenotype as described above; deletion of sequences 1 and 2 yielded a partially repressed phenotype, and deletion of sequences 1, 2, and 3 yielded the constitutive phenotype.

How Can an Inhibitor of Protein Synthesis Induce by Inhibiting Translation?

Recalling the report of Pestka et al. (1976) that inhibitory and inducing activities in a set of erythromycin congeners were inseparable, it is pertinent to ask, Since induction requires an environment in which synthesis of the control peptide is inhibited, why is this same environment not equally inhibitory for synthesis of the methylase? To answer this question, we begin by assuming that ribosomes, during induction, fall under one of three descriptions: (1) sensitive ribosomes complexed with erythromycin, (2) sensitive ribosomes uncomplexed with erythromycin, and (3) resistant ribosomes containing specifically methylated 23S rRNA. Let us ask what might happen to ribosomes belonging to each of these classes during translation of the control peptide and of the methylase structural gene.

Sensitive ribosomes complexed with erythromycin would stall in the control region and induce conformational realignment. Eventually, they would either complete synthesis of the control peptide or release the partially completed control peptide. Sensitive ribosomes uncomplexed with erythromycin, as well as resistant ribosomes, would synthesize the control peptide without any marked effect on induction.

Following induction, sensitive ribosomes complexed with erythromycin would synthesize short peptides and release them. Sensitive ribosomes not complexed with erythromycin would synthesize methylase efficiently, as would resistant ribosomes. For induction to work effectively it is necessary that the concentration of erythromycin used should not saturate the ribosomes. This would require that subinhibitory concentrations be used for induction, in accord with experimental observation.

Erythromycin appears to inhibit protein synthesis in a manner consistent with its ability to act as inducer. Pestka (1974) and Tai et al. (1974) concluded that (1) inhibition by erythromycin is limited to an early stage of protein synthesis after initiation but before extensive elongation has occurred, and (2) inhibition results in polysome breakdown and release of short peptides (between 5 and 12 amino acids in length). Menninger and Otto (1982) have examined biochemical aspects of macrolide action, and they ascribe the inhibitory action of erythromycin, spiramycin, and carbomycin to their proposed ability to stimulate dissociation of peptidyl tRNA from ribosomes, rather than to direct inhibition of peptide bond formation. This type of effect would not only inhibit protein synthesis overall but would allow for synthesis of the methylase under inducing conditions, as described above.

Negative Feedback Components in the Regulation of MLS Resistance

Mechanisms of gene regulation that serve the cell optimally should contain negative feedback features that switch off gene expression when it is no longer required. The switch-off of gene expression can be in response to either reduction in the original inductive stimulus, or saturation of the cell with the products of induced synthesis. Optimal function of the latter mechanism should result in turning off gene expression even in the presence of continued inductive stimulation.

The translational attenuation model predicts that synthesis of methylase should eventually become self-limiting. Only sensitive ribosomes will stall in the control region, so when the level of resistant ribosomes reaches (an unspecified) critical level, the number of sensitive ribosomes remaining would not suffice to provide the amount of stall in sequence 1 needed to maintain a maximal level of methylase synthesis. Results that support this aspect of the model were obtained by Shivakumar et al. (1980), who reported that no induction was demonstrable in an oleandomycin-resistant *B. subtilis* carrying a chromosomal mutation that affects a ribosomal protein constituent of the 50S subunit. Moreover, these investigators showed that inactivation of the methylase structural gene by in vitro deletion of an undetermined small number of nucleotides yielded (in *B. subtilis*) an inactive methylase that appeared to be synthesized at an abnormally high rate following induction by erythromycin. This interesting phenomenon was ascribed to an intracellular maximal level of sensitive ribosomes resulting, in turn, in maximal expression of the induced phenotype.

Structure and Function of the pAM77 MLS Attenuator

Our results point to two types of functional similarity among the MLS antibiotics. First, only MLS antibiotics induce, and second, ribosomes that contain the specifically methylated adenine residue are resistant to MLS antibiotics exclusively. At this point, we feel that overlap of the ribosomal binding sites for different MLS antibiotics, rather than strict congruence at a single site, describes the interaction between MLS antibiotics and their receptor on the 50S ribosome subunit.

What connection is there, if any, between the amino acid sequence of the control peptide and the mission of the gene whose expression it controls? We can begin seeking an answer to this question by asking more generally, Can erythromycin switch on the tryptophan operon? Can histidine deficiency switch on MLS resistance? Part of the answer is supplied by studies of pAM77, a plasmid that specifies induced MLS resistance in *S. sanguis*.

The sequence of the putative pAM77 MLS methylase, including promoter and control region (shown in Figs. 5 and 6), has been reported by Horinouchi et al. (1983). Analysis of the DNA sequence shows the presence of an open

```
DdeI                          -35                  -10        +1
CTTAGAAGCAAACTTAAGAGTGTGTTGATAGTGCATTATCTTAAAATTTTGTATAATAGGAATTGAAGTTAAATTAGATGCTAAAAATTTGTAATTAAGA
---------+---------+---------+---------+---------+---------+---------+---------+---------+---------+    100
GAATCTTCGTTTGAATTCTCACACAACTATCACGTAATAGAATTTTAAAACATATTATCCTTAACTTCAATTTAATCTACGATTTTTAAACATTAATTCT

   SD-1 HinfI    MetLeuValPheGlnMetArgAsnValAspLysThrSerThrIluLeuLysGlnThrLysAsnSerAspTyrValAspLysTyrVa
AGGAGGGATTCGTCATGTGGTATTCCAAATGCGTAATGTAGATAAAACATCTACTATTTTGAAACAGACTAAAAACAGTGATTACGTAGATAAATACGT
---------+---------+---------+------●--+---------+---------+---------+---------+---------+---------+    200
TCCTCCCTAAGCAGTACAACCATAAGGTTTACGCATTACATCTATTTTGTAGATGATAAAACTTTGTCTGATTTTTGTCACTAATGCATCTATTTATGCA
                                                   (icrs I)
lArgLeuIluProThrSerAspEND
TAGATTAATTCCTACCAGTGACTAATCTTATGACTTTTTAAACAGATAACTAAAATTACAAACAAATCGTTTAACTTCTGTATTTGTTTATAGATGTATC
---------+----●----+---------+---------+---------+---------+---------+---------+---------+---------+    300
ATCTAATTAAGGATGGTCACTGATTAGAATACTGAAAAATTTGTCTATTGATTTTAATGTTTGTTTAGCAAATTGAAGACATAAACAAATATCTACATAG
                   (icrs 6)
                    MetLysLysAsnIluLysTyrSerGlnAsnPheLeuThrAsnGluLysValLeuAsnGlnIluIluLysGlnLeuAsnLeuLy
             SD-2                                              RsaI
ACTTCAGGAGTGATTACATGAAAAAAAATATAAAATATTCTCAAAACTTTTTAACGAATGAAAAGGTACTCAACCAAATAATAAAACAATTGAATTTAAA
-----●---+---------+---------+---------+---------+---------+---------+---------+---------+---------+    400
TGAAGTCCTCACTAATGTACTTTTTTTTATATTTTATAAGAGTTTTGAAAAATTGCTTACTTTTCCATGAGTTGGTTTATTATTTTGTTAACTTAAATTT
(icrs 14)
sGluThrAspThrValTyrGluIluGlyThrGlyLysGlyHisLeuThrThrLysLeuAlaLysIluSerLysLysGlnValThrSerIluGluLeuAspSer
AGAAACCGATACCGTTTACGAAATTGGAACAGGTAAAGGGCATTTAACGACGAAACTGGCTAAAATAAGTAAACAGGTAACGTCTATTGAATTAGACAGT
---------+---------+---------+---------+---------+---------+---------+---------+---------+---------+    500
TCTTTGGCTATGGCAAATGCTTTAACCTTGTCCATTTCCGTAAATTGCTGCTTTGACCGATTTTATTCATTTGTCCATTGCAGATAACTTAATCTGTCA

HisLeuPheAsnLeuSerSerGluLysLeuLysAsnIluArgValThrLeuIluHisGlnAspIluLeuGluGlnPheGlnPheProAsnLysGlnArgT
CATCTATTCAACTTATCTTCAGAAAAATTAAAACTGAACATTCGTGTCACTTTAATTCACCAAGATATTCTACAGTTTCAATTCCCTAACAAACAGAGGT
---------+---------+---------+---------+---------+---------+---------+---------+---------+---------+●   600
GTAGATAAGTTGAATAGAAGTCTTTTTAATTTTGACTTGTAAGCACAGTGAAATTAAGTGGTTCTATAAGATGTCAAAGTTAAGGGATTGTTTGTCTCCA

yrLysIluValGlySerIluProTyrHisLeuSerThrGlnIluIluIluLysLysValValPheGluSerHisAlaSerAspIluTyrLeuIluValGluGl
ATAAAATTGTTGGGAGTATTCCTTACCATTTAAGCACACAAATTATTAAAAAAGTGGTTTTTGAAAGCCATGCGTCTGACATCTATCTGATTGTTGAAGA
---------+---------+---------+---------+---------+---------+---------+---------+---------+---------+    700
TATTTTAACAACCCTCATAAGGAATGGTAAATTCGTGTGTTTAATAATTTTTTCACCAAAAACTTTCGGTACGCAGACTGTAGATAGACTAACAACTTCT

uGlyPheTyrLysArgThrLeuAspIluHisArgSerLeuGlyLeuLeuLeuHisThrGlnValSerIluGlnGlnLeuLeuLysLeuProAlaGluCys
          HinfI                                          TaqIHinfI              Fnu4HI    RsaI
AGGATTCTACAAGCGTACCTTGGATATTCACCGTTCACTAGGGTTGCTCTTGCACACTCAAGTCTCGATTCAGCAATTGCTTAAGCTGCCAGCGGAATGC
---------+---------+---------+---------+---------+---------+---------+---------+---------+---------+    800
TCCTAAGATGTTCGCATGGAACCTATAAGTGGCAAGTGATCCCAACGAGAACGTGTGAGTTCAGAGCTAAGTCGTTAACGAATTCGACGGTCGCCTTACG

PheHisProLysProLysValAsnSerValLeuIluLysLeuThrArgHisThrThrAspValProAspLysTyrTrpLysLeuThrThrTyrPheValS
                                                                                            RsaI
TTTCATCCTAAACCAAAGTAAACAGTGTCTTAATAAAACTTACCCGCCATACCACAGATGTTCCAGATAAATATTGGAAGCTATATACGTACTTTGTTT
---------+---------+---------+---------+---------+---------+---------+---------+---------+---------+    900
AAAGTAGGATTTGGTTTTCATTTGTCACAGAATTATTTTGAATGGGCGGTATGGTGTCTACAAGGTCTATTTATAACCTTCGATATATGCATGAAACAAA

erLysTrpValAsnArgGluTyrArgGlnLeuPheThrLysAsnGlnPheHisGlnAlaMetLysHisAlaLysValAsnAsnLeuSerThrValThrTy
        TaqI                                                              RsaI
CAAAATGGGTCAATCGAGAATATCGTCAACTGTTTACTAAAAATCAGTTTCATCAAGCAATGAAACACGCCAAAGTAAACAATTTAAGTACCGTTACTTA
---------+---------+---------+---------+---------+---------+---------+---------+---------+---------+    1000
GTTTTACCCAGTTAGCTCTTATAGCAGTTGACAAATGATTTTTAGTCAAAGTAGTTCGTTACTTTGTGCGGTTTCATTTGTTAAATTCATGCAATGAAT

rGluGlnValLeuSerIluPheAsnSerTyrLeuLeuPheAsnGlyArgLysEND     MetSerArgPheCysLysPheGlyLysLeuHisValThrLy
                                          SD-3             HinfI
TGAGCAAGTATTGTCTATTTTTAATAGTTATCTATTATTTAACGGGAGGAAATAATTCTATGAGTCGCTTTTGTAAATTTGGAAAGTTACACGTTACTAA
---------+---------+---------+---------+---------+---------+---------+---------+---------+---------+    1100
ACTCGTTCATAACAGATAAAAATTATCAATAGATAATAAATTGCCCTCCTTTATTAAGATACTCAGCGAAAACATTTAAACCTTTCAATGTGCAATGATT

sGlyAsnValAspLysLeuLeuGlyIluLeuLeuThrAlaSerLysLysLeuLysArgSerLeuAlaProThrGlyAsnLeuTyrArgEND
                                                   AvaII                        ClaI    RsaI
AGGGAATGTAGATAAATTATTAGGTATACTACTGACAGCTTTCCAAGAAGCTAAAGAGGTCCCTAGCGCCTACGGGGAATTTGTATCGATAAGGGGTACAA
---------+---------+---------+---------+---------+---------+---------+---------+---------●-+---------+    1200
TCCCTTACATCTATTTAATAATCCATATGATGACTGTCGAAGGTTCTTCGATTTCTCCAGGGATCGCGGATGCCCCTTAAACATAGCTATTCCCCATGTT

   DdeI     AvaII            DdeI                    RsaI
ATTCCCACTAAGCGCTCGGGACCCCTTGTAGGAAATTGTCCTAAGTGTGGCAACAATATTGGGGTACATGGAACATAGAAAACAAAGTGAAGTATCTTTC
---------+---------+---------+---------+---------+---------+---------+---------+---------●-+---------+    1300
TAAGGGTGATTCGCGAGCCCTGGGGAACATCCTTTAACAGGATTCACACCGTTGTTATAACCCCATGTACCTTGTATCTTTTGTTTCACTTCATAGAAAG

                                                   DdeI
TAGGGTAAATATACCACCCCCACCTTTAAGGCTCTTAAAATACGTTTTAGAGCCTTAGAAAACGACACAAAAAGCAAGAGCATTTTTGACCTTGCTTTT
---------+---------+---------+------+●●--+---------+---------+---------+---------+----●●---+---------+    1400
ATCCCATTTATATGGTGGGGGTGGAAATTCCGAGAATTTTATGCAAAATCTCGGAATCTTTTGCTGTGTTTTTTCGTTCTCGTAAAAACTGGAACGAAAA

TTTATTGTCGTGCAATCCGATACACTTTATACGAAGCTCTAAA
---------+---------+---------+---------+---   1443
AAATAACAGCACGTTAGGCTATGTGAAAATATGCTTCGAGATTT
```

Figure 5 (See facing page for legend.)

112

reading frame containing 245 amino acids, which corresponds to a protein with a calculated molecular mass of 29,400 daltons, assuming an average molecular mass of 120 daltons per amino acid. The putative promoter and control regions for the pAM77 methylase resemble those of pE194 in general outline but with notable differences.

Similarities between pAM77 and pE194. A sequence in the methylase mRNA capable of encoding a 36-amino-acid control peptide begins at nucleotide 115 preceded by its Shine and Dalgarno sequence GGAGG (SD-1). The putative methylase structural gene sequence begins at residue 318 preceded by its Shine and Dalgarno sequence GGAGT (SD-2). SD-2 falls within the loop region of the inverted complementary repeat sequence numbered 14 (Fig. 6). We postulate that induction unmasks SD-2 by ultimately disrupting or preventing the formation of the secondary structure of inverted complementary repeat sequence 14, permitting the synthesis of methylase. We note that both the pAM77 and pE194 control peptide structural gene sequences use GGAGG (as SD-1), the "strong" Shine and Dalgarno sequence proposed by McLaughlin et al. (1981) as that preferentially used by gram-positive bacteria for initiation of protein synthesis. We also note the additional parallel between pAM77 and pE194 methylase structural genes that use GGAGU and AGAGG, respectively. Both of these sequences would be expected to pair more weakly with 16S rRNA than would SD-1, which is GGAGG in both cases.

Differences between pAM77 and pE194. The control region of pAM77 differs from that of pE194 with respect to the (predicted) length of the control peptide and the degree of complexity of inverted complementary repeat sequences whose secondary structure determines the rate of methylase synthesis. The pAM77 control peptide sequence codes for 36 amino acids, whereas the corresponding pE194 sequence codes for 19. More remarkable, we can identi-

Figure 5 The pAM77 methylase promoter, attenuator, and structural gene sequence. The promoter-attenuator region of the pAM77 MLS methylase is shown indicating putative locations of the −35, −10, and +1 sequences, associated with the promoter, and the two Shine and Dalgarno sequences, SD-1 and SD-2, associated with the attenuator. The location of the +1 site is putative. Three of fourteen possible symmetric configurations (numbers 1, 6, and 14) that span the control region are shown. (The entire set of 14 inverted complementary repeat sequences that we postulate span this region are shown in Fig. 6.) A potential Shine and Dalgarno sequence, labeled SD-3 at nucleotide 1043, might serve as loading site for the synthesis of a putative 43-amino-acid peptide beginning at nucleotide 1060. This hypothetical peptide terminates in the center of an unpaired loop, at nucleotide 1190, a site that is also located at the center of an inverted complementary repeat sequence. This inverted repeat sequence is the nearest such sequence to the 3′ end of the methylase structural gene capable of serving as transcription terminator and would have the interesting property that hindered translation of the hypothetical peptide could serve to link induced MLS resistance with other functions located farther downstream by transcriptional attenuation control. Part of this downstream sequence was determined (nucleotides 1200–1443).

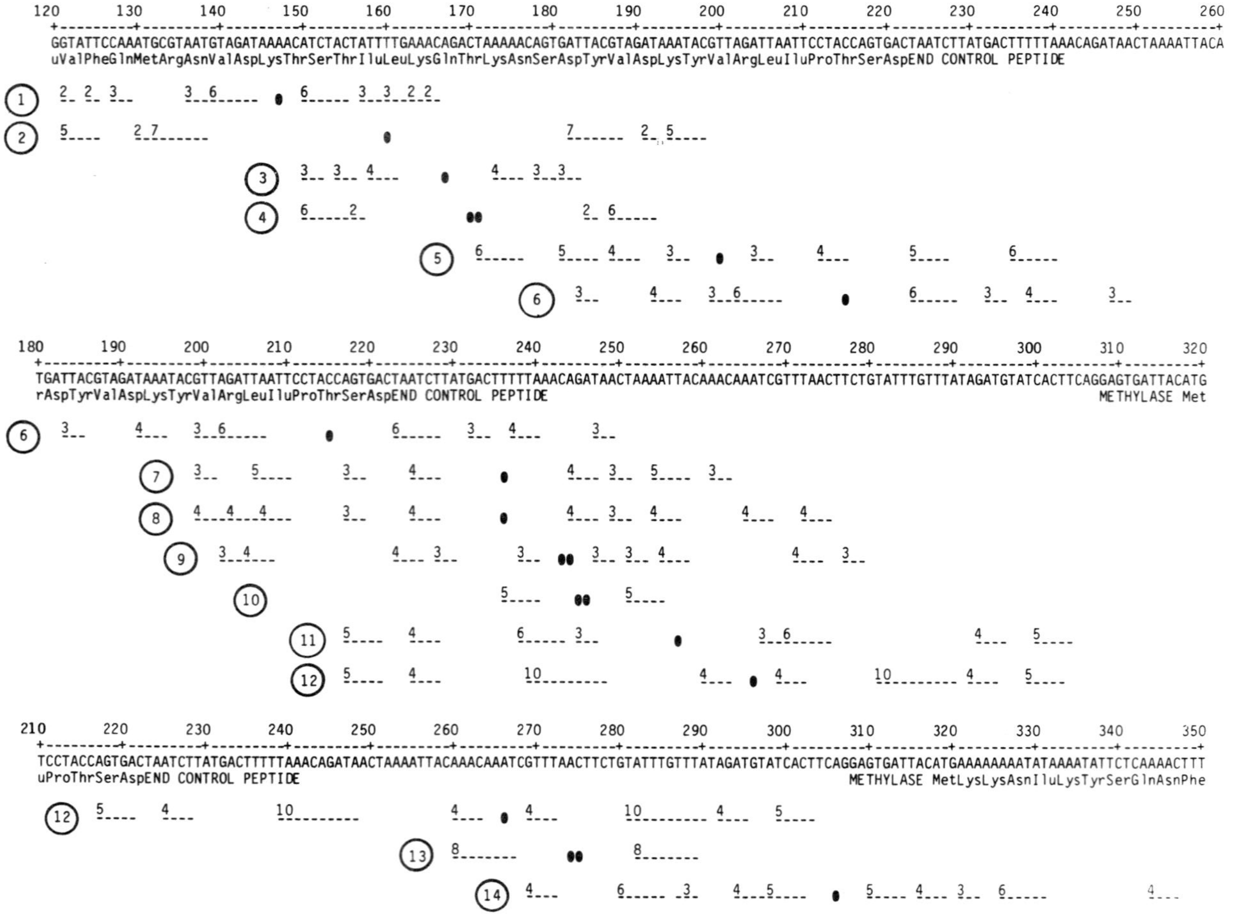

114

GGTATTCCAAATGCGTAATGTAGTAGATAAAACATCTACTATTTTGAAACAGACTAAAACAGTGATTACGTAGATAAATACGTTAGATAATTCCTACCAGTGACTAATTCTTATGACTTTTAAACAGATAACTAAAATTACA
uValPheIlnMetArgAsnValAspLysThrSerThrIluLeuLysGlnThrLysThrValAspSerAspThrLysSerTyrValAspLysLysTyrValArgLeulluProThrSerAspEND CONTROL PEPTIDE

TGATTACGTAGATAAATACGTTAGATAATTCCTACCAGTGACTAATTCTTATGACTTTTAAACAGATAACTAAAATTACAACAGATAACAAATGGTTTACTTCTGTATTTGTTTATAGATGTATCACTTCAGGAGTGATTACATG
rAspTyrValAspLysLysTyrValArgLeulluProThrSerAspEND CONTROL PEPTIDE
METHYLASE Met

TCCTACCAGTGACTAATTCTTATGACTTTTAAACAGATAACTAAAATTACAACAGATAACAAATGGTTTACTTCTGTATTTGTTTATAGATGTATCACTTCAGGAGGATTACTGAAAAAAATATAAAATATTCTCAAAACTTT
uProThrSerAspEND CONTROL PEPTIDE
METHYLASE MetLysLysAsnIluLysTyrSerGlnAsnPhe

Figure 6 (See facing page for legend.)

fy at least 13 successive centers of symmetry of inverted complementary repeat sequences in the pAM117 control regions as shown in Figure 5. These inverted complementary repeat sequences, collectively comprising 164 nucleotides, cover the coding region for the last 33 amino acids of the control peptide, as well as the coding sequence for the first 9 amino acids of the methylase. The 14 inverted complementary repeat sequences would enable the pAM77 control region to assume a correspondingly larger number of alternative paired conformational states than the pE194 control region. The corresponding pE194 region, in contrast, contains only three clearly defined centers of symmetry, which in turn could support six different conformational states falling into two classes (as discussed above and shown in Fig. 3). In addition, the inverted complementary repeat sequences in pE194 are 12−14 nucleotides in length, each perfectly paired except for a single nucleotide bulge.

The pAM77 inverted complementary repeat sequences all appear highly interrupted. We feel that this feature, together with the twofold greater number of codons at which ribosomes can potentially stall in the pAM77 control region, may account for the greater number of MLS antibiotics capable of induction in this system. We propose for the pAM77 attenuator that the inducing antibiotic can cause ribosomes to stall over the entire control peptide coding sequence (residues 115−222) during transcription of the pAM77 methylase mRNA. This stall effectively traps inverted complementary repeat sequences 1−6 in their entirety, as well as the 5′ arms of inverted complementary repeat sequences 7, 8, and 9. To the extent that inverted complementary repeat sequences 11, 12, and 13 form associated hairpin structures, inverted complementary repeat sequence 14 is prevented from forming, leaving SD-2 accessible for formation of initiation complexes with ribosomes and synthesis of methylase.

The large number of alternative conformations that the pAM77 control region is capable of assuming would tend to diminish the significance with which any single conformation, destabilized by mutation, can effectively result in constitutive expression. Consistent with this is the experimental finding that we have not been able to apply the conventional techniques successfully used in *S. aureus* to select spontaneous mutants of pAM77 constitutively resistant to MLS antibiotics, namely selection using MLS antibiotics that do not induce in the particular system. This temporarily precludes critical tests of the pAM77 attenuator model by analysis of spontaneous mutants similar to that described above for the pE194 attenuator.

Figure 6   The pAM77 methylase attenuator region. Fourteen alternative conformations at the 5′ end of the pAM77 methylase mRNA, together with their respective symmetric centers, are shown in relation to the control peptide and methylase structural gene sequences as proposed by Horinouchi et al. (1983). Inverted complementary repeat sequences are postulated to associate around the 13 symmetric centers (●). The region is subdivided into three sequential overlapping sections to facilitate presentation of the data and the model on a single page (see text for details).

Possible Host Contributions to the Specificity of Induction

In our study of pAM77 we show that two different MLS plasmids from different backgrounds with different induction specificities, introduced into *B. subtilis*, confer an induction phenotype that resembles that of the host strain in which the plasmid resided when first isolated. From this observation we infer that induction specificity is a function encoded by the plasmid DNA and that host contributions, if any, are general rather than specific.

Malke and Holm (1981), Barany et al. (1982), and Hardy and Haefeli (1982) have reported that MLS resistance is expressed in *Escherichia coli*. In their examination of induction specificity, Hardy and Haefeli noted that in *E. coli*, clindamycin appears to induce more effectively than erythromycin by the disk assay. The same determinant in two different host backgrounds can also express two different phenotypes. We feel that this latter observation may be related to systematic differences between gram-positive and gram-negative organisms in initiation of translation (McLaughlin et al. 1981).

Differences between the pE194 and pAM77 MLS Determinants at Their 3′ Ends

The two methylase structural genes show significant differences at their 3′ termini as well. The pAM77 methylase structural gene terminates at residue 1055. At residue 1045 we find a potential Shine and Dalgarno sequence, GGAGG (SD-3), preceding the sequence of 129 nucleotides beginning at 1055 capable of encoding a 43-amino-acid peptide. The hypothetical carboxyterminal peptide coding sequence terminates in an inverted complementary repeat sequence that we speculate could serve as a transcription terminator. If so, the rate of SD-3-related peptide synthesis could serve as a regulatory link to a function specified by a gene farther downstream. In contrast, the pE194 methylase structural gene appears to terminate at the beginning of an inverted complementary repeat sequence that resembles ρ-independent transcription terminators found in other systems (Rosenberg and Court 1980; Siebenlist et al. 1980).

SUMMARY

In his early paper describing the phenomenon that we have subsequently recognized as inducible MLS resistance, Garrod (1957) stated, "This is not an antagonism in the ordinary sense, but a kind of interaction of which, so far as I am aware, no previous example has been seen." Indeed, from a mechanistic point of view, posttranscriptional modification of the antibiotic receptor is relatively unique. However, as we have searched for additional examples of induced MLS resistance, we have learned that the phenotype is so pervasive in certain organisms (e.g., *Streptomyces*) that its absence in this group seems to be the exception.

Our understanding of the molecular mechanism of inducible MLS resistance is directly applicable to a wide range of medical, technological, as well as speculative questions. The work reviewed above helps us to understand how a *Staphylococcus* found resistant to erythromycin, but sensitive to other MLS antibiotics, can easily become constitutively resistant as a result of a single nucleotide change in the control region. On the basis of reports of Desmyter and Reybrouck (1964) and of Watanakunakorn (1976), there is ample evidence that this type of change has in fact occurred. Also provocative is the report of Tomich and Clewell (1980) that both transposition and MLS resistance are induced by erythromycin in *Streptococcus*. This finding opens the possibility that inducers of MLS resistance may switch on a wide range of inducible genes in addition to rRNA methylase.

Studies of induced MLS resistance in *Streptomyces*, particularly *S. lincolnensis*, have given us a possible clue and a model system with which to begin to unravel the series of adaptational events that occur when an antibiotic-producing organism goes into the producing phase of its life cycle. This organism has an inducible rRNA methylase, not expressed under conditions of active growth, for which lincomycin was shown to act as inducer (Fujisawa and Weisblum 1981). In the concluding remarks of his review of transcriptional attenuators that regulate amino acid biosynthesis, Yanofsky (1981) offered the suggestion that attenuation (in contrast to repression) mechanisms serve to link aminoacyl tRNA (and therefore reduce specific growth rate), rather than just amino acid levels, to expression of specific genes. This is precisely the type of metabolic linkage that *Streptomyces* might find advantageous during the stationary phase of their life cycle characterized by reduced growth rate and switchover to synthesis of antibiotics.

These considerations open up a wide range of considerations, including the possible role of MLS antibiotics as endogenous inducers acting on several regulons simultaneously, mediated by a corresponding number of attenuators all sharing the feature that they are required for stationary-phase differentiation. Such regulons could be distributed throughout the genome, either under transcriptional or translational control (or both), and would share the property that they respond to ribosome stall signals. In such systems the amplitude of response for any single unit could be finely modulated by the amino acid sequence of its associated control peptide.

ACKNOWLEDGMENTS

The work reviewed in this paper is based on the collaboration of many colleagues to whom I express my gratitude. Don Clewell generously provided *B. subtilis* transformed with pAM77, which allowed us to extend our molecular studies of the attenuator. I also thank the National Science Foundation, the National Institutes of Health, the Upjohn Company, and Eli Lilly and Company for support of this research.

REFERENCES

Allen, N.E. 1977. Macrolide resistance in *Staphylococcus aureus*: Inducers of macrolide resistance. *Antimicrob. Agents Chemother.* **11:** 669.

Barany, F., J.D. Boecke, and A. Tomasz. 1982. Staphylococcal plasmids that replicate and express erythromycin resistance in both *Streptococcus pneumoniae* and *Escherichia coli*. *Proc. Natl. Acad. Sci.* **79:** 2291.

Barber, M. and P.M. Waterworth. 1964. Antibacterial activity of lincomycin and pristinamycin: A comparison with erythromycin. *Br. Med. J.* **2:** 603.

Bourse, R, and J. Monier. 1968. Effet de l'erythromycin sur la croissance de *Staph. aureus* "resistant dissocie" en bacteriostase par un autre macrolide ou un antibiotique apparente. *Ann. Inst. Pasteur* **90:** 67.

Brefort, G., M. Magot, H. Ionesco, and M. Seybald. 1977. Characterization and transferability of *Clostridium perfringens* plasmids. *Plasmid* **1:** 32.

Chabbert, Y. 1956. Antagonisme in vitro entre l'erythromycine et la spiramycine. *Ann. Inst. Pasteur* **90:** 787.

Clewell, D.B. 1981. Plasmids, drug resistance, and gene transfer in the genus *Streptococcus*. *Microbiol. Rev.* **45:** 409.

Clewell, D.B., Y. Yagi, G.M. Dunny, and S.K. Schultz. 1974. Characterization of three plasmid deoxyribonucleic acid molecules in a strain of *Streptococcus faecalis*: Identification of a plasmid determining erythromycin resistance. *J. Bacteriol.* **117:** 283.

Coyle, M.B., B.H. Minshew, J.A. Bland, and P.C. Hsu. 1979. Erythromycin and clindamycin resistance in *Corynebacterium diphtheriae* from skin lesions. *Antimicrob. Agents Chemother.* **16:** 525.

Desmyter, J. and G. Reybrouck. 1964. Lincomycin sensitivity of erythromycin resistant staphylococci. *Chemotherapia* **9:** 183.

DeVriese, L.A. 1976. In vitro susceptibility and resistance of animal staphylococci to macrolide antibiotics and related compounds. *Ann. Rech. Vet.* **7:** 65.

Dixon, J. and A. Lipinski. 1974. Infections with beta-hemolytic streptococcus resistant to lincomycin and erythromycin and observations on zonal-pattern resistance to lincomycin. *J. Infect. Dis.* **130:** 351.

Docherty, A., G. Grandi, R. Grandi, T.J. Gryczan, A.G. Shivakumar, and D. Dubnau. 1981. Naturally occurring macrolide-lincosamide-streptogramin B resistance in *Bacillus licheniformis*. *J. Bacteriol.* **145:** 129.

Ehrlich, S.D. 1977. Replication and expression of plasmids from *Staphylococcus aureus* in *Bacillus subtilis*. *Proc. Natl. Acad. Sci.* **74:** 1680.

Engel, H.W., N. Soedirman, J.A. Rost, W.J. VanLeeuen, and J.D.A. VanEmbden. 1980. Transferability of macrolide, lincomycin, and streptogramin resistances between groups A, B, and D streptococci, *Streptococcus pneumoniae*, and *Staphylococcus aureus*. *J. Bacteriol.* **142:** 407.

Fujisawa, Y., and B. Weisblum. 1981. A family of r-determinants in Streptomyces spp. that specifies inducible resistance to macrolide, lincosamide, and streptogramin type B antibiotics. *J. Bacteriol.* **148:** 621.

Garrod, L.P. 1957. The erythromycin group of antibiotics. *Br. Med. J.* **2:** 57.

Gibson, E.M., N.M. Chace, S.B. London, and J. London. 1979. Transfer of plasmid-mediated antibiotic resistance from streptococci to lactobacilli. *J. Bacteriol.* **137:** 614.

Goldmann, S.F., and F. Heiss. 1971. Untersuchungen zum Phaenomen von Rezistenz und Rezistenzkopplung gegenueber Erythromycin, Lincomycin, und Staphylomycin Z. *Med. Mikrobiol. Immunobiol.* **156:** 168.

Graham, M.Y. and B. Weisblum. 1979a. 23S ribosomal ribonucleic acid of macrolide-producing streptomycetes contains methylated adenine. *J. Bacteriol.* **137:** 1464.

——————— . 1979b. Altered methylation of adenine in 23S ribosomal RNA associated with erythromycin resistance in *Streptomyces erythreus* and *Streptococcus fecalis*. *Contrib. Microbiol. Immunol.* **6:** 159.

Griffith, L.J., W.E. Ostrander, C.G. Mullins, and D.E. Beswick. 1965. Drug antagonism between lincomycin and erythromycin. *Science* **147:** 746.

Gryczan, T.J. and D. Dubnau. 1978. Construction and properties of chimeric plasmids in *Bacillus subtilis*. *Proc. Natl. Acad. Sci.* **75:** 1428.

Gryczan, T.J., G. Grandi, J. Hahn, and D. Dubnau. 1980. Conformational alteration of mRNA structure and the posttranscriptional regulation of erythromycin-induced drug resistance. *Nucleic Acids Res.* **8:** 6081.

Hahn, J., G. Grandi, T.J. Gryczan, and D. Dubnau. 1982. Translational attenuation of *ermC*: A deletion analysis. *Mol. Gen. Genet.* **186:** 204.

Hardy, K. and C. Haefeli. 1982. Expression in *Escherichia coli* of staphylococcal gene for resistance to macrolide, lincosamide, and streptogramin type B antibiotics. *J. Bacteriol.* **152:** 524.

Horinouchi, S. and B. Weisblum. 1980. Posttranscriptional modification of messenger RNA conformation: Mechanism of erythromycin inducible resistance. *Proc. Natl. Acad. Sci.* **77:** 7079.

—————. 1981. The control region for erythromycin resistance: Free energy changes related to induction and mutation to constitutive expression. *Mol. Gen. Genet.* **182:** 341.

—————. 1982a. Nucleotide sequence and map of pE194, a plasmid that specifies inducible resistance to macrolide, lincosamide, and streptogramin type B antibiotics. *J. Bacteriol.* **150:** 804.

—————. 1982b. Nucleotide sequence and map of pC194, a plasmid that specifies inducible resistance to chloramphenicol. *J. Bacteriol.* **150:** 815.

Horinouchi, S., W.-H. Byeon, and B. Weisblum. 1983. A complex attenuator regulates inducible resistance to macrolides, lincosamides, and streptogramin type B antibiotics in *Streptococcus sanguis*. *J. Bacteriol.* **154:** 1252.

Horodniceanu, T., L. Bougueleret, and G. Bieth. 1981. Conjugative transfer of multiple antibiotic resistance in beta hemolytic groups A, B, F, and G streptococci in the absence of extrachromosomal deoxyribose nucleic acid. *Plasmid* **5:** 127.

Hyder, S.L. and M.M. Streitfeld. 1973. Inducible and constitutive resistance to macrolide antibiotics and lincomycin in clinically isolated strains of *Streptococcus pyogenes*. *Antimicrob. Agents Chemother.* **4:** 327.

Iordanescu, S. 1976. Three distinct plasmids originating in the same *Staphylococcus aureus* strain. *Arch. Roum. Pathol. Exp. Microbiol.* **35:** 111.

Iordanescu, S. and M. Surdeanu. 1980. New compatibility groups for *Staphylococcus aureus* plasmids. *Plasmid* **4:** 256.

Jones, W.F., Jr., R.L. Nichols, and M. Finland. 1956. Development of resistance and cross-resistance in vitro to erythromycin, carbomycin, oleandomycin, and streptogramin. *Proc. Soc. Exp. Biol. Med.* **93:** 388.

Kasai, T. 1974. Regulation of the expression of the histidine operon in *Salmonella typhimurium*. *Nature* **249:** 523.

Kolter, R. and C. Yanofsky. 1982. Attenuation in amino acid biosynthetic operons. *Annu. Rev. Genet.* **16:** 113.

Kono, M., H. Hashimoto, and S. Mitsuhashi. 1966. Drug resistance of staphylococci. III. Resistance to some macrolide antibiotics and inducible system. *Jpn. J. Microbiol.* **10:** 59.

Lai, C.-J. and B. Weisblum. 1971. Altered methylation of ribosomal RNA in an erythromycin-resistant strain of *Staphylococcus aureus*. *Proc. Natl. Acad. Sci.* **68:** 856.

Lai, C.-J., J.E. Dahlberg, and B. Weisblum. 1973a. Structure of an inducibly methylatable nucleotide sequence in 23S ribosomal ribonucleic acid from erythromycin-resistant *Staphylococcus aureus*. *Biochemistry* **12:** 457.

Lai, C.-J., B. Weisblum, S.R. Fahnstock, and M. Nomura. 1973b. Alteration of 23S ribosomal RNA and erythromycin-induced resistance to lincomycin and spiramycin in *Staphylococcus aureus*. *J. Mol. Biol.* **74:** 67.

LeBlanc, D.J., J.J. Hawley, L.N. Lee, and E.J. St. Martin. 1978. "Conjugal" transfer of plasmid DNA among oral streptococci. *Proc. Natl. Acad. Sci.* **75:** 3484.

Malke, H. 1974. Genetics of resistance to macrolide antibiotics and lincomycin in natural isolates of *Streptococcus pyogenes*. *Mol. Gen. Genet.* **135:** 349.

——————. 1978. Zonal-pattern resistance to lincomycin in *Streptococcus pyogenes*: Genetic and physical studies. In *Microbiology—1978* (ed. D. Schlessinger), p. 142. American Society for Microbiology, Washington, D.C.

——————. 1979. Conjugal transfer of plasmids determining resistance to macrolides, lincosamides, and streptogramin-B type antibiotics among group A, B, D, and H streptococci. *FEMS Microbiol. Lett.* **5:** 335.

Malke, H. and S.E. Holm. 1981. Expression of streptococcal plasmid-determined resistance to erythromycin and lincomycin in *Escherichia coli*. *Mol. Gen. Genet.* **184:** 283.

Malke, H., W. Reichardt, M. Hartmann, and F. Walter. 1981. Genetic study of plasmid-associated zonal resistance to lincomycin in *Streptococcus pyogenes*. *Antimicrob. Agents Chemother.* **19:** 91.

McLaughlin, J.R., C.L. Murray, and J.C. Rabinowitz. 1981. Unique features in the ribosome binding site sequence of the gram positive *Staphylococcus aureus* gene. *J. Biol. Chem.* **256:** 11283.

Menninger, J.R. and D.P. Otto. 1982. Erythromycin, carbomycin, and spiramycin inhibit protein synthesis by stimulating the dissociation of peptidyl-tRNA from ribosomes. *Antimicrob. Agents Chemother.* **21:** 811.

Nakajima, Y., M. Inoue, Y. Oka, and S. Yamagishi. 1968. A mode of resistance to macrolide antibiotics in *Staphylococcus aureus*. *Jpn. J. Microbiol.* **12:** 248.

Niitu, Y., S. Hasegawa, and H. Kubota. 1974. In vitro development of resistance to erythromycin, other macrolide antibiotics, and lincomycin in *Mycoplasma pneumoniae*. *Antimicrob. Agents Chemother.* **5:** 513.

Parisi, J.T., J. Robbins, B.C. Lampson, and D.W. Hecht. 1981. Characterization of a macrolide, lincosamide, and streptogramin resistance plasmid in *Staphylococcus epidermidis*. *J. Bacteriol.* **148:** 559.

Pestka, S. 1974. Antibiotics as probes of ribosome structure: Binding of chloramphenicol and erythromycin to polyribosomes: Effect of other antibiotics. *Antimicrob. Agents Chemother.* **5:** 255.

Pestka, S., R. Vince, R. LeMahieu, F. Weiss, L. Fern, and J. Unowsky. 1976. Induction of erythromycin resistance by erythromycin derivatives. *Antimicrob. Agents Chemother.* **9:** 128.

Privitera, G., A. Dublanchet, and M. Sebald. 1979. Transfer of multiple antibiotic resistance between subspecies of *Bacteroides fragilis*. *J. Infect. Dis.* **139:** 97.

Rosenberg, M. and D. Court. 1980. Regulatory sequences involved in the promotion and termination of RNA transcription. *Annu. Rev. Genet.* **13:** 319.

Saito, T. and S. Mitsuhashi. 1980. Methylated adenine in the 23S ribosomal RNA of *Staphylococcus aureus* carrying temperature-inducible resistance to macrolide antibiotics. *Microbiol. Immunol.* **23:** 1127.

Saito, T., H. Hashimoto, and S. Mitsuhashi. 1969. Drug resistance of staphylococci. Decrease in the formation of erythromycin-ribosomes complex in erythromycin-resistant strains. *Jpn. J. Microbiol.* **13:** 119.

——————. 1970. Macrolide resistance in *Staphylococcus aureus*. Isolation of a mutant in which leucomycin is an active inducer. *Jpn. J. Microbiol.* **14:** 473.

Sanders, E., M.T. Foster, and D. Scott. 1968. Group A beta-hemolytic streptococci resistant to erythromycin and lincomycin. *N. Engl. J. Med.* **278:** 538.

Shine, J. and L. Dalgarno. 1974. The 3'-terminal sequence of *Escherichia coli* ribosomal RNA: Complementary to nonsense triplets and ribosome binding sites. *Proc. Natl. Acad. Sci.* **71:** 342.

Shivakumar, A.G. and D. Dubnau. 1981. Characterization of a plasmid-specified ribosome methylase associated with macrolide resistance. *Nucleic Acids Res.* **9:** 2549.

Shivakumar, A.G., J. Hahn, and D. Dubnau. 1979. Studies on the synthesis of plasmid-coded proteins and their control in *Bacillus subtilis* minicells. *Plasmid* **2:** 279.

Shivakumar, A.G., J. Hahn, G. Grandi, Y. Kozlov, and D. Dubnau. 1980. Posttranscriptional regulation of an erythromycin resistance protein specified by plasmid pE194. *Proc. Natl. Acad. Sci.* **77:** 3903.

Siebenlist, U., R.B. Simpson, and W. Gilbert. 1980. *E. coli* RNA polymerase interacts homologously with two different promoters. *Cell* **20:** 269.

Skinner, R.H. and E. Cundliffe. 1982. Dimethylation of adenine and resistance of *Streptomyces erythreus* to erythromycin. *J. Gen. Microbiol.* **128:** 2411.

Steitz, J.A. 1969. Polypeptide chain initiation: Nucleoside sequences of the three ribosomal binding sites in bacteriophage R17 RNA. *Nature* **224:** 957.

Tai, P.C., B.J. Wallace, and B.D. Davis. 1974. Selective action of erythromycin on initiating ribosomes. *Biochemistry* **13:** 4653.

Tanaka, T. and B. Weisblum. 1974. Mutant of *Staphylococcus aureus* with lincomycin- and carbomycin-inducible resistance to erythromycin. *Antimicrob. Agents Chemother.* **5:** 538.

Teraoka, H. and K. Tanaka. 1974. Properties of ribosomes from *Streptomyces erythreus* and *Streptomyces griseus*. *J. Bacteriol.* **120:** 316.

Tomich, P.K. and D.B. Clewell. 1980. Properties of erythromycin-inducible transposon Tn*917* in *Streptococcus faecalis*. *J. Bacteriol.* **141:** 1366.

Watanakunakorn, C. 1976. Clindamycin therapy of *Staphylococcus aureus* endocarditis. Clinical relapse and development of resistance to clindamycin, lincomycin, and erythromycin. *Am. J. Med.* **60:** 419.

Weaver, J.R. and P.A. Pattee. 1964. Inducible resistance to erythromycin in *Staphylococcus aureus*. *J. Bacteriol.* **88:** 574.

Weisblum, B. 1975. Altered methylation of ribosomal ribonucleic acid in erythromycin-resistant *Staphylococcus aureus*. In *Microbiology—1974* (ed. D. Schlessinger), p. 199. American Society for Microbiology, Washington, D.C.

Weisblum, B. and V. Demohn. 1969. Erythromycin-inducible resistance in *Staphylococcus aureus*: Survey of antibiotic classes involved. *J. Bacteriol.* **98:** 447.

Weisblum, B., S.B. Holder, and S.M. Halling. 1979a. Deoxyribonucleic acid sequence common to staphylococcal and streptococcal plasmids which specify erythromycin resistance. *J. Bacteriol.* **138:** 990.

Weisblum, B., M.Y. Graham, T. Gryczan, and D. Dubnau. 1979b. Plasmid copy number control: Isolation and characterization of high-copy-number mutants of plasmid pE194. *J. Bacteriol.* **137:** 635.

Weisblum, B., C. Siddhikol, C.-J. Lai, and V. Demohn. 1971. Erythromycin-inducible resistance in *Staphylococcus aureus*: Requirements for induction. *J. Bacteriol.* **106:** 835.

Yagi, Y., A.E. Franke, and D.B. Clewell. 1975. Plasmid determined resistance to erythromycin: Comparison of strains of *Streptococcus fecalis* and *Streptococcus pyogenes* with regard to plasmid homology and resistance inducibility. *Antimicrob. Agents Chemother.* **7:** 871.

Yagi, Y., T.S. McClellan, W.A. Frez, and D.B. Clewell. 1978. Characterization of a small plasmid determining resistance to erythromycin, lincomycin, and vernamycin B alpha in a strain of *Streptococcus sanguis* isolated from dental plaque. *Antimicrob. Agents Chemother.* **13:** 884.

Yanofsky, C. 1981. Attenuation in the control of expression of bacterial operons. *Nature* **289:** 751.

Role of RNA Polymerase, ρ Factor, and Ribosomes in Transcription Termination

Terry Platt
Department of Molecular Biophysics and Biochemistry
Yale University
New Haven, Connecticut 06510

David G. Bear
Department of Cell Biology
University of New Mexico School of Medicine
Albuquerque, New Mexico 87131

INTRODUCTION

Efficient transcription of the *Escherichia coli* genome depends on the ability of the cell to regulate the type and amount of mRNA produced from each of its 3000 to 4000 genes. This not only requires the information to specify precise starting points for transcription, but stopping points as well. The operon, as the functional polycistronic unit of gene expression, may in fact be defined by the occurrence of an efficient termination site at its distal end. This simple housekeeping function prevents transcriptional readthrough by RNA polymerase into adjoining regions of the genome. In addition, however, a termination site can be utilized within a regulatory region to alter gene expression by controlling the ability of RNA polymerase to transcribe beyond that site.

Thus, sites that specify termination may have several purposes in the cell. (1) As efficient punctuation signals, they permit the differential regulation of adjacent gene clusters. (2) As modulating elements (e.g., in attenuation), they permit differential control of expression within operons. (3) As conditional abortive elements (e.g., in mutational polarity), they prevent wasteful depletion of cellular metabolites. (4) As barriers to elongation, they minimize sequestering of important enzymes (e.g., RNA polymerase or other RNA-binding macromolecules) that might be engaging in unnecessary transcription or binding to excess unusable products.

It has been generally expected that all termination sites, regardless of their function in regulation, would contain certain common structural features required for the termination response of RNA polymerase at that site. The primary structures (sequences) of a large number of regions specifying termination of transcription have now been analyzed, and some general similarities are apparent, as is discussed below. In many cases, secondary structure in the RNA also plays an important role. However, despite our knowledge of

123

termination sequences and our belief that the DNA sequence must ultimately encode the termination signal(s), deciphering the functional attributes of termination regions is proving to be a complex task.

Numerous parameters impinge on the termination response, including not only the nucleotide sequence but the character of the transcribing RNA polymerase complex, the presence of secondary structure in the RNA transcript, and the requirement for (or participation of) additional protein factors. In turn, these parameters permit control of the termination event in many different ways and permit an integrated response to the overall needs of the cell. In this paper, we attempt to provide an overview of transcription termination with particular emphasis on what is known about the signals involved in specifying termination and the interactions among the macromolecules that govern its efficiency and specificity.

DISCUSSION

Signal for Termination

Early observations revealed that many bacterial transcripts have a series of uridine residues at the 3′ end, preceded by a GC-rich region of self-complementarity (Adhya and Gottesman 1978). A consistent story of how these features probably function in termination gradually emerged, confirming most aspects of the initial hypotheses (Rosenberg and Court 1979; Farnham and Platt 1980; Platt 1981). More recently, this view has been complicated by the discovery and characterization of numerous termination regions that cannot be accommodated by the simple model. In this section we compare and contrast different classes of termination sites at the structural level, and in later sections we focus on more recent developments regarding their functional requirements and organization.

To focus on the features of termination sites that are moderately well understood, we have divided them into two admittedly arbitrary categories, on the basis of their behavior in vitro. Even this classification must be considered loosely, because the response at termination sites in vitro may be profoundly different from the response in vivo, complicating interpretations and predictions of their behavior. The adoption of this pragmatic approach is necessitated primarily by our limited understanding of termination mechanisms.

Simple Terminators

We have defined a termination site as "simple" if it can function efficiently in vitro and does not require factors other than RNA polymerase. All known simple terminators consist of a GC-rich segment of dyad symmetry centered about 20 nucleotides before the stop site, which is followed by a region encoding 4−8 uridine residues located near the 3′ terminus of the transcript (Fig. 1). Sites having this structure appear to catalyze efficient termination in vitro, whereas no sites lacking the hairpin or uridine-encoding regions have yet

125

a   trpC(RI)   b   trpC(Bam)   c   trp t

```
                              UCC
                  A U         U   G
      A U         G   C       G-C
      A   U       G-C         A-U
      G-C         C-G         C-G
    _ G-C _     _ C-G _       C-G
      G-C         G-C         G-C
      G-C         G-C         C-G
      U-A         U-A         C-G
      C-GUUUU     C-GUUUU     G-CAUUUU
```

rpo+   0%   15%   25%

rpo203   20%   30%   45%

d   trp a1419   e   trp a135   f   trp a

```
      AAU             AAU             AAU
      U   G           U   G           U   G
      C   A           C   A           C   A
      C-G             C-G             C-G
      G-C             G-C             G-C
      C-G             C-G             C-G
      C-G             C-G             C-G
      C-G             C-G             C-G
      G-C             G-C             G-C
      A-UUUU(GCAA)    A-U             A-UUUUUUUU
                      C-GUUUUUU
```

rpo+   3%   65%   95%

rpo203   35%   80%   98%

Figure 1   Simple terminators. The sequence and presumed secondary structure of transcripts corresponding to some termination sites derived from the tryptophan operon. (*a,b*) Sites made synthetically; (----) contributions from linker sequences inserted during terminator construction. The efficiency of wild-type (*rpo⁺*) and the hyperterminating *rpoB* 203 polymerase (Guarente and Beckwith 1978) are given as molar percentage. (Data from Christie et al. 1981.)

been shown to do so. Models for termination based on these common features propose that (1) the dyad symmetry is responsible for the formation of intramolecular hairpin structures in the RNA transcript (Adhya and Gottesman 1978; Rosenberg and Court 1979; Farnham and Platt 1980), and (2) the uridines facilitate dissociation of the transcript from the template (Farnham and Platt 1980; Martin and Tinoco 1980), since RNA-DNA hybrids with rU-dA pairing are exceptionally unstable (Chamberlin 1965; Riley et al. 1966).

This model is supported by observations that point mutations in the inverted repeat usually affect termination: Changes that strengthen the potential RNA hairpin encoded by the dyad symmetry increase termination, whereas those that weaken it decrease termination (Rosenberg and Court 1979; Zurawski and Yanofsky 1980). Moreover, mutations that reduce the number of terminal uridines also weaken the termination response (Bertrand et al. 1977; Zurawski and Yanofsky 1980). Additional evidence from studies in vitro with ribonucleotide analogs supports these indications that the transcript itself carries at least part of the signal for termination: At simple sites, ITP reduces termination, iodo-CTP enhances termination, and BrUTP enhances readthrough transcription (Lee and Yanofsky 1977; Neff and Chamberlin 1978; Adhya et al. 1979; Farnham and Platt 1980).

Studies with the *trp* operon attenuator region as a model termination system have demonstrated that the dramatic effects of transcription termination seen with base analog incorporation into the RNA do not occur when similar analogs are incorporated into the DNA (Farnham and Platt 1982). These observations showed that only analog incorporation that affects transcript-template pairing in the rU-dA region elicits a significant alteration in the termination response. This finding further supports the model described above, i.e., RNA-DNA rather than DNA-DNA interactions are the important ones in this region. Transcription experiments with hybrid *trp* templates carrying single base mismatches in the attenuator region provide additional evidence that DNA-DNA and DNA-RNA interactions in the region of dyad symmetry do not contribute to the termination event (Ryan and Chamberlin 1983).

Several functions for the RNA hairpin are possible. It might serve to (1) reduce the extent of the transcript-template interaction, which would otherwise be particularly unfavorable for RNA release in the GC-rich region; (2) provide a structurally recognizable feature that causes RNA polymerase to pause; (3) introduce a regulatory pivot, where slight changes in the rate of hairpin formation may have dramatic effects on the extent of termination; and (4) stabilize the completed transcript from degradation past a certain point by 3' exonucleases. These functions are not necessarily mutually exclusive and may all participate to some extent. Direct biochemical support for these ideas has been provided by the construction and characterization of synthetic termination sites and analysis of several naturally occurring termination sites (Christie et al. 1981).

One unusual finding is the occurrence of terminators that have a run of adenines preceding the GC-rich region, which will provide a symmetric counterpart to the uridine-encoding region. Such terminators would be predicted to function in both directions and include the *leu*, *thr*, and *ilvB* attenuators (Keller and Calvo 1979; Friden et al. 1982; Gardner 1982), the ribosomal RNA (rRNA) terminators (Young 1979; Brosius et al. 1981), the *ompA* terminator (Pirtle et al. 1980), the *atp* operon terminator (Saraste et al. 1981), and the termination site of the ρ gene (Pinkham and Platt 1983). Some of these termination sites have been shown to function in both orientations, in vitro and in vivo (W.M. Holmes et al.; J.L. Pinkham et al.; both in prep.). Thus, the transcript produced by a polymerase molecule coming from either direction will incorporate adenines on the proximal side, which can potentially base-pair with the uridines immediately following the GC-rich hairpin. The particular contributions of this additional interaction to termination have not been determined, though a potential use of bidirectional termination sites for the cell could be to provide important genes with a protective barrier to convergent transcription from distally placed promoters.

Complex Terminators

By definition, our second category of termination sites encompasses all those not included above. Such sites (in general) lack either dyad symmetry or uridine-encoding segments, and extensive sequence or structural homologies are not obvious. The salient characteristic of these termination sites (beyond unusual AT-richness) is the frequent dependence of RNA polymerase on additional protein factors for termination in vitro. Table 1 summarizes the properties of several such proteins; the best studied of the *E. coli* factors are ρ (Roberts 1969) and NusA protein, also known as L factor (Kung et al. 1975; Greenblatt et al. 1980; Greenblatt and Li 1981b), both of which will be discussed later in detail. An excellent summary of all the *nus*-gene products is provided by Friedman et al. (1983).

The three best-characterized termination sites of this class are λ t_{R1} (Roberts 1969; Rosenberg et al. 1978; Court et al. 1980; Lau et al. 1982; Morgan et al. 1983a), the *tyrT* terminator (Küpper et al. 1978; Rossi et al. 1981), and the *trp* operon terminator *trp t'* (Wu et al. 1981; Sharp and Platt 1984). The nucleotide sequences of these regions are shown in Figure 2, along with those of several other sites. Though some sequence homologies exist, such as the CAATCAA common to λ t_{R1} and *tyrT* (see Rossi et al. 1981) and the similar ATCAACAA in λ t_{R0} (Calva and Burgess 1980), this homology has not been preserved among other sequences (Lau et al. 1982; Morgan et al. 1983a). Some other possibilities also exist for homology near the termination site (see Platt et al. 1983).

The problems involved in attempting to understand the functional aspects of these terminators are illustrated by three examples. The λ 4S termination site has GC-rich dyad symmetry, followed by a uridine-encoding region, and is 80% efficient in the minimal transcription system. It would therefore normally be classified among the simple terminators. However, since the inclusion of ρ in the reaction increases termination efficiency to 100% (Howard et al. 1977), it is not clear whether or not the 4S site should be considered a ρ-dependent one. Similarly, the first *trp* operon terminator (*trp t*) has a classic simple terminator structure, but it is only 25% functional in vitro and does not respond to ρ under normal transcription conditions (Wu et al. 1981). However,

Table 1 Protein factors affecting transcription termination

Protein	Monomer size (kD)	Activities/requirements	Possible function	Synonomous genes and names
ρ	46	RNA-dependent NTPase	site-specific RNA release factor	SuA, NusD, NitA, Psu
NusA	69	binds core polymerase	termination fidelity factor	L factor
NusB	16	?	antitermination factor	GroNB
NusE	12	?	translational coupling factor (?)	ribosomal protein S10
λN	11	binds to NusA; also requires NusB and NusE	antitermination factor	
λO	23	requires NusA	antitermination factor	

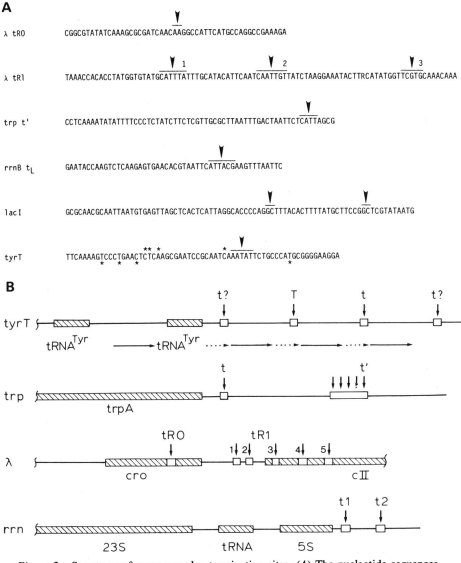

Figure 2 Structures of some complex termination sites. (*A*) The nucleotide sequences are shown for λ t_{R0} (Calva and Burgess 1980), the λ t_{R1} cluster (Rosenberg et al. 1978; Lau et al. 1982; Morgan et al. 1983a), the repeated regions of *tyrT* (Küpper et al. 1978; Rossi et al. 1981), the *trp t'* terminator (Wu et al. 1981), the site early in *rrnB* (Kingston and Chamberlin 1981), and the *lacI* terminator (Horowitz and Platt 1982; Cone et al. 1983). Arrows and overbars indicate observed regions of termination. For *tyrT*, the sequence shown is the second repeat and the major functional terminator. (*) Nucleotides that differ in the first or third repeat unit (see *B*; Rossi et al. 1981). (*B*) Schematic representations of some tandemly occurring, repeated, and clustered termination sites (see text for discussion).

it can be pushed to nearly 100% efficiency by using an appropriate combination of nucleotide concentrations, ρ, and NusA protein (Farnham et al. 1982). Finally, the sequence at the end of the *lacI* gene contains neither a GC-rich dyad symmetry nor a run of T residues but, in fact, encodes a termination signal that is 20−25% efficient in a minimal transcription system and does not respond to ρ and NusA protein added singly or in combination (Horowitz and Platt 1982). This termination is 80−90% efficient in assays in both *rho*[+] and several *rho*[−] strains in vivo, and the presence of bound *lac* repressor protein is not required for the termination event (Cone et al. 1983).

Since examination of sequences has not revealed common features of primary or secondary structure among these complex terminators, perhaps the relevant regions have not been compared. Increasing circumstantial evidence (Bektesh and Richardson 1980) suggests that perhaps (at least in the case of ρ-dependent termination) the crucial recognition regions may lie far upstream from the point of termination (see below). Moreover, as discussed by von Hippel et al. (1982), in the case of autogeneous regulation of the T4 gene-*32* protein, "specific" sequences are not necessarily required for recognition, and it may be that the termination sites discussed here have in common the "absence" of particular features (specific nucleotides or secondary structures). The characteristics distinguishing these terminators from DNA that does not specify termination have not been determined.

Although, as with simple terminators, important functional signals for termination appear to reside in the RNA (Sharp and Platt 1984), at present further generalizations cannot be drawn. The apparent variety and complexity of these multifactor interactions raise the challenge of considering new explanations that may eventually lead to an increased understanding of these phenomena.

Higher Order Structure

An intriguing aspect of several termination regions is the occurrence of multiple tandem termination sites, revealing modular organization at a higher level (Fig. 2B). The individual sites in these clusters may differ in strength or factor requirements. Examples include the λ t_{R1} terminator (Rosenberg et al. 1978; Lau et al. 1982; Morgan et al. 1983a,b), the tryptophan operon terminators in *E. coli* (Wu et al. 1981) and *Salmonella typhimurium* (Nichols et al. 1981), the tRNA[Tyr] terminators (Küpper et al. 1978; Rossi et al. 1981), and the rRNA (*rrn*) operons (Brosius et al. 1981; J. Li and C. Squires; in prep.). Termination in some of these regions, such as λ t_{R1} and *trp t'*, may be heterogeneous due to imperfect fidelity at one site, which can be improved by changing conditions or adding other factors (Platt 1981; Lau et al. 1982). In other cases, the independence of sites duplicated or repeated in tandem arrays must reflect other requirements not as yet understood.

Terminating the Transcript—A Dynamic Process

Termination must involve at least three conceptually discrete processes: cessation of transcription, release of the completed RNA molecule, and dissociation of the RNA polymerase complex. Though little is known about either the extent to which these are concerted events or the order in which they occur, recent studies have illuminated several aspects of the mechanism.

At the center of the process is the RNA polymerase molecule, and considerable evidence indicates that mutations in this enzyme can influence the extent of termination. For example, among revertants that can suppress defects in ρ, Guarente and Beckwith (1978) identified a mutation in RNA polymerase, $rpo203$, that restores efficient termination in the several different rho^- strains in which it was tested. This polymerase is believed to be deficient in its ability to stabilize the base-pairing between the uridines at the 3' end of the transcript and the DNA template and has been found to increase termination at several ρ-independent terminators (Christie et al. 1981). Interactions with ρ are not impaired, as judged by the enzyme's ability to terminate with normal efficiency at the ρ-dependent $trp\ t'$ site (T. Platt, unpubl.).

Other polymerase mutations, in which the ability to restore termination is specific for a particular rho allele, have also been isolated (Das et al. 1978; Guarente 1979). Two mutations, $nitB$ and $rif501$, which lie in the $rpoB$ gene (encoding the RNA polymerase β subunit) reduce termination at the $\lambda\ t_{L1}$ site or in the p_R operon (Inoko and Imai 1976; Lecocq and Dambly 1976). An extensive analysis of the effects of rifampin-resistant mutants on transcription termination at the trp attenuator has revealed RNA polymerase mutants that can either increase or decrease the efficiency of termination (Yanofsky and Horn 1981). Das et al. (1978) have found similar results with the termination site within the IS2 sequence. These findings indicate that the β subunit of RNA polymerase participates in the recognition of transcription termination signals and suggest that $rif^R rpoB$ mutations affect expression, at least in the trp operon, only under conditions where a termination structure is expected to form. Other rif^R mutations have been shown to affect simultaneously both termination and transcriptional pausing, suggesting that one region of the polymerase is probably involved in both events (Fisher and Yanofsky 1983).

Elongation Complex and NusA Protein

Until recently it was thought that after the initiation of transcription and the dissociation of σ factor, the remaining core polymerase molecule was the only macromolecule participating in elongation of the polynucleotide chain. Elegant experiments by Greenblatt and Li (1981a) demonstrated that the NusA protein (originally identified as a host factor required for the antitermination activity of the bacteriophage λN protein by Friedman [1971]) binds tightly to core RNA polymerase, but not to the holoenzyme, and can be displaced stoichiometrically by σ. These observations suggested that NusA might play an integral role in the elongation complex, possibly replacing σ just after initiation and serving to mediate the interactions of both termination and antitermination factors with RNA polymerase as a "fidelity factor" (Greenblatt 1981; see Fig. 3). This interpretation is consistent with experiments in coupled systems, showing that NusA is required for efficient synthesis of β-galactosidase, NADH dehydrogenase, and the β and β' subunits of polymerase (Kung et al. 1975; Greenblatt et al. 1980; Kung 1983).

NusA protein has been shown to be required for more efficient termination of transcription in vitro on λ DNA at the t_{R2} site (Greenblatt et al. 1981), in the leader region of the *rrnB* operon (Kingston and Chamberlin 1981), at the *trp t'* terminator (Platt 1981), and at the *trp* attentuator in vivo (Ward and Gottesman 1981). In addition, the combination of ρ and NusA proteins enhances termination at both terminators at the end of the *trp* operon (Farnham et al. 1982). At other sites, NusA appears to have an antitermination effect (see Friedman 1971; Greenblatt 1981; Greenblatt et al. 1981; Ward and Gottesman 1982).

Pauses during Transcription

It is now evident that RNA polymerase does not elongate RNA chains at a constant rate, and it has been found to hesitate at specific sites in vitro. The common feature of many of these sites is a significant stretch of dyad symmetry, permitting hairpin formation in the transcript. At the λ t_{R1} terminator, polymerase pauses in the absence of ρ at the site where it would normally cease elongation, before continuing readthrough polymerization (Rosenberg et al. 1978). RNA polymerase has also been found to hesitate at other ρ-

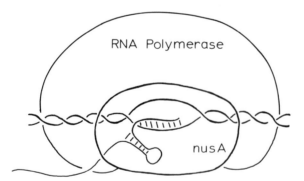

Figure 3   The elongation complex. RNA polymerase is depicted bound to the double-stranded DNA template, covering a region of 40–60 bp, with a bubble of about 17 bp (Gamper and Hearst 1982). The nascent transcript is thought to maintain hydrogen bonding across 6–10 nucleotides, and this diagram portrays the situation just as polymerase has arrived at a ρ-independent termination site. The hairpin is proposed to interact with the polymerase or the accompanying *nusA* protein, converting the elongation complex into a paused configuration.

dependent termination sites in the absence of ρ (Adhya and Gottesman 1978; Kassavetis and Chamberlin 1981; Morgan et al. 1983b). Sites other than ρ-dependent terminators can elicit a pause from RNA polymerase (Farnham and Platt 1981; Winkler and Yanofsky 1981; Gardner 1982). At two mutant *trp* attenuator sites that are ρ-independent (*trp a*135 and *trp a*1419; see Fig. 1), termination is incomplete or absent in vitro, but a substantial pause is observed; likewise, a pause occurs at *trp t*, the termination site at the end of the *trp* operon (Farnham and Platt 1981).

Hairpins other than those normally involved in termination can also elicit a pause just past a region of dyad symmetry. These include the early (1:2) hairpin in the *trp* leader region (Farnham and Platt 1981; Winkler and Yanofsky 1981), the analogous hairpin in the *thr* leader region (Gardner 1982), and several sites in the *c*II region of the λ genome (Morgan et al. 1983b). As the hairpin is strengthened, through analog incorporation into the RNA or by increasing the length of the dyad symmetry in the DNA, the pause increases. This suggests that the strength of the RNA-RNA interactions in the hairpin may control the time the polymerase remains paused.

A more puzzling situation exists with some complex termination sites. A substantial pause is observed at the *lacI* terminator site in vitro (Horowitz and Platt 1982), and similar experiments suggest pausing at the *trp t'* site (in the absence of ρ factor) as well (P.J. Farnham and T. Platt, unpubl.). In neither of these cases is there a thermodynamically significant potential hairpin in the transcript, and the particular signals that cause RNA polymerase to hesitate remain unknown.

Pausing has been postulated to be advantageous in the attenuation response, in slowing the polymerase molecule and permitting the translating ribosome to remain close behind (Winkler and Yanofsky 1981; Yanofsky 1981). In one instance, the *his* operon leader region (Fig. 4A) displays an even more extravagant structure than the leader regions of other biosynthetic operons. Only the first and last of the four potential structures would be expected to be required for the attenuation response (see below, Role of the Ribosome). Whether the intervening two hairpins serve as additional latent pausing sites to couple translation to transcription, as a means to open the region to other cellular controls or to mediate termination at the attenuator site in some other way, remains to be answered. The *rrn* termination region (Fig. 4B) also displays the potential for tandem structures, possibly reflecting special requirements for terminating rRNA transcription (Brosius et al. 1980). The necessity for organization in these repeated units is not obvious and may be related to factors as yet not understood, such as a backup function, RNA processing requirements, or retroregulation (see below).

If pausing by RNA polymerase is an obligatory prelude to termination, the length of time that a polymerase hesitates at a potential termination site may be directly related to the probability that termination (rather than readthrough transcription) will occur. To the extent that recognition of an RNA hairpin is

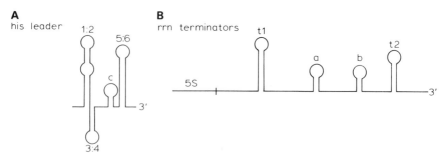

Figure 4 Tandemly repeated hairpins and termination. (*A*) The leader region of *his* operon mRNA. Hairpin 5:6 formation is required for termination, and competitive base-pairing is provided by formation of 2:3 and 4:5 (Johnston et al. 1980). These leader-region hairpins have been proposed to elicit pausing by RNA polymerase to insure that the ribosome remains close behind on the nascent transcript and to permit the coupling between transcription and translation that leads to the correct attenuation response (see Kolter and Yanofsky 1982). (*B*) A schematic of the termination region of the *rrn* operons of *E. coli*. The major site appears to be t_1, but t_2 has some function if t_1 is absent. The orientation of the region relative to the direction of the transcription also seems to be important, though t_1 would be predicted to function equally well in both orientations. The contribution of the a and b hairpins is unknown.

crucial to termination, the rate of formation of the hairpin, relative to the forward rate of RNA polymerase movement, is likely to be critical. Though it is difficult to separate cause from effect, factors reducing elongation rate seem to increase pausing. For instance, when ppGpp is included in vitro, the overall elongation rate of T7 RNA chains is significantly lowered, due to enhanced pausing of the transcriptional elongation complex at specific sites on the T7 DNA template (Kingston et al. 1981). In addition, ppGpp reduces the overall elongation rate and increases pausing on the leader region of the *rrnB* operon in vitro (Kingston and Chamberlin 1981). Transcription at low UTP concentrations also reduces the rate of elongation, and this has a striking effect on increasing termination at the *trp t* site (Farnham et al. 1982). Unexpectedly, this response is not only nucleotide-specific, but terminator-specific. The unique features of *trp t* responsible for these effects are not known.

The NusA protein also has an effect on pausing (Greenblatt et al. 1981; Kassavetis and Chamberlin 1981). NusA is required for termination at t_L in the *rrnB* operon, and its presence increases pausing at other sites in the same region (Kingston and Chamberlin 1981). In the *trp* operon, NusA dramatically increases the pause at stem 1:2 in the *trp* attenuator region (Farnham and Platt 1982). In general, the presence of NusA in a purified system results in about a twofold decrease in the rate of elongation; hence, the enhancement of termination by NusA may simply be a consequence of slower transcription. There is further evidence that pausing is essential for NusA-mediated termination: The *rif*501 mutation in RNA polymerase prevents termination in the p_R operon

(Lecocq and Dambly 1976) and also greatly decreases the extent of pausing during transcription of this operon in vitro (J. Greenblatt and J. Li, in prep.).

RNA Release and Polymerase Dissociation

RNA polymerase probably undergoes a conformational change (transient or otherwise) as a prerequisite for termination when it encounters a pause site (Farnham and Platt 1980; Platt 1981), which renders it susceptible to the action of other factors. The action of the antitermination factor, λN protein (see below) may then be explained by postulating that the N-polymerase elongation complex is resistant to this change, as in the "juggernaut hypothesis" (see Adhya and Gottesman 1978). NusA may also be involved in altering the conformational equilibrium of RNA polymerase in addition to simply slowing it down. Circumstantial evidence is consistent with this possibility: First, in the presence of NusA, the β' and α subunits of polymerase become more resistant to trypsin digestion, and second, the "antipausing" $rif501$ mutation makes the β' subunit more susceptible to trypsin, whether or not NusA is present (A. Dulhanty and J. Greenblatt, in prep.).

Martin and Tinoco (1980) suggested that after cessation of elongation, release of the transcript is facilitated by the instability of base-pairing between the commonly found stretch of uridines and the complementary adenines in the template DNA strand. It is not known in general how transcript release is related to dissociation of the polymerase-DNA complex, though in the case of the ρ-independent t_1 terminator of T7, the RNA is released with a half-time of less than 3 minutes, and polymerase then dissociates slowly from the DNA (half-time of ~ 12 min) (O'Hare and Hayward 1981).

Previous results have indicated that ρ is involved as a release factor for termination in vitro, in ρ-independent, as well as ρ-dependent systems (Howard et al. 1977; Fuller and Platt 1978; Richardson and Conaway 1980; Shigesada and Wu 1980). Though recent work does not support this conclusion for the trp attenuator site (Berlin and Yanofsky 1983), some ternary complexes do appear to be stable (see below). RNA release is thus not necessarily concomitant with the stopping of elongation, and a primary function of ρ may be to increase the rate of dissociation of termination complexes that have ceased to elongate.

Role of ρ

ρ was discovered by Roberts (1969) as an *E. coli* factor that enhanced transcription termination on bacteriophage λ DNA. The *rho* gene maps at about 83 minutes on the *E. coli* chromosome (Ratner 1976) and encodes a basic polypeptide of 419 amino acids and about 46,000 daltons (Finger and Richardson 1982; Pinkham and Platt 1983). The role of ρ protein in the termination of transcription at specific sites has been demonstrated in vitro and in vivo for a wide variety of DNA templates from *E. coli* and various bacteriophages (for reviews, see Roberts 1976; Fujimura and Hayashi 1978; Konings and Schoenmakers 1978; Herskowitz and Hagen 1980). Genetic studies involving ρ mutants confirm the requirement for ρ as a transcription termination factor in vivo (Richardson et al. 1975; Das et al. 1976). ρ can bind to single-stranded nucleic acids (Fig. 5) and has an RNA-dependent nucleoside triphosphatase (NTPase) activity that appears to be necessary for ρ-dependent termination (see below). Much evidence suggests that ρ interacts directly with the RNA transcript, and probably with polymerase as well, to function as a site-specific RNA release factor, but the molecular basis for this activity is poorly understood.

Examination of the small number of 3'-terminal sequences of ρ-dependent termination sites sequenced thus far (see Fig. 2) reveals few apparent structural homologies. Although sequencing additional ρ-dependent termination sites may further define common features among possible subclasses of complex terminators that are not yet evident, determination of the ρ interaction sites on the ternary transcription complex is also an important approach to gain further insights into the function of ρ. Here we focus on the interactions of ρ with the various components of the transcription apparatus, the involvement of the ρ-NTPase activity in termination, and our present understanding of the molecular organization of the entire transcription termination complex.

ρ-RNA Polymerase Interactions

One possible mechanism for ρ action involves binding to RNA polymerase, to reduce the affinity of polymerase for the template at a termination site (Roberts 1969). Although there is no definitive evidence that ρ makes direct physical contact with RNA polymerase during termination, ρ (in conjunction with the RNA transcript) must at some level cause RNA polymerase to modify its own interaction with template and transcript. Genetic studies also show the ability of particular RNA polymerase mutations (lesions located in the *rpoB* locus) to compensate for ρ mutants that are defective in termination and ATPase activity (Das et al. 1978; Guarente and Beckwith 1978; Guarente 1979). In addition, some biochemical experiments suggest a functional interaction between ρ and RNA polymerase (Das et al. 1978). Moreover, isolated RNA polymerase and ρ factor can form a stable complex (Goldberg and Hurwitz 1972; Ratner 1976), though such complexes may be due to nonspecific electrostatic interactions between the highly basic ρ protein and acidic regions on the RNA polymerase. Some ρ-polymerase complexes appear to be dependent on the presence of contaminants in the protein preparations (Ratner 1976).

NTPase Activity of ρ

Many studies have focused on the "uncoupled" NTPase activity, i.e., the NTPase activity stimulated by RNA alone in the absence of other transcriptional components (Lowery-Goldhammer and Richardson 1974; Lowery and

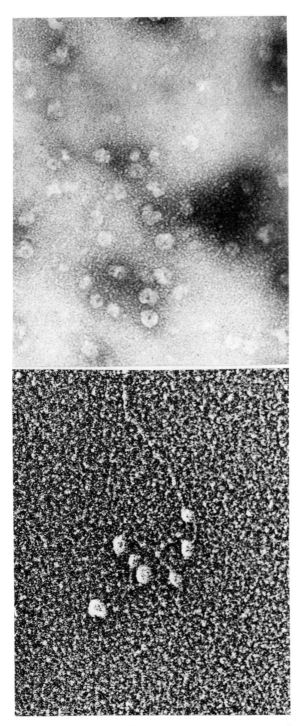

Figure 5 Electron micrographs of ρ. (*Left*) ρ molecules bound to poly(rC), mounted with spermidine and shadowed with tungsten as described by Griffith and Christiansen (1978). (*Right*) Purified ρ at higher magnification, stained with uranyl acetate; the diameter of the ρ molecule under these conditions is about 115 Å (Oda and Takanami 1972). Several good examples illustrating the hexameric structure are evident. (Courtesy of J.D. Griffith and D.G. Bear).

Richardson 1977a,b; Galluppi and Richardson 1980; Kent and Guterman 1981). The stimulation of the uncoupled NTPase activity requires a ribopolynucleotide containing cytosine (the amount of cytosine needed depends on the proportions of the other bases present) and chain length greater than about 50 nucleotides (Lowery and Richardson 1977a,b). The binding to high-molecular-weight poly(rC) appears to stimulate the highest activity (poly[dC] will not). Poly(rU) can also act as a weak cofactor, but no other polynucleotides lacking cytosine can elicit any NTPase activity. Recently, however, the combination of high-molecular-weight cytosine-containing single-stranded DNA and short oligoribonucleotides of any composition has been shown to stimulate the ρ NTPase (Richardson 1982). The precise significance of this observation is unclear at present but may be related to the small regions of single-stranded DNA generated at the growing point where RNA polymerase opens the DNA for access to the template strand.

Kinetic studies on the uncoupled NTPase activity utilizing ATP as substrate have provided some insight into the mechanism of this reaction (Sharp et al. 1983). The results of competition patterns are consistent with an ordered bi-ter reaction scheme (Fig. 6), in which ATP binds prior to the binding of the synthetic poly(C) substrate. After hydrolysis, the product ADP is released from the enzyme complex, followed by inorganic phosphate. It is not known whether poly(C) is released with each cycle of this uncoupled reaction in vitro.

Studies with blocked β-γ derivatives of NTPs (utilized by RNA polymerase in elongation but not cleavable by ρ) established the requisite role of the ρ-RNA-dependent NTPase activity in termination (Galluppi et al. 1976; Howard and de Crombrugghe 1976). Since a mutant RNA polymerase can stimulate the

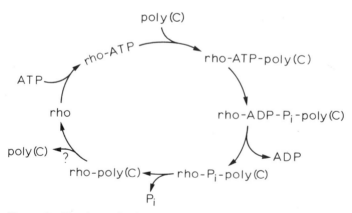

Figure 6   Kinetic mechanism of the poly(C)-dependent ATPase activity of ρ. The bi-ter reaction illustrated here is consistent with current observations on the uncoupled ATP hydrolysis; it has not been possible to determine whether the poly(C) actually dissociates from the ρ molecule at the end of each cycle. (Adapted from Sharp et al. 1983.)

NTPase activity of a mutant ρ factor, and increased termination activity results, the RNA polymerase molecule itself may also have some role in termination-linked NTPase activity (Das et al. 1978). RNA appears to be an essential cofactor in both the termination-coupled and -uncoupled NTPase activities, although other components may be involved. We presume that when hydrolysis of ATP is coupled to termination of transcription, the newly finished transcript, though possibly transiently bound to ρ, is ultimately released. Correlations have been demonstrated between the uncoupled ATPase activity and termination with both wild-type (Richardson and Macy 1981) and mutant ρ factors (Richardson and Carey 1982). Nevertheless, the significance of the uncoupled reaction is unclear, and its precise relationship to the coupled ATPase activity and the termination event has not been determined.

A possible role for the RNA-dependent ATPase activity is in translocation of a ρ molecule along mRNA in a search for a polymerase molecule paused at a termination site. The experiments devised to test this hypothesis have not yielded a definitive answer. Nuclease protection studies of ρ-poly(rC) complexes show enhanced protection during ATP hydrolysis from some ribonucleases and not from others (Galluppi and Richardson 1980). Studies designed to measure the exchange of ρ from radioactive poly(rC) onto cold poly(rC) do not demonstrate significantly shorter half-lives for ρ-poly(rC) complexes during ATP hydrolysis (Galluppi and Richardson 1980).

The NTPase activity may provide a mechanical mechanism for stripping the transcript from the transcription termination complex. The ATP-dependent release of RNA from isolated ternary complexes in the absence of NTP substrates seems to require ATP hydrolysis by ρ factor, though release occurs nonspecifically at various sites that are not ρ-dependent terminators (Richardson and Conaway 1980; Shigesada and Wu 1980). Though the only apparent requirement in these cases is that the transcript be at least 300 nucleotides in length (Richardson and Conaway 1980; Kassevetis and Chamberlin 1981), the real role of NTP hydrolysis in termination is not understood.

Interaction of ρ with DNA and RNA

ρ might function by binding to either a specific site on DNA (physically hindering the movement of RNA polymerase) or to a site on the nascent mRNA, subsequently stimulating polymerase to terminate. To distinguish between these possibilities, a number of groups have examined the DNA-binding properties of ρ. However, the binding of ρ to duplex DNA (even DNA containing ρ-dependent termination sites) is extremely weak. At moderate salt concentrations it is insignificant, although complexes can be isolated at very low salt concentrations (Beckmann et al. 1971; Darlix et al. 1971; Goldberg and Hurwitz 1972; Oda and Takanami 1972). In contrast, the single-stranded DNA-binding properties of ρ are not negligible, and this very strong interaction has revived interest in the possible role of ρ-DNA interactions during termination (Galluppi and Richardson 1980; Richardson 1982; D.G. Bear et al., in prep.). The nontemplate strand (presumably single-stranded at the transcription site) might be accessible to ρ factor during termination—this could play a part in the stability, as well as the activity, of the termination complex (Richardson 1982).

The most promising binding studies involve ρ interactions with RNA molecules. Work with synthetic polynucleotides demonstrates that ρ has a strong affinity for cytosine-containing RNA, with the highest affinity for the synthetic polyribonucleotide poly(rC) (Carmichael 1975; Galluppi and Richardson 1980). This appears to correlate with the strong stimulation of the NTPase activity by poly(C). Polynucleotides with increasing amounts of U or A weaken the relative binding constants and allow the ρ-polynucleotide complexes to be dissociated at lower monovalent ionic strengths (D.G. Bear et al., in prep.). However, strong polynucleotide binding appears to be a necessary but not sufficient condition for NTPase activity. Poly(dC) binds tightly to ρ but does not stimulate the hydrolytic activity at all (Galluppi and Richardson 1980).

In addition to base composition, the relative affinity of ρ for different RNAs is also dependent on the polynucleotide secondary structure (Richardson and Macy 1981). ρ displays no apparent binding to completely double-stranded structures like poly(A)·poly(U) and poly(I)·poly(C) (D.G. Bear et al., in prep.). Random RNA copolymers can be saturated with protein in proportion to the overall instability of secondary structure, as well as the amount of cytosine in single-stranded regions. ρ is able to melt only weak RNA helices and binds to only selected regions of highly double-stranded phage RNAs (D.G. Bear et al., in prep.).

The importance of the transcript was first indicated by Darlix (1973), showing that the inclusion of ribonuclease during transcription blocked termination at ρ-dependent sites. Along these same lines, current studies of the ρ-dependent termination site in the *trp* operon, *trp t'*, reveal several interesting features. It is located about 300 nucleotides past the last structural gene, in a region extremely rich in AT base pairs having little potential for secondary structure (Wu et al. 1981). In vitro, termination at *trp t'* is tightly coupled to the ATPase activity of ρ factor: A nonhydrolyzable analog, ATP-γ-S, that blocks hydrolysis also prevents termination. Moreover, ATPase activity depends on sequences in the *trp t'* region—transcription carried out with templates lacking *trp t'* (but otherwise similar, including ones carrying the ρ-independent site *trp t*) fails to elicit any significant ATPase activity from ρ factor. Termination at *trp t'* can also be eliminated by performing the transcription reactions in the presence of T1 RNase, though transcription remains unaffected (and can be quantitated by the yield of identifiable oligonucleotide fragments). This coincides with the absence of detectable ATPase activity. These observations support the conclusion that a segment of the RNA transcript (susceptible to T1 digestion) is required for termination to occur at a ρ-dependent site (Sharp and Platt 1984).

Preliminary results of the effect of deletions generated within this region both in vivo and in vitro indicate that the sequences required for the ρ-dependent termination event are located some distance upstream from the point(s) of termination (J.L. Galloway et al., in prep.). This is consistent with observations that transcripts must have a minimum length of a few hundred nuleotides before ρ-dependent termination can occur (Galluppi and Richardson 1980; Kassavetis and Chamberlin 1981; Richardson 1982), although rigorous tests of this hypothesis have yet to be performed.

Unlike the situation with ρ-independent termination, where enhancement of secondary structure in the RNA hairpin results in increased termination, in the ρ-dependent case the opposite seems to hold (see Table 2). That is, factors reducing secondary structure potential are correlated with an increase in the ATPase activity and termination efficiency (Das et al. 1978; Richardson and Macy 1981). This is consistent with our observations that *trp t'* is very low in G content (8%), as is the possible autoregulatory site within the leader region of the ρ gene itself (Pinkham and Platt 1983). We speculate that ρ may require only a region of RNA that is both untranslated and unstructured (see below) to catalyze termination of transcription. This could account for its participation in polarity, as well as its function at some of the ρ-dependent termination sites that have been characterized.

Structure of ρ and ρ Complexes

Electron microscopy, sedimentation, and chemical cross-linking studies have demonstrated that ρ is primarily a hexamer in solution, with an apparent sixfold axis of symmetry (Oda and Takanami 1972; Finger and Richardson 1982; D.G. Bear et al., in prep.). The hexameric structure also persists in the presence of poly(C) and ATP during hydrolysis, and binding of ρ to RNA [poly(rC)] appears to stabilize this form of the protein molecule (Fig. 5). At low protein concentrations, significant amounts of dissociated species can exist (Minkley 1973; Finger and Richardson 1982), and some mutant ρ factors

Table 2   Comparison of requirements for ρ-independent and ρ-dependent termination of transcription

	Termination of transcription	
	ρ-independent	ρ-dependent
Nucleic acid signal	RNA	RNA
Location	upstream (≈ 20)	upstream (>50)
Character	hairpin + uridines	lack of Gs
RNA-RNA interactions	yes	not known
RNA-DNA interactions	yes	not known
DNA-DNA interactions	no	probably not
NTP hydrolysis	no	yes

sediment primarily as dimers when free in solution, even though they retain ATPase activity (Shigesada and Imai 1978). The functional unit of wild-type ρ may thus be a dimer, and the symmetry could actually be 3-2 rather than a true sixfold.

The oligomeric form of ρ predominates in RNA binding (Finger and Richardson 1982), and spectroscopic measurements show that the protein saturates poly(rC) at a ratio of 1 monomer/12−14 nucleotides (or 1 hexamer/ 70−80 nucleotides) (D.G. Bear et al., in prep.). Extensive digestion of fully saturated ρ-poly(rC) complexes yields polynucleotide fragments with a maximum length of 60−80 nucleotides (Galluppi and Richardson 1980; D.G. Bear et. al., in prep.), and intermediate-size fragments are also generated (D.G. Bear et al., in prep.). This suggests that multiple interactions between hexamer and the poly(rC) may occur in units of monomer-binding-site sizes.

Proteolytic digestion of ρ-poly(rC) complexes in the presence of ATP yields primarily two fragments of about 31,000 daltons and 15,000 daltons (D.G. Bear et al., in prep.). The larger fragment is derived from the amino terminus and binds RNA, whereas the smaller fragment comes from the carboxyl terminus (D.G. Bear et al., in prep.). In the absence of ATP, proteolytic digestion occurs at several other sites as well, suggesting that there may be a conformational change in the structure of ρ, dependent upon the simultaneous binding (but not hydrolysis) of ATP and poly(C). Thus, the ρ polypeptide seems to be organized into two protease-resistant domains with an accessible hinge region, and this domain structure is dependent on the interaction of ρ with RNA and ATP. These conclusions are supported by studies with ρ temperature-sensitive mutants that suggest a conformational interaction between the ATP and RNA-binding sites (Kent and Guterman 1981).

A Model for ρ

The work of Richardson and his collaborators is largely responsible for our current ideas about how ρ functions (see Richardson 1983). The accumulated evidence for the participation of ρ in termination of transcription is consistent with the type of model presented in Figure 7. A specific recognition region in particular segments of RNA molecules provides a relatively high-affinity binding site for the ρ protein. The abundance of ATP in the cell ensures that the enzyme will be saturated prior to its encounter with the mRNA. Interaction with the transcript stimulates the protein to wrap up the remainder of the RNA until all of its sites are occupied and induces a conformational change in the protein. As a consequence of these events, ρ is now in close proximity to, if not directly in contact with, RNA polymerase and the accompanying NusA protein. We infer that multiple interactions between these various macromolecules then result in termination of transcription.

It is plausible, on the basis of the suggestions of von Hippel et al. (1982) in their study of the T4 gene-*32* autogeneous regulation, that ρ recognition may be provided by the *lack* of a specific structure or sequences favoring such

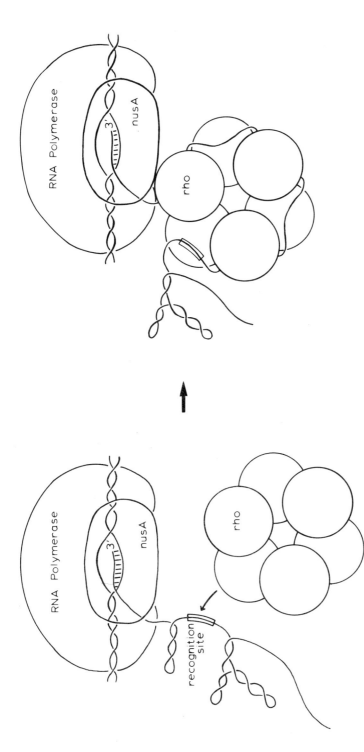

Figure 7 The participation of ρ in transcription termination. (*Left*) The postulated ρ recognition site has become exposed after synthesis, possibly defined as an unstructured region bounded by regions of some secondary structure. With a translated gene, the 5' region of the RNA would be masked by ribosomes. (*Right*) After ρ binding, the RNA is wrapped around the hexamer (which could melt distal secondary structure) and moves into contact with the elongation complex, probably interacting directly with RNA polymerase and/or NusA, and causing termination.

143

structures. Preliminary evidence for such a site in λ*cro* mRNA and further details of this aspect of ρ binding are discussed by Morgan et al. (1983a) and J.L. Galloway et al. (in prep.). Thus, ρ requires two conditions to function. First, the recognition site is in the RNA, located a considerable distance upstream of the (eventual) end of the transcript and must not be masked by the translational apparatus. Second, this site probably has little or no secondary structure (perhaps due to a predominance of AT base pairs) and requires only boundedness by structures to the 5′ side. Such a preference for lack of structure would allow a variety of sites to be utilized and would explain the lack of identifiable homology found within the current array of ρ-dependent termination sites.

Role of the Ribosome

In the preceding sections we have considered some of the interactions and recognition signals involved in causing termination of transcription. Yet, the real usefulness of termination as a regulatory mechanism depends on modulating the extent of termination in response to the particular requirements of the cell. Here we briefly discuss three phenomena for which this is true: attenuation, antitermination, and polarity. In each case, the sequences or structures specifying termination are masked or disrupted by competing factors whose influence in termination is dependent on the metabolic state of the cell. Reviews of attentuation (Yanofsky 1981; Kolter and Yanofsky 1982), polarity (Franklin and Yanofsky 1976; Adhya and Gottesman 1978), and antitermination (Greenblatt 1981; Ward and Gottesman 1982) may be recommended for those interested in pursuing these topics in greater depth.

Attenuation

The expression of many amino acid biosynthetic operons, including *his*, *ilv*, *leu*, *phe*, *thr*, and *trp*, is now known to be regulated by a phenomenon known as attenuation (Kasai 1974; Kolter and Yanofsky 1982). The attenuator site has the features of a simple terminator and is located in a leader region of about 200 nucleotides between the promoter and the first structural gene of the operon (Fig. 8a). Because of their structure, most attenuators function very efficiently in vitro to terminate transcription with RNA polymerase alone, and this finding would appear to preclude expression of the operon. However, other factors come into play in vivo to modulate the termination response and permit a certain fraction of the transcribing polymerase molecules to continue on into the structural genes. The major premise of attenuation is that formation of the stable base-paired GC-rich hairpin in the RNA transcript (which is followed by several consecutive uridine residues) is sufficient to cause termination of transcription. The extent of termination can therefore be controlled by interfer-

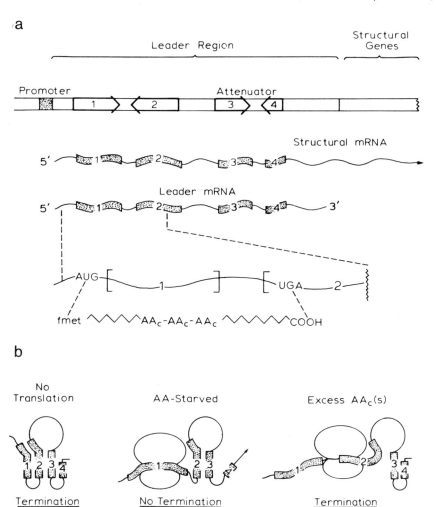

Figure 8   Attenuation in amino acid biosynthetic operons. (*a*) Schematic view of a typical operon, with a leader region of ~200 nucleotides preceding the structural genes. Dyad symmetry in the DNA is indicated by the large arrows; in the transcripts, this provides the potential for hairpin formation. Full translation of the leader peptide usually leads into region 2, allowing formation of a 3:4 structure, which leads to termination (see text). (*b*) The consequences of amino acid starvation or elimination of translation are illustrated. If the ribosome stalls on region 1 (starvation condition) early enough so that it cannot disrupt or mask region 2, termination will occur. (Adapted from Oxender et al. 1979.)

ing with hairpin formation to reduce or eliminate the ability of RNA polymerase to terminate at that site, and continuation of mRNA synthesis into distal regions will be enhanced.

Considerable evidence supports a mechanism in which formation of this distal hairpin (complementary to the template strand at the attenuator site itself) may be prevented by the formation of an alternate secondary structure in the transcript. The link to intracellular levels of the cognate amino acid(s) is provided by the ribosome, since the leader region in the biosynthetic operons mentioned above contains a ribosome-binding site that initiates translation of a short polypeptide rich in the end-product amino acid residue(s) of the enzymatic pathway. The movement of the ribosome across this region is governed by the efficiency of translation of these tandem codons and is thus coupled to the behavior of RNA polymerase at the attenuator site, since which of these mutually exclusive secondary structures actually forms is directed by the position of the ribosome on the leader mRNA (see Fig. 8b).

For example, if the end-product amino acid is abundant, translation of the leader peptide will be easily completed, and the ribosome will disrupt not only the self-complementary pairing between regions 1 and 2 but will also prevent 2:3 pairing. This, in turn, leaves 3:4 pairing available to form, which specifies termination. At the other extreme, if there is a deficiency of the amino acid in the cell, or of its charged cognate tRNA, the ribosome would stall at the tandem codons in the leader peptide at a relatively early position. In this case, only region 1 is masked, preventing 1:2 but allowing 2:3 formation, which precludes 3:4 pairing. The absence of this 3:4 termination signal at the attenuator site permits polymerase to transcribe completely through the leader region into the structural genes of the operon. Studies with a cell-free coupled system in the *trp* operon have demonstrated that these essential features and responses of attenuation can be reproduced in vitro (Das et al. 1982).

Attenuation appears to play a significant role in the regulation of expression in several other gene clusters. These include the *rplJ-rpoBC* operon, in which genes for several ribosomal proteins precede those for the β and β' subunits of RNA polymerase (Barry et al. 1980), the rRNA operon *rrnB* and possibly others (Kingston and Chamberlin 1981; M. Cashel, pers. comm.), and the *dnaG-rpoD* operon encoding primase and the σ subunit of RNA polymerase (Wold and MacMacken 1982). Bacteriophage λ also possesses numerous places in its genome where similar modes of regulation are employed (see Herskowitz and Hagen 1980; also see below).

Polarity

The strong coupling between transcription and translation in prokaryotic systems also manifests itself in the phenomenon of polarity, where eliminating translation within a polycistronic mRNA (by introduction of a nonsense codon, deletion, frameshift mutation, or insertion sequence) can reduce distal gene expression. In bacterial strains carrying nonsense suppressor mutations (or tRNAs), translation may be permitted to continue, and the polar effect is often alleviated. It has been proposed that RNA polymerase encounters previously latent sites for termination of transcription that become active due to the lack of translation (de Crombrugghe et al. 1973; Shimizu and Hayashi 1974).

Models to account for polarity include two major premises: (1) There are ρ-dependent sites of transcription termination within operons, and (2) these sites are not susceptible to the influence of ρ as long as the mRNA is concurrently translated (Adhya and Gottesman 1978). Support for these features is provided by the discovery of a class of polarity suppressors that do not restore translation, and these mutations have been shown to affect ρ factor (Richardson et al. 1975; Korn and Yanofsky 1976; Ratner 1976; Das et al. 1977). Polar effects can be suppressed not only by mutations in ρ factor but by the presence of other factors such as the N- and Q-gene products of bacteriophage λ (Adhya et al. 1974; Franklin 1974; Greenblatt 1981; Forbes and Herskowitz 1982; Ward and Gottesman 1982).

The nucleotide sequences in regions specifying ρ-dependent polar sites have not yet been precisely identified, but Ciampi et al. (1982) have reported the insertion of a Tn10 element just upstream of such a site in the *his* operon of *S. typhimurium*. The mRNA composition resembles *trp t′*: A segment of about 100 nucleotides has only 15% G but is flanked by short (30-nucleotide) regions containing 50% G residues (Bossi and Ciampi 1981; M.S. Ciampi, unpubl.).

Another transposable element, IS2, has been shown to severely reduce the expression of genes downstream when inserted in one orientation within an operon (Saedler et al. 1972; de Crombrugghe et al. 1973). This effect is apparently due to the presence of a ρ-dependent termination site within the IS2 sequence, since reduction in promoter-distal gene expression is suppressed in ρ mutants (Das et al. 1976; Besemer and Herpers 1977). The polarity induced by IS2 does not occur when the insertion is in the opposite orientation (Saedler et al. 1974). The insertion sequence IS1 also shows strong polarity effects upon insertion into operons and is apparently due to a strong ρ-dependent termination site (Das et al. 1977). The mechanism of polarity in the insertion sequence IS4 may not be entirely explained by the presence of ρ-dependent termination sites (Besemer and Herpers 1977).

In summary, the linking of polarity with ρ-dependent termination requires only that the polar sites within translated genes be segments of the transcript that have relatively high affinity for ρ (compared to neighboring regions). Only if translation is uncoupled from transcription will these sites become exposed at the requisite distance from RNA polymerase, allowing termination to occur. We would predict by analogy that these sites may be AT-rich and/or possess little secondary structure, and it seems likely that a molecular understanding of this phenomenon is not far off.

Antitermination

Both attenuation and polarity rely on the ribosome to block the termination response. In the former case, the ribosome can prevent the formation of a "simple" terminator structure (indirectly, via competing RNA-RNA interactions), and in the latter case, it can mask (presumably) ρ-specific recognition regions on the transcript. In this sense, the ribosome is an "antitermination" factor, but more generally any competing factor that prevents recognition of termination sites on the RNA might fulfill this function.

One of the best characterized systems for studying antitermination is provided, again, by bacteriophage λ. The two major antitermination factors have been identified as the products of the N and Q genes (see Greenblatt 1981; Forbes and Herskowitz 1982; Grayhack and Roberts 1982; Ward and Gottesman 1982), and their function is critical for the lytic/lysogenic decision in the life cycle of the phage (Herskowitz and Hagen 1980).

The requirements and responses of these antitermination factors are unusual. First, N protein can overcome termination at both ρ-dependent and ρ-independent sites, implying that N-mediated antitermination does not simply involve inactivation of ρ (see Gottesman et al. 1980). Second, suppression of termination is unique to transcripts initiated at the p_L and p_R promoters, due to the nearby *nut* (N *ut*ilization) sites contained in the transcripts that are essential to N function and independent of the promoter itself (Adhya et al. 1974, 1976; de Crombrugghe et al. 1979). A molecular analysis of the *nut* sites reveals dyad symmetry and some adjacent segments that are crucial for full function (see Fig. 9; Salstrom and Szybalski 1978; Drahos and Szybalski 1982; Friedman and Olson 1983).

The N protein is thought to function directly on those RNA polymerases that pass through the *nut* site, altering them in such a way that they fail to terminate at distal transcription termination sites (Adhya and Gottesman 1978). The host factor NusA has been shown to bind tightly to N and is required for its action (see Greenblatt 1981; Greenblatt and Li 1981a,b). The observation that the antitermination function of N can be demonstrated in a system using an S100 extract indicates that the ribosome, and, hence, translation of the RNA is not a critical ingredient for N to work (Ishii et al. 1980). Analysis of Q function in vivo (Forbes and Herskowitz 1982) suggests that its properties and requirements may be similar. Unlike N, however, which has not yet been demonstrated to function in a purified transcription system, Q protein can exert its antitermination effect in a minimal system consisting only of RNA poly-

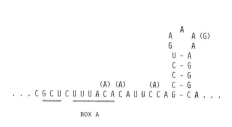

Figure 9   Antitermination and the *nut* sites. The probable structure in a transcript spanning the *nut*R (*nut*L) site is shown—the hairpin may well be a loading site for the N protein. A possible region for recognition by NusA, designated box A, has been identified and is indicated by the bars (Friedman and Olson 1983).

merase, a DNA template encoding the λ 6S transcript, purified Q protein, and NusA (Grayhack and Roberts 1982).

A more complex aspect of antitermination is probably involved in controlling the expression of rRNA operons in *E. coli*. ρ-dependent termination sites are thought to exist in these operons but are not manifested when transcription is driven by the normal *rrn* promoters (C.L. Squires, pers. comm.). Moreover, the strong polar effect normally exerted by the transposon Tn*10* is absent when this element is inserted into an *rrn* operon (Morgan 1980). Preliminary analysis suggests the presence of antitermination sites in the *rrn* leader region prior to the 16S RNA gene (M. Cashel, pers. comm.), which may be postulated to mediate the stringent response via the influence of ppGpp, though a direct relationship between these potential loading sites and regulation by ppGpp has not yet been established.

Several host factors have been identified as a requirement for λ*N* function by Friedman and co-workers (see Friedman et al. 1983), including NusA and NusB. The *nusE* locus, encoding ribosomal protein S10, has also been implicated as a requirement for the antitermination activity of *N* protein (Friedman et al. 1981), but this may be due to a reduced ability of the ribosome to translate the *N* mRNA efficiently (D. Steege et al., in prep.). Preliminary evidence that mutations in ribosomal protein L11 can affect *N*-mediated termination as well (Ward and Gottesman 1982) may be explained by similar considerations.

In general, antitermination events seem to require the presence of specific sites at which polymerase becomes modified and "resistant" to subsequent termination sites that it encounters. It is not known whether this resistance is due to a factor-stimulated conformational change in the polymerase or to the addition of a factor itself to the elongation complex. Moreover, not all termination sites are responsive—several sites in the *b2* region of λ do not permit significant readthrough by RNA polymerase in the presence of either *N* or *Q* proteins (Burt and Brammar 1982). Thus, an additional way to view termination sites, aside from their requirements (or lack thereof) for added factors, is according to their response to an antiterminating RNA polymerase molecule. Sites resistant to such readthrough, though rare, will be of paramount importance to cellular metabolism in those operons (such as the *rrn* group) where antitermination is a crucial feature of regulation.

Retroregulation and Processing

Control mechanisms utilizing feedback loops have been known to play a ubiquitous role in many kinds of biological systems. End-product inhibition of an early enzyme in a biosynthetic pathway and operator-repressor regulation of transcription in response to a metabolic cofactor are two classic examples of such mechanisms. We have already discussed attenuation, in which feedback control regulates termination of transcription at a specific site to control the expression of distal genes. Recent developments reveal a surprising additional observation: Effects at a terminator can also influence the expression of upstream genes that have already been transcribed. Moreover, in the case of λ "retroregulation," the particular regulatory site in the RNA serves overlapping dual functions that are mutually exclusive.

Retroregulation in λ

In bacteriophage λ, several mutations affecting the levels of the integrase activity (product of the *int* gene) map in the *b* region of the λ chromosome, several hundred base pairs downstream from the *int* gene itself (Guarneros and Galindo 1979). These mutations define a regulatory element designated *sib* (*s*ite of *i*nhibition in *b*), which acts in *cis* and whose DNA does not encode a protein. Point mutations inactivating the inhibitory function of *sib* have now been localized, and accumulated evidence indicates that the action of *sib*$^+$ is posttranscriptional, most probably affecting the levels of integrase that can be effectively translated (Schindler and Echols 1981; Guarneros et al. 1982).

The proposed mechanism for this retroregulation is outlined in Figure 10a, which shows the two modes of *int* expression. Under conditions favoring lysogenization of the phage, relatively high levels of integrase are required, whereas levels of excise protein, whose gene (*xis*) is located just proximal to *int*, must remain low to prevent excision once prophage integration has occurred. Transcription of *int* is dependent upon positive activation by the *c*II-gene product at the *int* promoter, located just upstream of the gene (and overlapping the distal portion of *xis*). Transcription proceeds beyond the *int* gene, through the attachment site (*att*), and into the *b* region of the phage genome. About 150 bp from the center of *att*, the polymerase will encounter a GC-rich region of dyad symmetry followed by a uridine-encoding region, which causes termination of transcription (Rosenberg and Schmeissner 1982).

A different pathway is utilized when λ embarks upon the lytic cycle. Under such conditions, transcription is initiated at the p_L promoter and continues directly through the *N* gene and across the *nut*L site, where modification by *N* renders RNA polymerase resistant to the distal termination sites. The altered elongation complex proceeds through the rest of the early region of the genome, which include *xis* and *int*, past *att*, and into the *b* region. As in the previous case, polymerase encounters the potential termination site in the *sib* region, but it is now unable to terminate there and reads through into distal sequences. Unexpectedly, the added sequence permits the formation of an RNase III recognition site, whose upper portion corresponds to the hairpin at the end of the p_{int} transcript. Only in the case of readthrough (from p_L) can cleavage by RNase III occur, however, which leads to lability of the *int* mRNA.

Recent evidence that the *sib* site can be moved distal to other genes and reduce their expression (Rosenberg and Schmeissner 1982) supports the observations mentioned above, and the following model appears to account for all

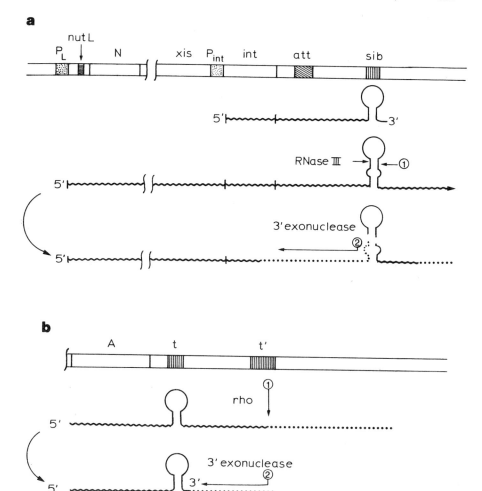

Figure 10   Retroregulation and processing. (*a*) Region of λ DNA transcribed from the p_L promoter (reversed from the usual orientation, to display transcription proceeding from left to right). The consequences of initiation at p_L versus p_{int} are shown. In the former case, the action of *N* prevents termination in the *sib* region, but subsequent cleavage by RNase III is proposed to result in the rapid 3'-exonucleolytic degradation of the *int* mRNA (see text). (*b*) The end of the tryptophan operon. The diagram indicates how previous observations may be accounted for that the end of *trp* mRNA occurred at *trp t*, immediately beyond *trpA*. We postulate that the major termination event occurs at *trp t'*, which is followed by rapid exonucleolytic degradation through the AT-rich region until it is impeded by the stable hairpin structure of the *trp t* region (see text).

of the data thus far. After cleavage by RNase III, the *int* mRNA is now lacking a 3'-hairpin structure and becomes extremely sensitive to degradation by 3' exonucleases, which in turn reduces the amount of integrase that can be produced from the p_L-promoted transcript. Although no specific ribonuclease has yet been implicated in this degradative event, the most likely candidate is RNase II, a 3' exonuclease that proceeds rapidly until it encounters regions of secondary structure that hinder its progress (Gupta et al. 1977).

Retroregulation in trp?

A similar explanation may account for the long-standing paradox concerning the tandem termination sites at the end of the tryptophan operon (Wu et al. 1980, 1981). In vivo, termination in this region is dependent upon ρ factor, and the mRNA has a 3' terminus located 36 nucleotides beyond the end of *trpA* (the last structural gene). This site, designated *trp t*, has GC-rich dyad symmetry, and the RNA terminates with 4 uridines immediately following the potential hairpin. Termination at this site exhibits little response to ρ factor in vitro, however (Wu et al. 1981; Farnham et al. 1982). Mutations enhancing readthrough into distally placed lactose genes (Guarente and Beckwith 1978) included several deletions localized distal to *trp t* (Guarente et al. 1979; Wu et al. 1980). This led to the discovery of a second termination site, *trp t'*, located about 250 bp beyond *trp t* and entirely dependent on the presence of ρ (Wu et al. 1981). Since *trp t* has the characteristics of a simple terminator, it has never been clear why readthrough transcription occurs at *trp t* in vivo in *rho⁻* strains, or when *trp t'* has been deleted.

A model we may now propose to explain these puzzling results requires a new assumption: that *trp t'* is used as the "real" termination site for transcription of the *trp* operon in vivo (Fig. 10b). If so, when a mutation in ρ prevents termination at *trp t'*, or if the *trp t'* region is deleted, transcription would proceed through this portion of the genome into distal sequences. In the wild-type cell, however, transcription would terminate efficiently at *trp t'*, but the lack of secondary structure in the final few hundred nucleotides of the transcript would make it very sensitive to degradation by one of the exoribonucleases of *E. coli* (Gupta et al. 1977; Kasai et al. 1977). Degradation proceeding rapidly to the strong secondary structure corresponding to the *trp t* site would finally yield the observed 3' terminus. If current experiments support this hypothesis, this case will provide another example of "retroregulation," though there is no obvious function for the sequences between *trp t* and *trp t'*, and no effect on levels of gene product are detectable. It will also strengthen arguments that processing events are of great importance for controlling mRNA levels in the cell.

CONCLUDING REMARKS

Genetic and biochemical evidence supports a simple model for ρ-independent termination of transcription (Rosenberg and Court 1979; Farnham and Platt 1980), in which RNA-RNA base-pairing in the transcript and RNA-DNA interactions between the transcript and the template have important but discrete functional contributions. The pausing of polymerase at the RNA hairpin may increase the probability of termination, at which point the instability of the rU-dA pairing causes dissociation of the transcript from the template. The termination efficiency of the transcription complex can be increased by the participation of NusA protein, which may slow elongation so that any required interactions between polymerase and the RNA or DNA can occur more accurately. If the formation of a termination complex requires the polymerase to undergo a conformational change, then a slower elongation rate may also enhance the opportunity for this isomerization. Much of the data on the currently known simple terminators is consistent with this model, and this general mechanism may be common to other transcription systems (Mills et al. 1978; Wiggs et al. 1979).

The importance of transcription termination as a regulatory phenomenon is now established, but many questions remain. For example, several sites lack the classical features of terminators (Fig. 2). In general, these are demonstrably factor-dependent, but some, like the *lacI* terminator (Horowitz and Platt 1982), do not respond to known factors under in vitro conditions. Moreover, in the other cases, we have little clue as to what the specific signals specifying ρ-dependent or NusA-catalyzed termination may be. It is curious that several terminators in this class may overlap processing sites, and the lack of sequence similarities may reflect the constraints of having multiple regulatory signals within the same region of DNA.

One common feature to both factor-independent and factor-dependent termination that does emerge is the participation of the RNA transcript in this event. There is yet no concrete evidence that specific sequences are involved; rather it is the structures assumed by the RNA itself that become crucial. This assertion is strongest with the hairpins of simple terminators and seems likely to be true in an opposite sense (i.e., a requirement for lack of structure) for the ρ-dependent cases. An interplay between alternative structures mediates the attenuation response, and antitermination mechanisms will probably invoke recognition of RNA secondary structures as well.

It seems certain that coordination of the many complex regulatory systems in *E. coli* will utilize numerous additional cellular components in vivo. For example, factors such as *nusB*- and *nusE*-gene products (Friedman et al. 1981) and the *N* and *Q* proteins of bacteriophage λ (Greenblatt 1981; Forbes and Herskowitz 1982; Ward and Gottesman 1982) are insufficiently characterized to understand their role in catalyzing and controlling termination and antitermination. Whether these protein factors themselves can recognize particular sequences, rather than altering the specificity of RNA polymerase, is also unknown. The role of coupling between transcription and translation in mediating termination is even more intricate, playing an integral function in polarity (Adhya et al. 1976; Franklin and Yanofsky 1976; Adhya and Gottesman 1978) and attenuation (Kolter and Yanofsky 1982). We can only surmise that recognition and response in these different systems must entail subtle but significant structural differences superimposed upon apparent sequence similarities to specify the signal strength, factor dependence, and response to metabolic requirements of the wide variety of termination regions that have now been identified.

ACKNOWLEDGMENTS

We thank Molly Schmid, Deborah Steege, Malcolm Winkler, and Charles Yanofsky for valuable discussions, and Sankar Adhya, Michael Chamberlin, Jack Greenblatt, and John Richardson for comments on the manuscript. We are especially grateful to Peter von Hippel for enthusiastic support, imaginative insights, and stimulating contributions to ideas in the manuscript, and the generous use of his facilities. We are also indebted to our laboratory colleagues and to the participants in the Airlie House Symposium on Termination (September 1982) for many fruitful interactions and thoughtful discussions. Jean Gross and Bobbie Squires provided expert and patient typing through many revisions, and Mary Gilland was a great help with the artwork. This was written, in part, while D.G.B. was funded by U.S. Public Health Service postdoctoral fellowship GM-06676 and T.P. was on sabbatical leave in the laboratory of P.H. von Hippel, Institute of Molecular Biology, University of Oregon. T.P. was supported by GM-22830 and a Senior Faculty Fellowship from Yale University.

REFERENCES

Adhya, S. and M. Gottesman. 1978. Control of transcription termination. *Annu. Rev. Biochem.* **47:** 967.

Adhya, S., M. Gottesman, and B. de Crombrugghe. 1974. Release of polarity in *E. coli* by gene N of phage λ. Termination and antitermination of transcription. *Proc. Natl. Acad. Sci.* **71:** 2534.

Adhya, S., M. Gottesman, B. de Crombrugghe, and D. Court. 1976. Transcription termination regulates gene expression. In *RNA polymerase* (ed. R. Losick and M. Chamberlin), p. 719. Cold Spring Harbor Laboratory, Cold Spring Harbor, New York.

Adhya, S., P. Sarkar, D. Valenzuela, and U. Maitra. 1979. Termination of transcription by *Escherichia coli* RNA polymerase: Influence of secondary structure of RNA transcripts on ρ-dependent termination. *Proc. Natl. Acad. Sci.* **76:** 1613.

Barry, G., C.L. Squires, and C. Squires. 1980. Attenuation and processing of the RNA from the transcription unit of *Escherichia coli. Proc. Natl. Acad. Sci.* **77:** 3331.

Beckmann, J.S., V. Daniel, Y. Tichauer, and U.Z. Littauer. 1971. Binding of termination factor rho to DNA. *Biochem. Biophys. Res. Commun.* **43:** 806.

Bektesh, S.L. and J.P. Richardson. 1980. A ρ-recognition site on phage λ cro gene mRNA. *Nature* **283:** 102.

Berlin, V. and C. Yanofsky. 1983. RNA polymerase spontaneously releases transcript and template during transcription termination of the *trp* operon attenuator. *J. Biol. Chem.* **258:** 1714.

Bertrand, K., L. Korn, F. Lee, and C. Yanofsky. 1977. The attenuator of the tryptophan operon of *Escherichia coli.* Heterogeneous 3'-OH termini in vivo and deletion mapping of functions. *J. Mol. Biol.* **117:** 227.

Besemer, J. and M. Herpers. 1977. Suppression of polarity of insertion mutations within the *gal* operon of *E. coli*. *Mol. Gen. Genet.* **151:** 295.

Bossi, L. and M.S. Ciampi. 1981. DNA sequences at the sites of 3 insertions of the transposable element Tn5 in the histidine operon of *S. typhimurium*. *Mol. Gen. Genet.* **183:** 406.

Brosius, J., T.J. Dull, D.D. Sleeter, and H.F. Noller. 1981. Gene organization and primary structure of a ribosomal RNA operon from *E. coli*. *J. Mol. Biol.* **148:** 107.

Burt, D.W. and W.J. Brammar. 1982. Transcriptional termination sites in the *b2* region of bacteriophage lambda that are unresponsive to antitermination. *Mol. Gen. Genet.* **185:** 462.

Calva, E. and R.R. Burgess. 1980. Characterization of a rho-dependent termination site within the *cro* gene of bacteriophage lambda. *J. Biol. Chem.* **255:** 11017.

Carmichael, G.G. 1975. Isolation of bacterial and phage proteins by homopolymer RNA-cellulose chromatgraphy. *J. Biol. Chem.* **250:** 6160.

Chamberlin, M. 1965. Comparative properties of DNA, RNA, and hybrid homopolymer pairs. *Fed. Proc.* **24:** 1446.

Christie, G.E., P.J. Farnham, and T. Platt. 1981. Rho-independent transcription termination: Comparison of natural and synthetic terminator sites *in vitro*. *Proc. Natl. Acad. Sci.* **78:** 4180.

Ciampi, M.S., M.B. Schmid, and J.R. Roth. 1982. Transposon Tn10 provides a promoter for transcription of adjacent sequences. *Proc. Natl. Acad. Sci.* **79:** 5016.

Cone, K.C., M.A. Sellitti, and D.A. Steege. 1983. *lac* repressor mRNA transcription terminates *in vivo* in the *lac* control region. *J. Biol. Chem.* **258:** 11296.

Court, D., C. Brady, M. Rosenberg, D.L. Wulff, M. Behr, M. Mahoney, and S. Izumi. 1980. Control of transcription termination: A rho-dependent termination site in bacteriophage λ. *J. Mol. Biol.* **138:** 231.

Darlix, J.L. 1973. The functions of rho in T7-DNA transcription *in vitro*. *Eur. J. Biochem.* **35:** 517.

Darlix, J.L., A. Sentenac, and P. Fromageot. 1971. Binding of termination factor rho to RNA polymerase and DNA. *FEBS Lett.* **13:** 165.

Das, A., D. Court, and S. Adhya. 1976. Isolation and characterization of conditional lethal mutants of *Escherichia coli* defective in transcription termination factor rho. *Proc. Natl. Acad. Sci.* **73:** 1959.

———. 1978. Interaction of RNA polymerase and rho in transcription termination: Coupled ATPase. *Proc. Natl. Acad. Sci.* **75:** 4828.

Das, A., I.P. Crawford, and C. Yanofsky. 1982. Regulation of tryptophan operon expression by attenuation in cell-free extracts of *Escherichia coli*. *J. Biol. Chem.* **257:** 8795

Das, A., D. Court, M. Gottesman, and S. Adhya. 1977. Polarity of insertion mutations is caused by rho-mediated termination of transcription. In *DNA insertion elements, plasmids, and episomes* (ed. A.I. Bukhari et al.), p. 93. Cold Spring Harbor Laboratory, Cold Spring Harbor, New York.

de Crombrugghe, B., S. Adhya, M. Gottesman, and I. Pastan. 1973. Effect of rho on transcription of bacterial operons. *Nat. New Biol.* **241:** 260.

de Crombrugghe, B., M. Mudryj, R. DiLauro, and M. Gottesman. 1979. Specificity of the bacteriophage lambda N gene product (pN): *nut* sequences are necessary and sufficient for antitermination by pN. *Cell* **18:** 1145.

Drahos, D., and W. Szybalski. 1982. Synthesis of the *nutL* DNA segments and analysis of antitermination and termination functions in coliphage lambda. *Gene* **18:** 343.

Farnham, P.J. and T. Platt. 1980. A model for transcription termination suggested by studies on the *trp* attenuator in vitro using base analogs. *Cell* **20:** 739.

———. 1981. Rho-independent termination dyad symmetry in DNA causes RNA polymerase to pause during transcription *in vitro*. *Nucleic Acids Res.* **9:** 563.

———. 1982. Effects of DNA base analogs on transcription termination of the tryptophan operon attenuator of *Escherichia coli*. *Proc. Natl. Acad. Sci.* **79:** 998.

Farnham, P.J., J. Greenblatt, and T. Platt. 1982. Effects of nusA protein on transcription termination in the tryptophan operon of *Escherichia coli*. *Cell* **29:** 941.

Finger, L.R. and J.P. Richardson. 1982. Stabilization of the hexameric form of *Escherichia coli* protein rho under ATP hydrolysis conditions. *J. Mol. Biol.* **156:** 203.

Fisher, R.F. and C. Yanofsky. 1983. Mutations of the β subunit of RNA polymerase alter both transcription pausing and transcription termination in the *trp* operon leader region *in vitro*. *J. Biol. Chem.* **258:** 8146.

Forbes, D. and I. Herskowitz. 1982. Polarity suppression dependent on the Q gene products of phage lambda. *J. Mol. Biol.* **160:** 549.

Franklin, N.C. 1974. Altered reading of genetic signals fused to the N operon by bacteriophage λ: Genetic evidence for modification of polymerase by the protein product of the N gene. *J. Mol. Biol.* **89:** 33.

Franklin, N.C. and C. Yanofsky. 1976. The N protein of λ: Evidence bearing on transcription termination, polarity and the alteration of *E. coli* RNA polymerase. In *RNA polymerase* (ed. R. Losick and M.J. Chamberlin), p. 693. Cold Spring Harbor Laboratory, Cold Spring Harbor, New York.

Friden, P., T. Newman, and M. Freundlich. 1982. Nucleotide sequence of the ilvB promoter-regulatory region: A biosynthetic operon controlled by attenuation and cyclic AMP. *Proc. Natl. Acad. Sci.* **79:** 6156.

Friedman, D.I. 1971. A bacterial mutant affecting lambda development. In *The bacteriophage lambda* (ed. A.D. Hershey), p. 773. Cold Spring Harbor Laboratory, Cold Spring Harbor, New York.

Friedman, D.I. and E.R. Olson. 1983. Evidence that a nucleotide sequence, "box A," is involved in the action of the NusA protein. *Cell* **34:** 143.

Friedman, D.I., A.T. Schauer, M.R. Baumann, L.S. Baron, and S.L. Adhya. 1981. Evidence that ribosomal protein S10 participates in control of transcription termination. *Proc. Natl. Acad. Sci.* **78:** 1115.

Friedman, D.I., A.T. Schauer, E.J. Mashni, E.R. Olson, and M.F. Baumann. 1983. *E. coli* factors involved in the action of the λ gene N antitermination function. In *Microbiology-1983* (ed. D. Schlessinger), p. 39. American Society for Microbiology, Washington, D.C.

Fujimura, F.K. and M. Hayashi. 1978. Transcription of isometric single-stranded DNA phage. In *The single-stranded DNA phages* (ed. D.T. Denhardt et al.), p. 485. Cold Spring Harbor Laboratory, Cold Spring Harbor, New York.

Fuller, R.S. and T. Platt. 1978. The attenuator of the tryptophan operon in *E. coli*: Rho-mediated release of RNA polymerase from a transcription termination complex in vitro. *Nucleic Acids Res.* **5:** 4613.

Galluppi, G. and J.P. Richardson. 1980. ATP-induced changes in the binding of RNA synthesis termination protein rho to RNA. *J. Mol. Biol.* **138:** 513.

Galluppi, G., C. Lowery, and J.P. Richardson. 1976. Nucleoside triphosphate requirement for termination of RNA synthesis by rho factor. In *RNA polymerase* (ed. R. Losick and M. Chamberlin), p. 657. Cold Spring Harbor Laboratory, Cold Spring Harbor, New York.

Gamper, H.B. and J.E. Hearst. 1982. A topological model for transcription based on unwinding angle analysis of *E. coli* RNA polymerase binary, initiation and ternary complexes. *Cell* **29:** 81.

Gardner, J.F. 1982. Initiation, pausing, and termination of transcription in the threonine operon regulatory region of *Escherichia coli*. *J. Biol. Chem.* **257:** 3896.

Goldberg, A.R. and J. Hurwitz. 1972. Studies on termination of in vitro ribonucleic acid synthesis by rho factor. *J. Biol. Chem.* **247:** 5637.

Gottesman, M.E., S. Adhya, and A. Das. 1980. Transcription antitermination by bacteriophage lambda N gene product. *J. Mol. Biol.* **140:** 57.

Grayhack, E.J. and J.W. Roberts. 1982. The phage λ Q gene product: Activity of a transcription antiterminator *in vitro*. *Cell* **30:** 637.

Greenblatt, J. 1981. Regulation of transcription termination by the N gene protein of bacteriophage lambda. *Cell* **24:** 8.

Greenblatt, J. and J. Li. 1981a. Interaction of the sigma factor and the *nusA* gene protein of *E. coli* with RNA polymerase in the initiation-termination cycle of transcription. *Cell* **24**: 421.

—————. 1981b. The *nusA* gene protein of *E. coli*: Its identification and a demonstration that it interacts with the gene N transcription antitermination protein of bacteriophage lambda. *J. Mol. Biol.* **147**: 11.

Greenblatt, J., M. McLimont, and S. Hanly. 1981. Termination of transcription by the *nusA* gene protein of *E. coli*. *Nature* **292**: 215.

Greenblatt, J., J. Li, S. Adhya, D.I. Friedman, L.S. Baron, B. Redfield, H. Kung, and H. Weissbach. 1980. L factor that is required for β-galactosidase synthesis is the *nusA* gene product involved in transcription termination. *Proc. Natl. Acad. Sci.* **77**: 1991.

Griffith, J.D. and G. Christiansen. 1978. Electron microscope visualization of chromatin and other DNA-protein complexes. *Annu. Rev. Biophys. Bioeng.* **7**: 19.

Guarente, L.P. 1979. Restoration of termination by RNA polymerase mutations is rho allele-specific. *J. Mol. Biol.* **129**: 295.

Guarente, L.P. and J. Beckwith. 1978. Mutant RNA polymerase of *Escherichia coli* terminates transcription in strains making defective rho factor. *Proc. Natl. Acad. Sci.* **75**: 294.

Guarente, L.P., J. Beckwith, A.M. Wu, and T. Platt. 1979. A mutation distal to the mRNA endpoint reduces transcription termination in the tryptophan operon of *E. coli*. *J. Mol. Biol.* **133**: 189.

Guarneros, G. and J.M. Galindo. 1979. The regulation of integrative recombination by the *b2* region and the cII gene of bacteriophage λ. *Virology* **95**: 119.

Guarneros, G., C. Montanez, T. Hernandez, and D. Court. 1982. Posttranscriptional control of bacteriophage λ int gene expression from a site distal to the gene. *Proc. Natl. Acad. Sci.* **79**: 238.

Gupta, R.S., T. Kasai, and D. Schlessinger. 1977. Purification and some novel properties of RNase II. *J. Biol. Chem.* **252**: 8945.

Herskowitz, I. and D. Hagen. 1980. The lysis-lysogeny decision of phage λ: Explicit programming and responsiveness. *Annu. Rev. Genet.* **14**: 399.

Horowitz, H. and T. Platt. 1982. A termination site for *lacI* transcription is between the CAP site and the *lac* promoter. *J. Biol. Chem.* **257**: 11740.

Howard, B.H. and B. de Crombrugghe. 1976. ATPase activity required for termination of transcription by the *E. coli* protein factor rho. *J. Biol. Chem.* **251**: 2520.

Howard, B.H., B. de Crombrugghe, and M. Rosenberg. 1977. Transcription in vitro of bacteriophage lambda 4S RNA: Studies on termination and rho protein. *Nucleic Acids Res.* **4**: 927.

Inoko, H. and M. Imai. 1976. Isolation and genetic characterization of the *nitA* mutants of *Escherichia coli* affecting termination factor rho. *Mol. Gen. Genet.* **143**: 211.

Ishii, S., K. Kuroki, Y. Sugino and F. Imamota. 1980. Purification and characterization of the N gene product of bacteriophage lambda. *Gene* **10**: 291.

Johnston, H.M., W.M. Barnes, F.G. Chumley, L. Bossi, and J.R. Roth. 1980. Model for regulation of the histidine operon of *Salmonella*. *Proc. Natl. Acad. Sci.* **77**: 508.

Kasai, T. 1974. Regulation of the expression of the histidine operon in *Salmonella typhimurium*. *Nature* **249**: 523.

Kasai, T., R.S. Gupta, and D. Schlessinger. 1977. Exoribonucleases in wild type *Escherichia coli* and RNase II-deficient mutants. *J. Biol. Chem.* **252**: 8950.

Kassavetis, G.A. and M.J. Chamberlin. 1981. Pausing and termination of transcription within the early region of bacteriophage T7 DNA in vitro. *J. Biol. Chem.* **256**: 2777.

Keller, E.B. and J.M. Calvo. 1979. Alternative secondary structures of leader RNAs and the regulation of the *trp*, *phe*, *his*, *thr*, and *leu* operons. *Proc. Natl. Acad. Sci.* **76**: 6186.

Kent, R.B. and S.K. Guterman. 1981. A mutant rho ATPase from *E. coli* that is temperature-sensitive in the presence of RNA. *Mol. Gen. Genet.* **181**: 367.

Kingston, R.E. and M.J. Chamberlin. 1981. Pausing and attenuation of *in vitro* transcription in the *rrnB* operon of *E. coli*. *Cell* **27**: 523.

Kingston, R.E., W.C. Nierman, and M.J. Chamberlin. 1981. A direct effect of guanosine tetraphosphate on pausing of *Escherichia coli* RNA polymerase during RNA chain elongation. *J. Biol. Chem.* **256**: 2787.

Kolter, R. and C. Yanofsky. 1982. Attenuation in amino acid biosynthetic operons. *Annu. Rev. Genet.* **16**: 113.

Konings, R.N.H. and J.G.G. Schoenmakers. 1978. Transcription of the filamentous phage genome. In *The single-stranded DNA phages* (ed. D.T. Denhardt et al.), p. 507. Cold Spring Harbor Laboratory, Cold Spring Harbor, New York.

Korn, L.J. and C. Yanofsky. 1976. Polarity suppressors defective in transcription termination at the attenuator of the tryptophan operon of *Escherichia coli* have altered rho factor. *J. Mol. Biol.* **106**: 231.

Kung, H. 1983. Role of nusA protein in coupled transcription and translation. In *Microbiology-1983* (ed. D. Schlessinger), p. 53. American Society for Microbiology, Washington, D.C.

Kung, H.F., C. Spears, and H.J. Weissbach. 1975. Purification and properties of a soluble factor required for the deoxyribonucleic acid-directed in vitro synthesis of β-galactosidase. *J. Biol. Chem.* **250**: 1556.

Küpper, H., T. Sekiya, M. Rosenberg, J. Egan, and A. Landy. 1978. A ρ-dependent termination site in the gene coding for tyrosine tRNA su₃ of *Escherichia coli*. *Nature* **272**: 423.

Lau, L.F., J.W. Roberts, and R. Wu. 1982. Transcription terminates at λ tR₁ in three clusters. *Proc. Natl. Acad. Sci.* **79**: 6171.

Lecocq, J.-P. and C. Dambly. 1976. A bacterial RNA polymerase mutant that renders λ growth independent of the *N* and *cro* functions at 42°C. *Mol. Gen. Genet.* **145**: 53.

Lee, F. and C. Yanofsky. 1977. Transcription termination at the trp operon attenuators of *Escherichia coli* and *Salmonella typhimurium*: RNA secondary structure and regulation of termination. *Proc. Natl. Acad. Sci.* **74**: 4365.

Lowery, C. and J.P. Richardson. 1977a. Characterization of the nucleoside triphosphate phosphohydrolase (ATPase) activity of RNA synthesis termination factor P. I. Enzymatic properties and effects of inhibitors. *J. Biol. Chem.* **252**: 1375.

————. 1977b. Characterization of the nucleoside triphosphate phosphohydrolase (ATPase) activity of RNA synthesis termination factor ρ. II. Influence of synthetic RNA homopolymers and random copolymers on the reaction. *J. Biol. Chem.* **252**: 1381.

Lowery-Goldhammer, C. and J.P. Richardson. 1974. An RNA-dependent nucleoside triphosphate phosphohydrolase (ATPase) associated with rho termination factor. *Proc. Natl. Acad. Sci.* **71**: 2003.

Martin, F. and I. Tinoco. 1980. DNA-RNA hybrid duplexes containing oligo (dA:rU) sequences are exceptionally unstable and may facilitate termination of transcription. *Nucleic Acids Res.* **8**: 2295.

Mills, D.R., C. Dubkin, and F.R. Kramer. 1978. Template-determined, variable rate of RNA chain elongation. *Cell* **15**: 541.

Minkley, E.G. Jr. 1973. Functional form of RNA synthesis termination factor rho. *J. Mol. Biol.* **78**: 577

Morgan, E.A. 1980. Insertions of Tn10 into an *E. coli* ribosomal RNA operon are incompletely polar. *Cell* **21**: 257.

Morgan, W.D., D.G. Bear, and P.H. von Hippel. 1983a. Rho-dependent termination of transcription. I. Identification and characterization of termination sites for transcription from the bacteriophage λ P_R promoter. *J. Biol. Chem.* **258**: 9553.

————. 1983b. Rho-dependent termination of transcription. II. Kinetics of mRNA elongation during transcription from the bacteriophage λ P_R promoter. *J. Biol. Chem.* **258**: 9565.

Neff, N.F. and M.J. Chamberlin. 1978. Termination of transcription by *Escherichia coli* RNA polymerase *in vitro* is affected by ribonucleoside triphosphate base analogs. *J. Biol. Chem.* **253**: 2455.

Nichols, B.P., M. Blumenberg, and C. Yanofsky. 1981. Comparison of the nucleotide sequence of *trp*A and sequences immediately beyond the *trp* operon of *Klebsiella aerogenes*, *Salmonella typhimurium*, and *Escherichia coli*. *Nucleic Acids Res.* **9**: 1743.

Oda, T. and M. Takanami. 1972. Observations on the structure of the termination factor rho and its attachment to DNA. *J. Mol. Biol.* **71**: 799.

O'Hare, K.M. and R.S. Hayward. 1981. Termination of transcription of the coliphage T7 "early" operon in vitro: Slowness of enzyme release, and lack of any role for sigma. *Nucleic Acids Res.* **9**: 4689.

Oxender, D., G. Zurawski, and C. Yanofsky. 1979. Attenuation in the *Escherichia coli* tryptophan operon: Role of RNA secondary structure involving the tryptophan codon region. *Proc. Natl. Acad. Sci.* **76**: 5524.

Pinkham, J. and T. Platt. 1983. Nucleotide sequence of the *E. coli* rho gene and its regulatory regions. *Nucleic Acids Res.* **11**: 3531.

Pirtle, R.M., I.L. Pirtle, and M. Inouye. 1980. Messenger ribonucleic acid of the lipoprotein of the *Escherichia coli* outer membrane. I. Nucleotide sequences at the 3' terminus and sequences of oligonucleotides derived from complete digest of the mRNA. *J. Biol. Chem.* **255**: 199.

Platt, T. 1981. Termination of transcription and its regulation in the tryptophan operon of *E. coli*. *Cell* **24**: 10.

Platt, T., H. Horowitz, and P.J. Farnham. 1983. Kinetic view of transcription termination. In *Microbiology-1983* (ed. D. Schlessinger), p. 21. American Society for Microbiology, Washington, D.C.

Ratner, D. 1976. Evidence that mutations in the *SuA* polarity suppressing gene directly affect termination factor rho. *Nature* **259**: 151.

Richardson, J.P. 1982. Activation of rho protein ATPase requires simultaneous interaction of two kinds of nucleic acid binding sites. *J. Biol. Chem.* **257**: 5760.

—————— . 1983. Involvement of a multi-step interaction between Rho protein and RNA in transcription termination. In *Microbiology-1983* (ed. D. Schlessinger), p. 31. American Society for Microbiology, Washington, D.C.

Richardson, J.P. and J.L. Carey III. 1982. Rho factors from polarity suppressor mutants with defects in their RNA interactions. *J. Biol. Chem.* **257**: 5767.

Richardson, J.P. and R. Conaway. 1980. RNA release activity of transcription termination protein ρ is dependent on the hydrolysis of nucleoside triphosphates. *Biochemistry* **19**: 4293.

Richardson, J.P. and M.R. Macy. 1981. RNA synthesis termination protein rho function: Effects of conditions that destabilize RNA secondary structure. *Biochemistry* **20**: 1133.

Richardson, J.P., C. Grimley, and C. Lowery. 1975. Transcription termination factor rho activity is altered in *Escherichia coli* with *su*A gene mutations. *Proc. Natl. Acad. Sci.* **72**: 1725.

Riley, M., B. Maling, and M.J. Chamberlin. 1966. Physical and chemical characterization of two- and three-stranded adenine-thymine and adenine-uracil homopolymer complexes. *J. Mol. Biol.* **20**: 359.

Roberts, J. 1969. Termination factor for RNA synthesis. *Nature* **224**: 1168.

—————— . 1976. Transcription termination and its control in *E. coli*. In *RNA polymerase* (ed. R. Losick and M. Chamberlin), p. 247. Cold Spring Harbor Laboratory, Cold Spring Harbor, New York.

Rosenberg, M. and D. Court. 1979. Regulatory sequences involved in the promotion and termination of RNA transcription. *Annu. Rev. Genet.* **13**: 319.

Rosenberg, M. and U. Schmeissner. 1982. Regulation of gene expression by transcription termination and RNA processing. In *Interactions of Translational and transcriptional controls in the regulation of gene expression* (ed. M. Grunberg-Manago and B. Safer), p.1. Elsevier, New York.

Rosenberg, M., D. Court, D.L. Wulff, H. Shimatake, and C. Brady. 1978. The relation between function and DNA sequence in an intercistronic regulatory region in the phage λ. *Nature* **272**: 414.

Rossi, J., J. Egan, L. Hudson, and A. Landy. 1981. The *tyrT* locus: Termination and processing of a complex transcript. *Cell* **26:** 305.

Ryan, T. and M.J. Chamberlin. 1983. Transcription analyses with heteroduplex *trp* attenuator templates indicate that the transcript stem and loop structure serves as the termination signal. *J. Biol. Chem.* **258:** 4690.

Saedler, H., H.J. Reif, S. Hu, and N. Davidson. 1974. IS2, A genetic element for turn-off and turn-on of gene activity in *E. coli*. *Mol. Gen. Genet.* **132:** 265.

Saedler, H., J. Besemer, B. Kemper, B. Rosenwirth, and P. Starlinger. 1972. Insertion mutations in the control region of the *gal* operon of *E. coli*. *Mol. Gen. Genet.* **115:** 258.

Salstrom, J.S. and W. Szybalski. 1978. Coliphage λ *nutL⁻*: A unique class of mutants defective in the site of gene N product utilization for antitermination of leftward transcription. *J. Mol. Biol.* **124:** 195.

Saraste, M., N.J. Gay, A. Eberle, M.J. Runswick, and J.E. Walker. 1981. The *atp* operon: Nucleotide sequence of the genes for the γ, β, and ε subunits of *Escherichia coli* ATP synthase. *Nucleic Acids Res.* **9:** 5287.

Schindler, D. and H. Echols. 1981. Retroregulation of the *int* gene of bacteriophage λ: Control of translation completion. *Proc. Natl. Acad. Sci.* **78:** 4475.

Sharp, J.A. and T. Platt. 1984. Rho-dependent termination and concomitant NTPase activity requires a specific intact RNA region. *J. Biol. Chem.* **259:** (in press).

Sharp, J.A., J.L. Galloway, and T. Platt. 1983. A kinetic mechanism for the poly(C)-dependent ATPase of the *E. coli* transcription termination protein, Rho. *J. Biol. Chem.* **258:** 3482.

Shigesada, K. and M. Imai. 1978. Studies on the altered rho factor in *nitA* mutants of *Escherichia coli* defective in transcription termination. *J. Mol. Biol.* **120:** 467.

Shigesada, K. and C.-W Wu. 1980. Studies of RNA release reaction catalyzed by *E. coli* transcription termination factor rho using isolated ternary transcription complexes. *Nucleic Acids Res.* **8:** 3355.

Shimizu, N. and M. Hayashi. 1974. In vitro transcription of the tryptophan operon integrated into a transducing phage genome. *J. Mol. Biol.* **84:** 315.

von Hippel, P.H., S.C. Kowalczykowski, N. Lonberg, J.W. Newport, L.S. Paul, G.D. Stormo, and L. Gold. 1982. Autoregulation of gene expression: Quantitative evaluation of the expression and function of the bacteriophage T4 gene 32 (single-stranded DNA binding) protein system. *J. Mol. Biol.* **162:** 795.

Ward, D.F. and M.E. Gottesman. 1981. The *nus* mutations affect transcription termination in *E. coli*. *Nature* **292:** 212.

————— . 1982. Suppression of transcription termination by phage lambda. *Science* **216:** 946.

Wiggs, J.L., J.W. Bush, and M.J. Chamberlin. 1979. Utilization of promoter and terminator sites on bacteriophage T7 DNA by RNA polymerases from a variety of bacterial orders. *Cell* **16:** 97.

Winkler, M.E. and C. Yanofsky. 1981. Pausing of RNA polymerase during in vitro transcription of the tryptophan operon leader region. *Biochemistry* **20:** 3738.

Wold, M.S. and R. McMacken. 1982. Regulation of expression of the *Escherichia coli dnaG* gene and amplification of the *dnaG* primase. *Proc. Natl. Acad. Sci.* **79:** 4907.

Wu, A.M., G.E. Christie, and T. Platt. 1981. Tandem termination sites at the end of the tryptophan operon in *E. coli*. *Proc. Natl. Acad. Sci.* **78:** 2913.

Wu, A.M., A.B. Chapman, T. Platt, L.P. Guarente, and J. Beckwith. 1980. Deletions of distal sequence affect termination of transcription at the end of the tryptophan operon in *Escherichia coli*. *Cell* **19:** 829.

Yanofsky, C. 1981. Attenuation in the control of tryptophan operon expression. *Nature* **289:** 751.

Yanofsky, C. and V. Horn. 1981. Rifampicin resistance mutations that alter the efficiency of transcription termination at the tryptophan operon attenuator. *J. Bacteriol.* **145:** 1334.

Young, R.A. 1979. Transcription termination in the *Escherichia coli* ribosomal RNA operon rrnC. *J. Biol. Chem.* **254:** 12725.

Zurawski, G. and C. Yanofsky. 1980. *Escherichia coli* tryptophan operon leader mutations which relieve transcription termination are *cis*-dominant to *trp* leader mutations which increase transcription termination. *J. Mol Biol.* **142:** 123.

Global Control Systems

Susan Gottesman
Laboratory of Molecular Biology
National Cancer Institute
National Institutes of Health
Bethesda, Maryland 20205

Frederick C. Neidhardt
Department of Microbiology and Immunology
The University of Michigan
Ann Arbor, Michigan 48109

INTRODUCTION

An efficient way to coordinate the synthesis of proteins that are required in fixed stoichiometric ratios under all environmental conditions is to place them in a single transcriptional unit. Prokaryotic cells make extensive use of this means of regulating the expression of genes that produce proteins of related function, such as those involved in the biosynthesis of histidine or of tryptophan. Most of the structural genes in species such as *Escherichia coli* are organized into operons consisting of two to eight individual cistrons (Bachman and Low 1980). A major portion of the work of molecular genetics in the past few decades has been devoted to unraveling the variety of ways in which the transcriptional activity of operons is adjusted by the interaction of specific protein regulators (activators and repressors), or by the process of attenuation at control sites (promoter-operator) near the beginning of each operon.

That single regulatory molecules regulating single operons is only the first step in understanding cell regulation was suggested by the long-standing observation that sugar utilization operons as a group respond to the presence of glucose in the medium (Magasanik 1962). This catabolite repression effect was demonstrated to be mediated through the cellular level of cAMP, although how the changes in cAMP are brought about is still not clear. The isolation of mutations in the cAMP synthesis (*cya*) and sensing (*crp*) systems was the first genetic approach to complex networks (Perlman and Pastan 1969; Emmer et al. 1970; Schwartz and Beckwith 1970). Two aspects of the genetic selections used have continued to be important in analyzing the regulation of networks: (1) By asking for mutations that simultaneously abolish utilization of two unrelated sugars, one can isolate mutations in regulators common to both sugar utilization operons and mapping far from either operon. (2) The use of indicator agar to screen for regulatory mutations simplifies the task of looking for negative (nonexpressing) mutations. The isolation of *crp* and *cya* mutations and the biochemical analysis of the purified components of the system demonstrated that the catabolite receptor protein, in the presence of cAMP, acts as a

163

positive regulator of transcription for the lactose operon, the maltose operon, and the arabinose operon, among others, and that this positive regulation can operate in addition to the sugar-specific regulators of transcription (de Crombrugghe and Pastan 1978; Botsford 1981).

Although the *cya/crp*-mediated catabolite repression system remains the premier example of the regulation of diverse genes, work in the last few years and the advent of new genetic and biochemical tools for the analysis of regulated genes have rapidly increased our recognition of complex and interlocking regulatory networks within the cell. These "global control systems" are characterized as a network of unlinked, separately controlled genes that share some common regulatory signals. These signals may or may not be expressed through common regulatory genes. Our understanding of how such systems work to allow the cell as a whole to respond to changes in its environment, and in particular to sudden stress, has already given us new insight into how physiological changes can be achieved. A better understanding of what networks exist in the cell and how they interact will undoubtedly be critical to an understanding of cell-division regulation and growth control, whether in bacteria or higher organisms.

In this brief paper we present some general considerations about the discovery and analysis of global control systems and then describe two such systems that have recently come into prominence.

APPROACHES TO GLOBAL CONTROL SYSTEMS

The unit of regulation in global control systems is the regulon. This term, first introduced by Maas and Clark (1964) to describe the organization of the genes encoding the arginine biosynthetic pathway in enteric bacteria, can be defined as two or more operons that share a common regulator molecule (e.g., a protein activator or repressor). These operons may be scattered on the genetic map and may be subject to separate individual controls.

The task of analyzing a regulon—the primary task in the study of global controls—has four components, usually pursued in the following order: (1) recognition of the genes and their protein products that comprise the regulon; (2) discovery of the regulatory elements (genes and proteins) that unite these genes into a single regulon; (3) elucidation of the molecular details of the activity of these controlling elements, including the nature of any metabolites ("alarmones"; Stephens et al. 1975) that may function as coeffectors; and (4) understanding the physiological significance of the regulon in the growth and survival of the cell.

Recognizing the Controlled Set of Functions

Two technical developments—one biochemical and one genetic—now make it possible to launch studies designed explicitly to discover regulons, i.e., to recognize genes and their protein products that are coregulated. The resolution of individual cellular proteins on two-dimensional polyacrylamide gels (O'Farrell 1975; O'Farrell et al. 1977), which has become one of the most useful analytical techniques in cell biology, is uniquely suited for discovering sets of coregulated gene products. Virtually all of the protein produced by a cell in a given medium (e.g., ~110 proteins in *E. coli*) can be resolved and then made visible by silver staining (Merril et al. 1980) or radiographic methods (MacConkey 1979), and various labeling protocols make it relatively easy to detect even modest changes in the rate of synthesis of individual polypeptides (Reeh et al. 1977). Recognizing coregulated gene products is done by direct observation of extracts of normal cells—mutants need not be isolated—and therefore even growth-essential proteins can be studied. Furthermore, the function of the gene products need not be known initially, because activity assays are unnecessary. Additional advantages to this approach in *E. coli* derive from the extensive characterization of the protein map of this organism (Pedersen et al. 1978; Herendeen et al. 1979), including the biochemical identification of approximately 180 polypeptide spots (Neidhardt et al. 1983b).

The second approach to the study of global control was made possible by the development of techniques for fusing genes (such as *lacZ*) randomly to chromosomal promoters (Casadaban and Cohen 1979; Beckwith 1981). After being treated with the Mu phage engineered to produce fusions, cells can be plated and large numbers of colonies screened for production of the fusion gene product (e.g., β-galactosidase) under whatever environmental condition is being studied. In this way promoters can be identified that respond to the stimulus, and they can then be mapped and their regulation analyzed. This approach is limited to discovering promoters of nonessential genes, but it is reasonable to assume that many genes that are strongly induced by a given stimulus are not essential under other conditions.

Both approaches have figured prominently in recent studies. The genetic approach using Mud*lac* fusions has been employed (Wanner and McSharry 1982) to reveal 20 promoters that are induced by phosphate starvation; the biochemical approach has led to the discovery of the high-temperature regulon of bacteria, as is described below.

Defining the Regulatory Elements

It is possible, even likely in some cases, that a group of proteins that are simultaneously induced (or repressed) by a particular environmental condition are really responding to a variety of internal signals through a variety of regulatory molecules. Genetic analysis is sine qua non for establishing that a set of proteins (recognized on gels) or promoters (recognized by Mud*lac* fusions) really constitute a regulon.

Genetic evidence that a group of genes constitute a regulon can be obtained by the isolation of a regulatory mutant. Each of the two regulons, to be described more fully later in this paper, was in fact discovered by isolation of pleiotropic mutants defective in the regulatory response. Regulatory mutants can be isolated by direct selection for an inability to respond to a particular stress. The *htpR* mutants affecting the heat-shock regulon of *E. coli* were isolated as high-temperature lethals. The Mud*lac* fusions provide another means for defining regulatory elements. After stabilizing the *lacZ* fusions to prevent Mu-promoted rearrangements or hops, constitutive mutants can be selected on lactose, and noninducibles can be selected by the toxic lactose analog, t*ONPG (Beckwith 1981). Once the mutants have been obtained, techniques that by now have become routine can be used to clone the regulatory gene and identify its product, as is illustrated in the analysis of the two regulons that follow. Isolation of the control elements can then permit in vitro studies to probe the molecular details of the regulation.

Metabolic Signalers

Discovery of the low-molecular-weight effectors that act as triggers or signals for a regulon is not as easy as finding the regulatory gene and its product. Some of the known metabolites that function in this way are nucleotides (e.g., cAMP and guanosine tetraphosphate). New techniques are now available to display all the nucleotides of a cell (Bochner and Ames 1982), and these should facilitate searches for alarmones. However, even in situations where the alarmone has been identified, it is not always clear how the cell regulates the alarmone level or how the alarmone works.

Cell Function

Finally, there remains the task of evaluating the role of the regulon in the life of the cell. Interestingly, even the existence of mutants with well-defined phenotypes and defective in the control of the regulon may not give the complete answer to the significance of the regulon. This situation is well illustrated by the high-temperature regulon. It was some time after analysis of this regulon that it was discovered that it is triggered equally well by the addition of ethanol to the medium as by raising the temperature. Furthermore, it is not at all easy to sort out which phenotypic consequences of a pleiotropic regulatory mutant are the consequence of which particular operons of the regulon. This point is also illustrated by the *htpR* mutant that dies at 42°C—it is not yet known which of the 14 known genes of the regulon is (are) responsible for preventing death at high temperature.

In this study, we examine in depth two systems in which we know a great deal about the members of the controlled set, the mechanisms of their control and, at least in one case, the functions served by the controlled genes. In the first case, the SOS system of *E. coli*, many and possibly all members of the controlled set are dispensable functions, needed only after the emergency stress of DNA damage. The operon fusion techniques used to study this system have depended on the dispensability of these genes. In the second case, the heat-shock genes of *E. coli*, a primarily biochemical approach was necessary to look at a set of genes that may play roles essential for cell growth.

SOS SYSTEM

DNA damage and the interruption of DNA replication can have disastrous consequences for the cell; in response, *E. coli* has developed a well-regulated and redundant system for repairing DNA damage. Lysogenic phage that inhabit the *E. coli* chromosome have utilized the system to ensure their ability to "jump ship" if the damage to the cell is too severe. Repair systems include excision of DNA damage products, such as pyrimidine dimers, recombination between damaged genomes to reconstitute intact copies, and error-prone DNA replication. Most of the enzymes necessary for these repair processes are synthesized de novo after DNA damage and are jointly controlled by two regulatory genes, *recA* and *lexA*, in a network known as the SOS system (Radman 1975; Witkin 1976; Little and Mount 1982). In addition to a rapid induction of new gene products in response to damage, the system has a mechanism for self-limitation, ensuring a rapid return to equilibrium after the damage has been repaired. Although some of the early steps in the induction of this network remain obscure, much of the process has been elegantly demonstrated genetically and biochemically.

Role of the RecA Protease

The critical and unique regulator for the SOS network is the RecA protein, a key recombination protein of *E. coli*. Mutations in the *recA* gene reduce the level of homologous recombination 1000-fold and cannot easily be overcome by secondary mutations (Clark 1973). Purified RecA protein has a number of DNA interaction properties that hint at its role in recombination. It stimulates DNA strand assimilation and, by variations on this reaction, can create a "recombinant" or hybrid molecule from a double-stranded DNA circle and a single-stranded homologous invading strand (Radding 1982).

RecA is also capable of carrying out a very different reaction, specific to the recovery from DNA damage. After DNA damage it acts as a protease and clips specific repressors of transcription, allowing induction of a set of DNA repair functions. Its substrates fall into two classes: The bacterial repressor LexA represents the sole known member of one class, and the lysogenic phage repressors, such as the λcI product, represent the second class. The proteolytic cleavage of these repressors initiates a cascade of inductions, both for the prophage, now able to enter lytic growth and escape its damaged host, and for the bacteria, which synthesizes the functions necessary for the repair of the damage.

The role of proteolysis in initiating this cascade was first recognized through study of the mechanism of induction in bacteriophage λ. Roberts and Roberts (1975) were able to demonstrate that the cI repressor is cleaved shortly after UV or mitomycin treatment of lysogenic cells. Mutant λ repressor that cannot be induced by UV (ind⁻) is not cleaved, and mutations in the host recA gene render the cell uninducible for λ prophage and unable to carry out the cleavage of repressor. Purified RecA protein can be demonstrated to cleave and inactivate cI repressor in vitro in a purified system; the reaction requires only RecA protein, repressor, ATP, dATP or an ATP analog such as ATP-γ-S, and a polynucleotide (Roberts et al. 1978; Craig and Roberts 1980; 1981).

Given these observations, it was quickly demonstrated that LexA protein is also a substrate for the RecA protease and that the dominant, UV-sensitive lexA mutations, which had long puzzled workers in this field, were ind⁻ (uncleavable) alleles of lexA (Little et al. 1980). If, in fact, LexA serves as a repressor for cellular repair functions in the same way that λ repressor acts for phage functions, RecA proteolysis of LexA might be sufficient to turn on expression of SOS functions. This has, in fact, proved to be the case (see Fig. 1 and discussion below.)

The most puzzling aspect of this induction mechanism remains the question of how the RecA protein, present in growing cells but inactive as a protease, is changed into an active protease within minutes after DNA damage. Increasing the level of RecA in the cell is not sufficient to cause induction (Ogawa et al.

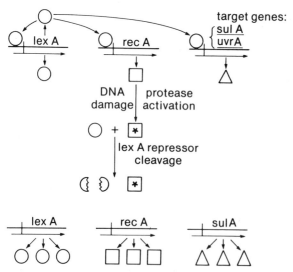

Figure 1   Model for control of the *E. coli* SOS functions. The product of the *recA* gene, after DNA damage, is activated to act as a protease and cleave *lexA* repressor. After DNA damage is repaired, LexA protein will accumulate and rerepress the SOS operons.

1979). No differences in the in vitro characteristics of RecA protein itself can be detected when isolated from SOS-induced cells, strongly suggesting that any modification of RecA that occurs is easily reversible during purification. The in vitro requirement for a polynucleotide for cleavage of λ repressor or LexA suggests that single-strand gaps in the DNA or polynucleotide pieces formed by an interruption of DNA synthesis may accumulate and activate RecA. In vitro, there does not seem to be any specificity for a particular polynucleotide, and even mutations in *recA* that have protease activity in vivo in the absence of DNA damage (*tif*) do not abolish the polynucleotide requirement in vitro. Purified *tif* protease, however, does interact more efficiently than wild-type RecA with polynucleotide and dATP to carry out proteolysis (Phizicky and Roberts 1981).

Regulated Set of Genes

The observation that many of the repair and mutagenic capacities of the cell appear only after UV treatment and do not appear in *recA* and *lexA* mutant cells led to the proposal that an SOS regulatory network existed. The in vivo definition of the extent of these regulated functions was one of the first and most elegant exploitations of the Mud*lac*-Amp fusion technique (Casadaban and Cohen 1979). Starting with the observation that new functions appear after UV treatment of cells, Kenyon and Walker (1980) defined genes whose transcription is responsive to DNA damage by isolating random insertions of Mud*lac*-Amp into the *E. coli* chromosome and replica-plating the isolated colonies to lactose indicator plates with and without mitomycin, an inducer of the SOS response. Using this method, they isolated ten independent fusions in genes they call *din* (*d*amage *in*ducible), which turn on β-galactosidase synthesis after SOS-induction treatment. Mapping and further characterization led to the finding of at least five separate genes, including the *uvrA* gene (Kenyon and Walker 1981). The functions of a number of these genes are not yet known, but *uvrB*, *sfiA*, *umuC*, and the *recA* and *lexA* genes themselves also form part of the SOS-induced gene set (see Table 1; Brent and Ptashne 1980;

Table 1   Functions of some genes regulated as part of the SOS system

Gene name	Map location	Function
recA	58'	general recombination; protease
lexA	91'	negative regulator of transcription
umuC,D	25'	mutagenesis
uvrA	92'	excision of UV damage
uvrB	17'	excision of UV damage
sulA	22'	inhibitor of septation[a]

[a]*sulA* = *sfiA*; *sulB* = *sfiB*.

TACTGTATGAGCATACAGTA   <u>rec</u>A operator
TGCTGTATATACTCACAGCA   <u>lex</u>A operator 1
AACTGTATATACACCCAGGG   <u>lex</u>A operator 2
TACTGTATATTCATTCAGGT   <u>uvr</u>A operator
AACTGTTTTTTTATCCAGTA   <u>uvr</u>B operator
TACTGTACATCCATACAGTA   <u>sul</u>A
taCTGTatat–cat–CAG–a   consensus

Figure 2 Consensus sequence for LexA binding. This sequence has been found in all LexA-repressed operons. Capital letters are conserved in all six sequences (Beck and Bremer 1980; Little and Mount 1982).

Bagg et al. 1981; Fogliano and Schendel 1981; Huisman and D'Ari 1981; Kenyon and Walker 1981; Sancar et al. 1982).

In vitro, LexA can be shown to repress transcription of RNA from the *lexA* promoter, the *recA* promoter, as well as *sfiA*, *uvrA*, and *uvrB*, and a number of the *din* genes (Brent and Ptashne 1981; Little et al. 1981; Kenyon et al. 1982; Sancar et al. 1982; Mizusawa et al. 1983). These operons have in common a sequence of about 20 bp, in the promoter region, of which 7 are invariant in the eight cases now examined (Fig. 2). DNase protection studies have demonstrated LexA binding to this region of the *recA*, *lexA*, and *uvrB* genes. These results all support the hypothesis that LexA itself directly represses these genes.

The genetic definition of regulation by LexA is provided by examining the inducibility of a given gene in a *recA* mutant that is also lacking LexA repressor through mutation in the *lexA* gene (*spr* or *tsl*). These cells are missing RecA protease and therefore cannot initiate the induction of the SOS system after DNA damage. However, because they are also missing the LexA repressor, they should constitutively express functions that are directly repressed by LexA. Thus far, all SOS-inducible functions tested are, in fact, constitutive in such a *recA lexA⁰* cell, suggesting that LexA may be the only cellular repressor cleaved by RecA (Huisman and D'Ari 1981; Kenyon et al. 1982). Therefore, the SOS induction system is tightly linked by the RecA cleavage of LexA. Variability in response of given operons to different UV doses may reflect different affinities of the LexA-binding sites in these operons (Kenyon et al. 1982). λ repressor is a much poorer substrate for the RecA protease than LexA, so that λ prophage induction may not occur unless the cell is severely damaged by UV or other inducing treatments (Little et al. 1981).

The method of defining regulated genes by insertion of Mud*lac*-Amp does not, of course, detect genes that are essential for cell survival or genes that are themselves involved in induction of the system. Therefore, it is not surprising that Mud*lac*-Amp insertions in the *lexA* and *recA* genes were not detected in the experiments of Kenyon and Walker, and there may exist an additional set of essential functions that are induced after DNA damage. It is certainly clear from the examples already examined that there is nothing inconsistent about constitutive expression of a gene that may be further induced by the SOS system. RecA itself plays an important role as a recombination enzyme and must be present in the cell when damage occurs to initiate the inducing cycle. *uvrB* has three promoters, all active in vitro and only two of which are repressed by LexA (Sancar et al. 1982). *lexA* of course, is also present in uninduced cells, where it serves as a repressor.

Reestablishing Equilibrium

At first glance, it seems odd that LexA should regulate its own synthesis, since it will accumulate in just that situation when it is not needed, when it is being actively cleaved by RecA. However, if one makes the assumption that LexA affinity for its own operator will be somewhat less than its affinity for the *recA* operator, then this self-regulatory circuit appears as an efficient mechanism for increasing the synthesis rate of LexA such that it will be in plentiful supply to rerepress the *recA* operon, as well as the other SOS genes, as soon as RecA protease activity declines. Since RecA is apparently not covalently modified to become a protease, the disappearance of the still mysterious polynucleotide activator from the cell will inactivate remaining RecA, and repression will return. In fact, LexA does bind more tightly to the *recA* operator than to either the *uvrB* or *lexA* operator (Brent and Ptashne 1981). Measurements of LexA half-life indicate that it changes from a few minutes immediately after DNA damage induction to 30−40 minutes within a few hours of induction if repair has taken place (Little and Mount 1982). Studies of the induction and disappearance of UV-induced mutagenesis and repair capabilities suggest similar kinetics.

The return to equilibrium includes more than simply the shutoff of transcription of the induced set of genes. A network that responds to emergencies and solves them by whatever means are available may include functions that are not compatible with normal growth. For the SOS system, such lethal functions are suggested by the finding that mutations that constitutively induce the system (either the *tif* mutation or the *tsl* or *spr* alleles of *lexA*) are lethal unless the cell carries an additional mutation in the *sulA* or *sulB* genes (George et al. 1975).

SulA and SulB seem to be involved in the arrest of cell septation during DNA repair; cells that are constitutively expressing the SOS functions form long, nonseptated filaments. *sulA* or *sulB* mutations abolish this filamentation without affecting any of the other SOS-induced functions. *sulA* is apparently a typical example of an SOS-regulated gene. Fusions of Mud*lac*-Amp into *sulA* express β-galactosidase after SOS-inducing treatments (Huisman and D'Ari 1981), the *sulA* gene carries a typical LexA-binding site in the promoter, and transcription of *sulA* is repressed by LexA in vitro (Mizusawa et al. 1983). *sulB*, on the other hand, has not been shown to be inducible after SOS-inducing treatments, and recent evidence suggests that *sulB* mutations are alleles of the essential cell-division function *ftsZ* (Lutkenhaus 1983). Therefore, SulA may act by inhibiting the action of SulB, thereby preventing septation.

Although the inhibition of cell division during DNA repair may be useful to the cell, this is clearly not a function that can persist once cell growth has resumed. The cell has developed an additional safeguard for returning to equilibrium and resuming cell division. The SulA product is extremely unstable, with a half-life of only 1.2 minutes (Mizusawa and Gottesman 1983). Therefore, any SulA synthesized during SOS induction will disappear rapidly when repression of the *sulA* gene is returned. The importance of this instability in recovering from DNA damage is demonstrated in *E. coli lon* cells, which lack the protease activity necessary for SulA degradation. In these cells, SulA has a half-life of 19 minutes (Mizusawa and Gottesman 1983). As predicted, SOS induction in *lon* cells is lethal, unless the cells carry an additional mutation in *sulA* or *sulB*. *lon* cells express their DNA repair functions normally but filament lethally even after mild UV doses (Huisman et al. 1980; Gottesman et al. 1981).

The SOS regulatory system comprises a fascinating collection of reversible and irreversible processes that interact to provide new functions to the cell after DNA damage and a rapid return to equilibrium when the functions are no longer needed. The initiating event is an apparently reversible change in the activity of RecA protein to convert it into a highly specific protease; the cleavage of LexA substrate by the protease initiates an irreversible induction of the LexA-repressed functions. This irreversibility is countered by the self-regulation of *lexA*, so that high levels of LexA rapidly accumulate in induced cells, poising the system for a return to repression whenever the RecA protease activity drops. Finally, at least one of the induced functions, the potentially lethal SulA septation inhibitor, is itself subject to rapid proteolysis, not by the RecA protease, but by an apparently constitutive protease, the product of the *E. coli lon* gene. Thus, both classical repressors and regulation through proteolysis combine to increase the responsiveness of the system beyond that which could be provided by either mechanism alone.

HIGH-TEMPERATURE RESPONSE OF *E. coli*

Effect of Temperature on Growth and Gene Expression

The response of *E. coli* to shifts in temperature is both dramatic and complex (Lemaux et al. 1978). When wild-type cells are shifted from one temperature to another within the normal range of growth temperature (20−40°C), the overall rate of protein synthesis quickly (within a minute or two) changes to its new steady-state rate, but the differential rate of synthesis of most individual proteins either increases or decreases transiently as much as 50-fold. Within approximately 20 minutes these transient changes abate, and the new steady-state rates of synthesis are achieved. In most cases these steady-state differential rates are independent of the temperature. The magnitude of the transient responses depends on the extent of the temperature shift, and a shift downward in temperature generates a response opposite to a shift upward. In a few cases, these transient changes accomplish a rapid attainment of the new steady-state level and therefore may be active regulatory responses to temperature shifts. For a large number of proteins, these transient changes are of unknown biological significance. They may represent passive effects of temperature on promoter structure followed by active reestablishment of the preshift level of function. In any event, these changes are of sufficient magnitude that they must be taken into account when interpreting experiments that involve temperature shifts with mutant or normal cells.

When these results were first obtained, they underscored how little was known of how *E. coli* adjusts its affairs to maximize growth rate in the normal temperature range (20–37°C) and also what prevents this optimization during restricted growth at the lower (9–20°C) and upper (37–49°C) ranges for growth. The first comprehensive study of the adjustment of gene activity at different temperatures was undertaken using the two-dimensional gel technique. It was completed in 1979 (Herendeen et al. 1979), with the following conclusions: (1) There is little variation in the levels of most proteins in the midrange of temperature; (2) the restricted rate of growth outside this range is accompanied by marked changes in levels of many proteins; (3) a few proteins are thermometerlike in varying simply over the whole temperature range irrespective of cellular growth rate; and (4) proteins in certain metabolic regulation classes (Pedersen et al. 1978) display a homogeneous, class-specific response to temperature.

Discovery of the High-temperature Regulon of *E. coli*

Out of these studies emerged a group of 14 polypeptides that seemed to merit further study. They shared a common response to temperature shifts. All were dramatically, but transiently, derepressed during a shift up in temperature and repressed following a shift down. All but one or two of them also displayed a permanent elevation in level at temperatures above 37°C. In one case this steady-state derepression was particularly noteworthy: Protein B56.5 is one of the major proteins of the cell at normal temperatures (comprising ~1% of the cellular protein at 37°C), and its steady-state rate of synthesis at 46°C accounted for over 12% of the total protein synthesized (Neidhardt et al. 1981). Furthermore, the cellular levels of all of these polypeptides varied no more than twofold at any one temperature in ordinary media, no matter what the growth rate of the cells.

The major breakthrough in the study of this group of coregulated proteins came with the recognition that a mutant existed in which their high-temperature induction was abolished. Cooper and Ruettinger (1975) isolated a mutant that was temperature-sensitive lethal (at 42°C). The parental strain, SC122, possessed a temperature-sensitive nonsense suppressor, and it could be shown that the conditional lethal mutation in the mutant, K165, was a nonsense mutation in a chromosomal gene mapping between 60 minutes and 80 minutes. Our study of this mutant quickly revealed the exciting finding that heat induction of the entire set of 14 polypeptides was abolished in the mutant. Upon a shift from 28°C (permissive) to 42°C (restrictive) the mutant accelerates its growth in an apparently normal fashion for approximately a mass doubling but without induction of the 14 high-temperature proteins. The cells then die and lyse (Neidhardt and VanBogelen 1981).

Given that the mutation in strain K165 was a nonsense mutation, it appeared that the gene must code for a protein that in some way was essential for growth at high temperature and for the heat induction of the 14 polypeptides. We called this gene *htpR*, and the set of proteins it controls, the HTP (for *high-temperature production*) regulon. Essentially the same conclusion was drawn from similar observations by Yamamori and Yura (1982), who called the putative regulatory gene *hin*.

Two plasmids of the Clarke-Carbon recombinant ColE1 bank (Clarke and Carbon 1976), pLC31-16 and pLC31-32, were found to yield transconjugants with strain K165 that grew well at high temperature and had recovered heat inducibility for the HTP regulon (Neidhardt and VanBogelen 1981). Both plasmids were known to carry the *ftsE* gene (mapped at ~73−74 min), and we were able to show that they also carried the *livJ* and *livK* genes, which map at 75.4 minutes. We therefore assigned the *htpR* gene a position at approximately 75 minutes, consistent with the earlier finding by Cooper and Ruettinger (1975) that the essential gene(s) mutated in strain K165 were complemented by F' 41, covering the region from 63 minutes to 76 minutes.

A few of the genes for the 14 polypeptides of the HTP regulon were identified and mapped (see below) at scattered positions distant from *htpR*. Since the *htpR* product functions in *trans* (as shown by plasmid complementation) and is blocked by a nonsense mutation, we postulated that it codes for a positive regulatory protein that stimulates expression of the HTP genes upon a shift up in temperature (Neidhardt and VanBogelen 1981).

Molecular Cloning and Expression of the Regulatory Gene

Plasmid pLC31-16 was used to construct a restriction map of the region containing *htpR* (Fig. 3). Below the circular map in Figure 3, restriction fragments are drawn that were subsequently cloned into derivatives of pBR322. Analysis of these fragments showed that (1) the *htpR* gene lies within the 1.69-kb segment subcloned in pFN97, and (2) the mutation in the *htpR* allele of strain K165 lies within the 0.75-kb segment that is shared by pFN82 and pFN92 (Neidhardt et al. 1983a).

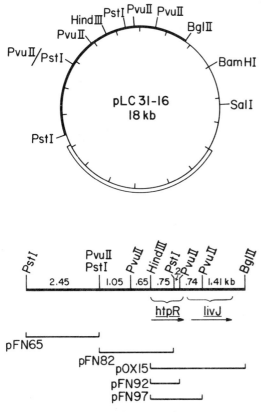

Figure 3   Restriction map of the *E. coli* insertion in plasmid pLC31-16. The double
line in the circular map represents ColE1 DNA. The portion of the insertion indicated
by the heavy line is drawn in more detail as the horizontal segment immediately below
the circular map. The approximate locations of *htpR* and *livJ* (R. Landrick and D.
Oxender, unpubl.) are shown, together with the probable direction of transcription (→)
of *htpR* inferred from the properties of the subcloned fragments shown by the other
horizontal lines (Neidhardt et al. 1983a).

The results in Figure 4 illustrate the normal induction of the HTP regulon in
an *htpR⁺* strain, the absence of such an induction in an *htpR* mutant, and the
resumption of the response in this mutant carrying a normal allele of the
regulatory gene on the plasmid pFN97. Plasmid pFN92 fails to complement
the mutant allele, but in a *recA⁺* host it yields wild-type recombinants.

Identification of the *htpR*-gene Product

The results shown in Figure 4 suggested a candidate for the product of *htpR*.
The arrow in Figure 4A indicates a polypeptide with a molecular weight of

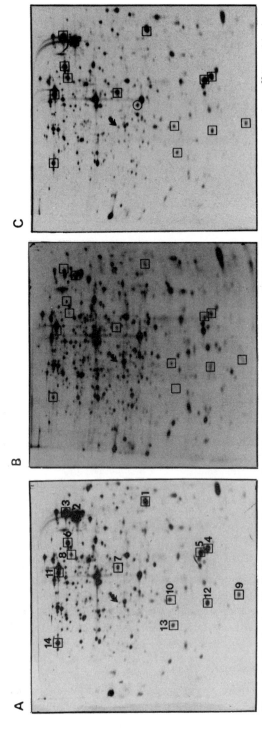

Figure 4 Synthesis of individual proteins in various *E. coli* strains after a shift from 28°C to 42°C. The autoradiographs show [³⁵S]methionine incorporation from 4 min to 7 min after the shift. The numbered boxes indicate polypeptide spots that were induced only in cells carrying *htpR⁺*. (→) The position of F33.4. (*A*) Strain SC122 (*htpR⁺*). (*B*) Strain K165 (*htpR*). (*C*) Strain K166 (*htpR*) with plasmid pFN97 (*htpR⁺*). The circled protein is the plasmid-encoded β-lactamase (Neidhardt et al. 1983a).

approximately 33,000. It is visible in the normal strain and in the mutant strain that also carries the *htpR*$^+$ allele on a plasmid (Fig. 4C) but is not visible in the *htpR* mutant (Fig. 4B). On the basis of its isoelectric point and its apparent size, this polypeptide is designated F33.4 (Pedersen et al. 1978). Each of the plasmids containing cloned restriction fragments on pLC31-16 was transferred into minicells (Reeve 1979) or maxicells (Sancar et al. 1979) to study the expression of the genes carried by the plasmids. Each DNA segment that complemented *htpR* made a polypeptide that migrated similarly to F33.4 on two-dimensional gels. In addition, plasmid pFN92, which fails to complement (but does yield recombinants), coded for a very similar polypeptide. Subsequent testing, however, revealed that the polypeptide encoded by this plasmid is smaller than that produced by the complementing plasmids.

From the size of F33.4 (equivalent to ~0.9 kb of coding capacity), the *htpR* gene must occupy most of the pFN92 segment (0.95 kb) plus a small stretch extending into the pFN97 segment. As a corollary, the *htpR* gene must be read in the direction shown by the arrow in Figure 4, since pFN92 must be missing only some carboxyterminal amino acids rather than its promoter and initiation sites. The entire sequence of this region of DNA has been determined (R. Landrick, pers. comm.), and the results are completely consistent with this interpretation.

The fact that the slightly truncated version of F33.4 made by plasmid pFN92 is physiologically inactive provides strong support for the notion that F33.4 is the product of the regulatory gene *htpR*. Confirmation of the prediction that it is a "positive" regulatory protein (postulated from the information about the role of the *htpR* gene in the heat-shock response) awaits the results of experiments currently in progress. A fascinating possibility arises from the work of Yamamori et al. (1982), who showed that there is enhanced synthesis of HTP polypeptides when cellular levels of the σ subunit of RNA polymerase are lowered. This result might reflect only an indirect effect, but it might indicate that the HtpR protein interacts not with HTP promoters directly but with polymerase core enzyme.

Response of the HTP Regulon to Temperatures Above the Growth Range

Normal strains of *E. coli* are capable of sustained (balanced) growth in rich medium up to a temperature of 46–48°C. It is clear, therefore, that expression of all growth essential genes occurs along with the induced expression of the HTP regulon under these conditions. What happens at temperatures too high for growth? A shift to 50°C halts growth quite abruptly, but there is residual protein synthesis at an easily detected rate. Virtually all this synthesis is accounted for by the activity of the HTP regulon; there is a virtual cessation of translation from other genes. In a general sense the pattern resembles protein synthesis in any eukaryotic cell exposed to growth-restricting temperatures, but there is extensive production of multiple species of the few polypeptides that are produced at 50°C. These species appear to be the result of large-scale mistranslation at high temperature (R.A. VanBogelen and F.C. Neidhardt, unpubl.).

Response of the HTP Regulon to Ethanol

In *Drosophila* a large number of conditions or chemical agents induce the heat-shock proteins or a subset of these proteins (Ashburner and Bonner 1979). In yeast, ethanol induces the heat-shock proteins and also confers thermotolerance (Plesset et al. 1982). We have recently found that the proteins of the HTP regulation of *E. coli* are induced when cells are grown in the presence of ethanol and that this response of the regulon is controlled by the protein product of the *htpR* gene (R.A. VanBogelen et al., in prep.).

Addition of 4% ethanol to cells growing at 28°C had a similar effect to a shift to 42°C. Growth slowed, and the HTP proteins were induced. With 10% ethanol the induction was even greater and was comparable to a shift from 28°C to 50°C. The *htpR* mutant that is defective in the synthesis of F33.4 displayed a diminished induction and was hypersensitive to ethanol. In 10% ethanol the heat-shock polypeptides were made almost exclusively and most appeared as multiple spots on two-dimensional gels, suggestive of high-level mistranslation (R.A. VanBogelen et al., in prep.)

Heat and ethanol apparently affect a common cellular process or component which, acting through the *htpR*-gene product, induces the same set of polypeptides even under conditions where net growth and the expression of most cellular genes are inhibited.

Proteins and Genes of the HTP Regulon

Current work directed at constructing a gene-protein index of *E. coli* (Neidhardt et al. 1983b) has helped identify some of the 14 polypeptides of the HTP regulon. In Table 2 we list the HTP polypeptides together with current information about their biochemical identity and the map locations of their structural genes.

Unlike the situation with the SOS regulon, one cannot divine the biological function of the HTP regulon from this partial knowledge of its components. The products of the *groEL* and *groES* genes are necessary for steps in the morphogenesis of most of the phages of *E. coli* (Tilly and Georgopoulos 1982) and seem to be essential for bacterial growth as well, based on the fact that most *groE* mutants are temperature-sensitive for growth (R.A. VanBogelen and F.C. Neidhardt, unpubl.). The *dnaK*-gene product is thought to be necessary for some step in initiation of phage DNA synthesis, but its function in the uninfected cell is unknown. Finally, there is a peculiar situation that one of the proteins is an auxiliary lysyl-tRNA synthetase, the product of the *lysU* gene (Hirshfield et al. 1981; VanBogelen et al. 1983). The genes for these products are not linked to the regulatory gene *htpR*. The identities of the

Table 2   Heat-shock polypeptides of *E. coli* K12

Protein number[a]	Biochemical identification	Genetic map location (min)
1 (B25.3)	—	—
2 (B56.5)	*groEL*-gene product	93.5
3 (B66.0)	*dnaK*-gene product	0.5
4 (C14.7)	—	—
5 (C15.4)	*groES*-gene product	93.5
6 (C62.5)	—	—
7 (D33.4)	—	—
8 (D60.5)	*lysU*-gene product (second form of lysyl-tRNA synthetase)	93.0
9 (F10.1)	—	—
10 (F21.5)	—	—
11 (F84.1)	—	—
12 (G13.5)	—	—
13 (G21.0)	—	—
14 (H94.0)	—	—

[a]The numbers refer to Fig. 4A. The alphanumeric designations (in parentheses) follow the convention described by Pedersen et al. (1978).

remaining ten HTP polypeptides are unknown, as are the map locations of their structural genes; but it is known that they are not tightly linked to the *htpR* gene, because they are not encoded by plasmids that include this gene.

Biological Function of the HTP Regulon

One would have hoped to learn from the nature of the HTP proteins what role the regulon plays in the life of the cell. Unfortunately, we are left only with the primary information that induction of this regulon confers thermal tolerance on the cell (Yamamori and Yura 1982) and resistance to the deleterious effects of ethanol as well. We know neither the nature of the presumed damage to the cell that triggers the response, nor in what way the response may be advantageous. Although membrane damage may be thought to constitute a likely common factor in the cellular encounter with heat and ethanol, none of the HTP proteins so far identified seem to be integral membrane proteins (R.A. VanBogelen and F.C. Neidhardt, unpubl.), and there are a multitude of other possibilities. Hopefully, an answer will be forthcoming from the extensive ongoing work with this system. If so, the answer is likely to be of considerable general significance, because not only is the heat-shock response similar in general terms from bacteria to man, but the similarity even includes the extraordinary finding that the *dnaK* gene of *E. coli* is 45–50% homologous (by DNA sequence analysis) to the hsp70 heat-shock gene of *Drosophila* (Craig et al. 1982).

CONCLUSION

The two global control systems reviewed here share some similarities. Each fits the definition of a regulon, each involves a dramatic cellular response to potentially lethal situations, and each has its counterpart in cellular systems throughout the biological world.

Luigi Gorini, next to whom one of us (F.C.N.) was privileged to work closely in the late 1950s (in the Department of Bacteriology and Immunology at Harvard) studied the first recognized regulon of *E. coli*. S.G. was a graduate student in the same department 10 years later and was enriched by Gorini's presence. We believe he would have been deeply interested in this aspect of bacterial physiology as it is currently emerging, and we wish we had the benefit of his good ideas and intuition today, as well as the pleasure of sharing his joy in exploration.

ACKNOWLEDGMENT

The experimental work on the high-temperature response and part of the preparation of this review were supported by U.S. Public Health Service grant GM-17892 from the National Institute of General Medical Sciences to F.C.N.

REFERENCES

Ashburner, M and J.J. E   .er. 1979. The induction of gene activity in *Drosophila* by heat shock. *Cell* **17**: 241.

Bachman, B.J. and K.B. Low. 1980. Linkage map of *Eschericia coli* K-12, edition 6. *Microbiol. Rev.* **44**: 1.

Bagg, A., C.J. Kenyon, and G.C. Walker. 1981. Inducibility of a gene product required for UV and chemical mutagenesis in *Escherichia coli*. *Proc. Natl. Acad. Sci.* **78**: 5749.

Beck, E. and E. Bremer. 1980. Nucleotide sequence of the gene *ompA* coding the outer membrane protein II* of *Escherichia coli* K-12. *Nucleic Acids Res.* **8**: 3011.

Beckwith, J. 1981. A genetic approach to characterizing complex promoters in *E. coli*. *Cell* **23**: 307.

Bochner, B.R. and B.N. Ames. 1982. Complete analysis of cellular nucleotides by two-dimensional thin layer chromatography. *J. Biol. Chem.* **257**: 9759.

Botsford, J.L. 1981. Cyclic nucleotides in procaryotes. *Microbiol. Rev.* **45**: 620.

Brent, R. and M. Ptashne. 1980. The *lexA* gene product represses its own promoter. *Proc. Natl. Acad. Sci.* **77**: 1932.

————. 1981. Mechanism of action of the *lexA* gene product. *Proc. Natl. Acad. Sci.* **78**: 4204.

Casadaban, M.J. and S. Cohen. 1979. Lactose genes fused to exogenous promoters in one step using a Mu-*lac* bacteriophage: *In vivo* probe for transcriptional control sequences. *Proc. Natl. Acad. Sci.* **76**: 4530.

Clark, A.J. 1973. Recombination deficient mutants of *E. coli* and other bacteria. *Annu. Rev. Genet.* **7**: 69.

Clarke, L. and J. Carbon. 1976. A colony bank containing synthetic ColE1 hybrid plasmids representative of the entire *E. coli* genome. *Cell* **9**: 91.

Cooper, S. and T. Ruettinger. 1975. A temperature sensitive nonsense mutation affecting the synthesis of a major protein of *Escherichia coli* K-12. *Mol. Gen. Genet.* **139**: 167.

Craig, E., T. Ingolia, M. Slater, L. Manseau, and J. Bardwell. 1982. *Drosophila*, yeast and *E. coli* genes related to the *Drosophila* heat-shock genes. In *Heat shock from bacteria to man* (ed. M.J. Schlessinger et al.), p. 11. Cold Spring Harbor Laboratory, Cold Spring Harbor, New York.

Craig, N.L. and J.W. Roberts. 1980. *E. coli recA* protein-directed cleavage of phage λ repressor requires polynucleotide. *Nature* **283**: 26.

_____. 1981. Function of nucleoside triphosphate and polynucleotide in *Escherichia coli recA* protein-directed cleavage of phage λ repressor. *J. Biol. Chem.* **256**: 8039.

de Crombrugghe, B. and I. Pastan. 1978. Cyclic AMP, the cyclic AMP receptor protein, and their dual control of the galactose operon. In *The operon* (ed. J.H. Miller and W.S. Reznikoff), p. 303. Cold Spring Harbor Laboratory, Cold Spring Harbor, New York.

Emmer, M., B. de Crombrugghe, I. Pastan, and R. Perlman. 1970. Cyclic AMP receptor protein of *E. coli*: Its role in the synthesis of inducible enzymes. *Proc. Natl. Acad. Sci.* **66**: 480.

Fogliano, M. and P.R. Schendel. 1981. Evidence of the inducibility of the *uvrB* operon. *Nature* **289**: 196.

George, J., M. Castellazzi, and G. Buttin. 1975. Prophage induction and cell division in *E. coli*. III. Mutations *sfiA* and *sfiB* restore division in *tif* and *lon* strains and permit the expression of mutator properties of *tif*. *Mol. Gen. Genet.* **140**: 308.

Gottesman, S., E. Halpern, and P. Trisler. 1981. Role of *sulA* and *sulB* in filamentation by *lon* mutants of *Escherichia coli* K-12. *J. Bacteriol.* **148**: 265.

Herendeen, S.L., R.A. VanBogelen, and F.C. Neidhardt. 1979. Levels of major proteins of *Escherichia coli* during growth at different temperatures. *J. Bacteriol.* **139**: 185.

Hirshfield, I.N., P.L. Bloch, R.A. VanBogelen, and F.C. Neidhardt. 1981. Multiple forms of lysyl-transfer ribonucleic acid synthetase in *Escherichia coli*. *J. Bacteriol.* **146**: 345.

Huisman, O. and R. D'Ari. 1981. An inducible DNA replication-cell division coupling mechanism in *E. coli*. *Nature* **290**: 797.

Huisman, O., R. D'Ari, and J. George. 1980. Further characterization of *sfiA* and *sfiB* mutations in *Escherichia coli*. *J. Bacteriol.* **144**: 185,

Kenyon, C.J. and G.C. Walker. 1980. DNA-damaging agents stimulate gene expression at specific loci in *Escherichia coli*. *Proc. Natl. Acad. Sci.* **77**: 2819.

_____. 1981. Expression of the *E. coli uvrA* gene is inducible. *Nature* **289**: 808.

Kenyon, C.J., R. Brent, M. Ptashne, and G.C. Walker. 1982. Regulation of damage-inducible genes in *Escherichia coli*. *J. Mol. Biol.* **160**: 445.

Lemaux, P.G., S.L. Herendeen, P.L. Bloch, and F.C. Neidhardt. 1978. Transient rates of synthesis of individual polypeptides in *E. coli* following temperature shifts. *Cell* **13**: 427.

Little, J.W. and D.W. Mount. 1982. The SOS regulatory system of *Escherichia coli*. *Cell* **29**: 11.

Little, J.W., D.W. Mount, and C.R. Yanisch-Perron. 1981. Purified lexA protein is a repressor of the *recA* and *lexA* genes. *Proc. Natl. Acad. Sci.* **78**: 4199.

Little, J.W., S.H. Edmiston, L.Z. Pacelli, and D.W. Mount. 1980. Cleavage of the *Escherichia coli lexA* protein by the *recA* protease. *Proc. Natl. Acad. Sci.* **77**: 3225.

Lutkenhaus, J.F. 1983. Coupling of DNA replication and cell division: *sulB* is an allele of *ftsZ*. *J. Bacteriol.* **154**: 1339.

Maas, W.K. and A.J. Clark. 1964. Studies on the mechanism of repression of arginine biosynthesis in *Escherichia coli*. II. Dominance of repressibility in diploids. *J. Mol. Biol.* **8**: 365.

MacConkey, E.H. 1979. Double-label autoradiography for comparison of complex protein mixtures after gel electrophoresis. *Anal. Biochem.* **96**: 39.

Magasanik, B. 1962. Catabolite repression. *Cold Spring Harbor Symp. Quant. Biol.* **26**: 249.

Merril, C., D. Goldman, S.A. Sedman, and M.H. Ebert. 1980. Ultrasensitive stain for proteins in polyacrylamide gels shows regional variation in cerebrospinal fluid proteins. *Science* **211**: 1437.

Mizusawa, S. and S. Gottesman. 1983. Protein degradation in *Escherichia coli*: The *lon* gene controls the stability of *sulA* protein. *Proc. Natl. Acad. Sci.* **80**: 358.

Mizusawa, S., D. Court, and S. Gottesman. 1983. Transcription of the *sulA* gene and repression by LexA. *J. Mol. Biol.* **171**: (in press).

Neidhardt, F.C. and R.A. VanBogelen. 1981. Positive regulatory gene for temperature-controlled proteins in *Escherichia coli*. *Biochem. Biophys. Res. Commun.* **100:** 894.

Neidhardt, F.C., R.A. VanBogelen, and E.T. Lau. 1983a. Molecular cloning and expression of a gene that controls the high-temperature regulon of *Escherichia coli*. *J. Bacteriol.* **153:** 597.

Neidhardt, F.C., V. Vaughn, T.A. Phillips, and P.L. Bloch. 1983b. Gene-protein index of *Escherichia coli* K-12. *Microbiol. Rev.* **47:** 231.

Neidhardt, F.C., T.A. Phillips, R.A. VanBogelen, M.W. Smith, Y. Georgalis, and A.R. Subramanian. 1981. Identification of the B56.5 protein, the A-protein, and the *groE* gene product of *Escherichia coli*. *J. Bacteriol.* **145:** 513.

O'Farrell, P.H. 1975. High-resolution two-dimensional electrophoresis of proteins. *J. Biol. Chem.* **250:** 4007.

O'Farrell, P.Z., H.M. Goodman, and P.H. O'Farrell. 1977. High resolution two-dimensional electrophoresis of basic as well as acidic proteins. *Cell* **12:** 1133.

Ogawa, T., H. Wabiko, T. Tsurimoto, T. Horii, H. Masukato, and H. Ogawa. 1979. Characteristics of purified *recA* protein and the regulation of its synthesis *in vivo*. *Cold Spring Harbor Symp. Quant. Biol.* **43:** 909.

Pedersen, S., P.L. Bloch, S. Reeh, and F.C. Neidhardt. 1978. Pattern of protein synthesis in *E. coli*: A catalog of the amount of 140 individual proteins at different growth rates. *Cell.* **14:** 179.

Perlman, R.L. and I. Pastan. 1969. Pleiotropic deficiency of carbohydrate utilization in an adenylate cyclase deficient mutant of *Escherichia coli*. *Biophys. Biochem Res. Commun.* **37:** 151.

Phizicky, E.M. and J.W. Roberts. 1981. Induction of SOS functions: Regulation of proteolytic activity of *E. coli recA* protein by interaction with DNA and nucleoside triphosphates. *Cell* **25:** 259.

Plesset, J., C. Palm, and C.S. McLaughlin. 1982. Induction of heat shock proteins and thermotolerance by ethanol in *Saccharomyces cerevisiae*. *Biochem. Biophys. Res. Commun.* **108:** 1340.

Radding, C.M. 1982. Homologous pairing and strand exchange in genetic recombination. *Annu. Rev. Genet.* **16:** 405.

Radman, M. 1975. SOS repair hypothesis: Phenomenology of an inducible DNA repair which is accompanied by mutagenesis. In *Molecular mechanisms for repair of DNA* (ed. P. Hanawalt and R.B. Setlow), p. 355. Plenum Press, New York.

Reeh, S., S. Pedersen, and F.C. Neidhardt. 1977. Transient rates of synthesis of five aminoacyl-transfer ribonucleic acid synthetases during a shift-up of *Escherichia coli*. *J. Bacteriol.* **129:** 702.

Reeve, J. 1979. Use of minicells for bacteriophage-directed polypeptide synthesis. *Methods Enzymol.* **68:** 493.

Roberts, J.W. and C.W. Roberts. 1975. Proteolytic cleavage of bacteriophage lambda repressor in induction. *Proc. Natl. Acad. Sci.* **72:** 147.

Roberts, J.W., C.W. Roberts, and N.L. Craig. 1978. *Escherichia coli recA* gene product inactivates phage repressor. *Proc. Natl. Acad. Sci.* **75:** 4714.

Sancar, A., A.M. Hack, and W.D. Rupp. 1979. Simple method for identification of plasmid-coded protein. *J. Bacteriol.* **137:** 692.

Sancar, G.B., A. Sancar, J.W. Little, D.W. Mount, and W.D. Rupp. 1982. The *uvrB* gene of *E. coli* has both *lexA*-repressed and *lexA*-independent promoters. *Cell* **28:** 523.

Schwartz, D. and J.R. Beckwith. 1970. Mutants missing a factor necessary for the expression of catabolite sensitive operons in *E. coli*. In *The lactose operon* (ed. J.R. Beckwith and D. Zipser), p. 417. Cold Spring Harbor Laboratory, Cold Spring Harbor, New York.

Stephens, J.C., S.W. Artz, and B.N. Ames. 1975. Guanosine 5'-diphosphate 3'-diphosphate (ppGpp): Positive effector for histidine operon transcription and general signal for amino acid deficiency. *Proc. Natl. Acad. Sci.* **72:** 4389.

Tilly, K. and C. Georgopoulos. 1982. Evidence that the two *Escherichia coli groE* morphogenetic gene products interact *in vivo*. *J. Bacteriol.* **149:** 1082.

VanBogelen, R.A., V. Vaughn, and F.C. Neidhardt. 1983. Gene for heat-inducible lysyl-tRNA synthetase (*lysU*) maps near *cadA* in *Escherichia coli*. *J. Bacteriol*. **153:** 1066.

Wanner, B.L. and R. McSharry. 1982. Phosphate-controlled gene expression in *Escherichia coli* K-12 using Mud*l*-directed *lacZ* fusions. *J. Mol. Biol*. **158:** 347.

Witkin, E.M. 1976. Ultraviolet mutagenesis and inducible DNA repair in *Escherichia coli*. *Bacteriol. Rev*. **40:** 869.

Yamamori, T. and T. Yura. 1982. Genetic control of heat shock protein synthesis and its bearing on growth and thermal resistance in *Escherichia coli* K-12. *Proc. Natl. Acad. Sci*. **79:** 860.

Yamamori, T., T. Osawa, T. Tobe, K. Ito, and T. Yura. 1982. *Escherichia coli* gene (*hin*) controls transcription of heat-shock operons and cell growth at high temperature. In *Heat shock from bacteria to man* (ed. M.J. Schlessinger et al.), p. 131. Cold Spring Harbor Laboratory, Cold Spring Harbor, New York.

Protein-mediated Translational Repression

Kristine M. Campbell, Gary D. Stormo, and Larry Gold
Department of Molecular, Cellular, and Developmental Biology
University of Colorado
Boulder, Colorado 80309

INTRODUCTION

Regulation of protein synthesis at the level of translation is one of the many ways cells control the concentrations of gene products. We define (for this review) translational regulation to be the variation from "constitutive" levels of translation caused by the action of specific proteins on specific mRNAs. The constitutive efficiencies of translation of mRNAs are determined by both the sequence and the structure of their initiation regions (Steitz 1979; Gold et al. 1981; Stormo et al. 1982). mRNAs are in excess of ribosomes; i.e., free ribosomes are less abundant than are open initiation regions on mRNA. Selection of messages for translation occurs according to the relative strengths of their initiation domains. Primary sequence features important for recognition by ribosomes include the polypurine Shine and Dalgarno region, the initiation codon, and other nearby nucleotides (Gold et al. 1981). Initiation regions possess varying degrees of structure, from the unstructured T4 gene-*32* message to the obviously structured MS2 replicase message (see below).

The features of mRNAs recognized by translational regulatory proteins include both sequence and structure. The role of structure in mRNA function is often overlooked but is of prime importance to translational regulation. Work on tRNA structure and function (Altman 1978), the autocatalytic splicing of *Tetrahymena* ribosomal (r)RNA (Kruger et al. 1982), mRNA attenuation regions (see Bauer et al., this volume), and the role of RNA in RNase P activity (Reed et al. 1982) suggest that the relationship between RNA structure and function is complex.

In this paper we discuss several well-studied examples of prokaryotic translational control. We also discuss some ideas about the evolution of these regulatory systems that allow us to predict systems that would be best served by regulation at the translational level.

DISCUSSION

RNA Phages

The RNA phages are the simplest viruses known to infect *Escherichia coli*. Both MS2[1] and the similar but nonidentical phage, $Q\beta$, are plus-strand RNA

---

[1]We use the term MS2 everywhere that data exist for R17, MS2, or f2.

viruses encoding four polypeptides (Fig. 1) (Horiuchi 1975; Beremand and Blumenthal 1979). Adsorption and injection of the single-stranded genome, a molecule containing more than 3000 nucleotides, places the virus at great risk. A single nucleolytic break at any moment between adsorption and completion of a minus strand, several minutes after infection, will be lethal to the infecting virus. Any regulation of expression of the RNA genome must occur in the absence of transcriptional control.

After injection, the plus strand is attacked by ribosomes; the strongest ribosome-binding site is for the coat protein, which is ultimately the most abundantly expressed viral gene product. However, viral progeny cannot be made without expression of the other genes. Initiation at the coat-ribosome-binding site and subsequent elongation leads, by a complex mechanism of antipolarity (MinJou et al. 1972), to initiation by other ribosomes at the

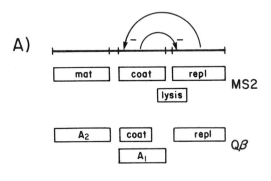

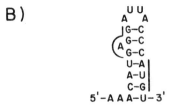

Figure 1 Arrangement of information in the genomes of the RNA phages. (*A*) A hybrid genome, represented as the plus-strand RNA, is shown in the direction 5'−3'. Although *Qβ* is about 20% larger than MS2, both phages encode four proteins. MS2 encodes a lysis gene that overlaps both the coat and replicase cistrons. *Qβ* uses its maturation protein (A_2) as a lysis protein (R. Winter and L. Gold, in prep.). The important point is that both *Qβ* and MS2 use translational repression; the coat protein of each can repress its own replicase translational initiation, and the replicase (of at least *Qβ*) can repress coat translation. (*B*) The site at which the MS2 coat protein binds to block replicase initiation; a similar structure can be drawn for the *Qβ* replicase initiation domain. For the MS2 replicase initiation site, the Shine and Dalgarno domain and the initiation codon are overlined.

replicase gene. After a few moments in an infected cell, the plus strand has directed the synthesis of a small number of coat protein monomers and replicase molecules. The replicase protein captures the *E. coli* elongation factors EF-Tu and EF-Ts and ribosomal protein S1 to build the tetramer that is the complete replicase enzyme (Blumenthal and Carmichael 1979). The preferred binding site for the first molecules of complete replicase appears to be the site overlapping the coat initiation domain (Fig. 1; Weber et al. 1972; Meyers et al. 1981). The binding of replicase to the coat-ribosome-binding site represses coat translation. For successful infection to occur, the binding constant for replicase at the coat initiation region must be greater than the binding constant for the sites used for replication. Repression prevents translation from the coat initiation codon, eliminating the possibility of collisions between elongating ribosomes and elongating replicase (Kolakofsky and Weissmann 1971; Taniguchi and Weissmann 1979). It may be important that dissociation of replicase from the coat initiation site be slow, so that ribosomes do not have the chance to sneak onto a plus strand whose replication has begun. Amazingly, the secondary binding site for replicase is more than 1000 nucleotides from the 3' end, even though this site is thought to be crucial for replication (Meyers et al. 1981).

The binding site for coat protein is the replicase initiation domain (Fig. 1; Bernardi and Spahr 1972). Binding of coat to this site represses translation (Lodish and Zinder 1966). The K_d for coat binding to the replicase initiation region at physiological ionic strength and 37°C is in the micromolar range (J. Carey et al., in prep.). Repression of replicase expression in vivo will not occur until the concentration of coat protein corresponds to perhaps a few thousand molecules per cell. The reason that the coat protein represses the replicase gene is not understood, although the following observations suggest that binding of coat to this region is a necessary event in phage infection. A cell synthesizing $Q\beta$ coat protein from a plasmid cannot be successfully infected with wild-type $Q\beta$ phage (R. Winter and L. Gold, in prep.). In vitro experiments with the MS2 coat-protein-binding site have shown that several single base substitutions in this region prevent coat binding (J. Carey et al., in prep.). However, no $Q\beta$ mutants able to form plaques on the coat-producing bacterial strain have been isolated, suggesting that simple mutations that destroy coat binding to the replicase gene are themselves lethal to $Q\beta$ (R. Winter and L. Gold, in prep.). This presupposes that the inability of $Q\beta$ to infect the coat-producing host is due to premature repression of replicase synthesis. The other possibility is immediate encapsidation of the incoming RNA. Further work is required to understand the significance of repression of replicase expression by coat protein.

T4 Gene-32 Protein

Gene *32* of bacteriophage T4 encodes an early protein that is essential for DNA replication, recombination, and repair (for review, see Doherty et al.

1983). Purified gene-*32* protein binds single-stranded DNA cooperatively (Alberts and Frey 1970) and, with a lower affinity, also binds unstructured RNA (von Hippel et al. 1982). At least two factors influence the amount of gene-*32* protein made during the infection. When little or no single-stranded DNA is made, small amounts of gene-*32* protein are produced (Gold et al. 1973, 1976; Krisch et al. 1974). Both gene-*32* nonsense fragments and temperature-sensitive missense proteins are overproduced (Gold et al. 1973, 1976; Krisch et al. 1974; M. Nelson et al., in prep.). These observations led to the suggestion that gene-*32* protein represses its own synthesis and that single-stranded DNA induces *32* synthesis (Gold et al. 1976; Russel et al. 1976). This model explicitly proposed that gene-*32* protein's primary ligand is single-stranded DNA and that only upon saturation of this ligand would the protein turn off its own production.

Gene-*32* mRNA is stable, and the repression of *32* synthesis was shown to be reversible in vivo, even in the absence of RNA synthesis (Russel et al. 1976). The repressed cells contained large amounts of gene-*32* mRNA that could be extracted and efficiently translated in vitro. These in vivo experiments showed that the regulation occurred at the level of translation. In vitro experiments demonstrated the specificity of repression and confirmed that repression occurred translationally (Lemaire et al. 1978). RNA isolated from cells infected with phage carrying an amber mutation in gene *32* was translated in vitro to produce a collection of T4 proteins, including the gene-*32* amber fragment. Addition of purified gene-*32* protein (to $3-4$ μM) specifically inhibited the synthesis of the amber fragment. This is approximately the concentration of protein calculated to exist in vivo with no single-stranded DNA present (Gold et al. 1977). The abruptness of the shutoff of expression with increasing amounts of protein suggested cooperative binding to the mRNA "operator" site. Addition of threefold to fourfold more gene-*32* protein diminished translation of other T4 genes, indicating only a quantitative difference in the sensitivity of the other T4 genes to translational repression. Derepression was produced by addition of single-stranded but not double-stranded DNA. The slope of the repression curve was decreased by the addition of poly[r(U)], as though poly[r(U)] were competing with the operator site on gene-*32* mRNA. It was proposed that the operator overlaps the ribosome-binding site; binding of gene-*32* protein would then prevent ribosome binding. It was also suggested that gene-*32* protein bound to its own initiator region preferentially because it is less structured than the initiator regions of other T4 mRNAs.

Three thermodynamic parameters describe the binding of a protein to a nucleic acid (McGhee and von Hippel 1974): the binding site size (n, in units of nucleotides per protein monomer), the intrinsic binding constant (K, in units of M^{-1}), and the cooperativity of the binding (ω, unitless). The binding of gene-*32* protein to a variety of polynucleotides has been characterized (for summary, see von Hippel et al. 1982). The site size is a constant of about 7 nucleotides; the cooperativity is constant at about 2×10^3. The intrinsic binding constant varies with salt concentration, temperature, and nucleotide composition (both sugar and base type). At 37°C and physiological ionic strength, binding to DNA of T4 base composition occurs with a net binding constant $(K\omega)$ of about 10^8. The $K\omega$ for gene-*32* protein binding to RNA is about 4×10^6. This 25-fold difference in binding constants for DNA versus RNA accounts for the preferential saturation of single-stranded DNA in the experiments of Lemaire et al. (1978) but cannot account for the specificity of repression. Since gene-*32* protein binds more strongly to guanosine than to other residues, one could account for the specificity if the gene-*32* operator were uniquely rich in guanosine. This is certainly not the case (see Fig. 3).

A sequence of nucleotides may bind to another sequence, forming some secondary structure, or it may bind to gene-*32* protein. These possibilities are in competition in the cell. The overall reaction and the reaction coordinate diagram are shown in Figure 2. The value ΔG_{net} determines whether a particular section of a polynucleotide is bound by gene-*32* protein or exists in a double-stranded nucleic acid structure without gene-*32* protein. ΔG_{net} is the

A)

B)

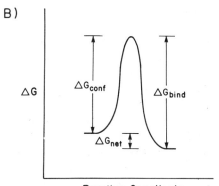

Figure 2 Alternative fates of single-stranded polynucleotides in the presence of gene-*32* protein. (*A*) Schematic diagram of the competition for polynucleotide binding. A single-stranded polynucleotide may either bind to itself, forming some secondary structure, or it may bind to gene-*32* protein. (*B*) The reaction coordinate diagram. The equilibrium distribution between the two states is determined by ΔG_{net}, the difference between the energy of gene-*32* binding (ΔG_{bind}), and the energy of secondary structure conformation (ΔG_{conf}).

difference between the energy of secondary conformation, ΔG_{conf} (a function of the nucleotide sequence), and the energy of gene-*32* protein binding, ΔG_{bind} (a function of the concentration of gene-*32* protein and the number of protein monomers that can be bound). Because of the cooperativity of binding of gene-*32* protein, the site size is particularly important in binding to small targets. Since the first (or last) monomer binds without the benefits of cooperativity, binding is better at sites where there are few ends per monomer bound, i.e., for long sites (Newport et al. 1981).

For a given nucleotide sequence, it is possible to calculate, using rules of secondary structure formation (Tinoco et al. 1973), the expected ΔG_{conf} (and also the K_{conf}) for that region. The fractional saturation of such sites by gene-*32* protein can then be described by

$$\theta = \frac{(K_{conf})\ (K_{bind})\ [P]^m}{1 + (K_{conf})\ (K_{bind})\ [P]^m}$$

where [P] is the concentration of gene-*32* protein, m is the target size (in monomers of gene-*32* protein) and K_{bind} is the total binding constant (from the intrinsic constant [K], cooperativity [ω], and the target size). The sequence of gene *32* (Krisch and Allet 1982) and the assumption that the operator overlaps the initiation region allow a calculation of the concentration of gene-*32* protein necessary for translational repression. In a 63-nucleotide-long stretch (m=9) extending from before the Shine-Dalgarno sequence into the ninth codon of the protein-coding region, only one weak structure would probably exist. About 1.5 μM gene-*32* protein would be sufficient to denature this structure and coat the 63 nucleotides of the operator (to 90% saturation, consistent with the level of in vitro repression) (Fig. 3).

To account for the specificity of repression, this large unstructured (or weakly structured) region should be unique among T4 ribosome-binding sites and nearly unique among all T4 mRNA sequences. Examination of ten other sequenced T4 gene beginnings showed that all have more potential structure than the gene-*32* start; these mRNAs, therefore, are not likely to be bound by gene-*32* protein at concentrations where autorepression occurs. Furthermore, looking at all T4 sequences available (~5% of the genome), we have noted that the gene-*32* ribosome-binding site is more unstructured than almost any T4 RNA sequence. Most importantly, at gene-*32* protein concentrations predicted for shutoff, which are very close to those measured in vivo and in vitro, about 99% of the single-stranded DNA would be covered by gene-*32* protein; nearly all of the mRNA would be left unbound (Fig. 4) (von Hippel et al. 1982).

Some details of the model for gene-*32* autogenous regulation remain to be experimentally verified. Direct binding studies between purified gene-*32* protein and gene-*32* mRNA are needed to prove that the mechanism of repression is as postulated (N. Guild and L. Gold, in prep.). The existence of a set of missense mutants that overproduce gene-*32* protein (Doherty et al. 1982; M.A.

met   lys   lys   thr   glu   ala   gln   ala   leu   gly(etc)
    phe   arg   ser   ala   leu   ala   met   lys   asn

```
          1            2            3            4            5
GCTCATGAGGTAAAGTGTCATAGCACCAACTGTTAATTAAATTAAAAAGGAAATAAAAATGTTAAACGTAACTACTGCTGACTCGCTGCACAAATGGCTAAACTGACTGGCAATAAAGGTTTTCTCTGAAGATAAAGGCGAGT
aaa bbbb       bbbbaaccd    ee fffff    fffff    eed         cc           gggggghhh ii        iihhh  jjj   jjj        ggggg
```

$$\Delta G^{\circ}_{conf(a-b)} = -5.2 \qquad \Delta G^{\circ}_{conf(c-f)} = -3.6 \qquad \Delta G^{\circ}_{conf(g-j)} = -14.6$$

line	nucleotide residues(N)	protein monomers(m)	ΔG°_{conf}
B	18	2	0
C	39	5	-2.4
D	65	9	-3.6
E	89	12	-8.8
F	130	18	-18.2

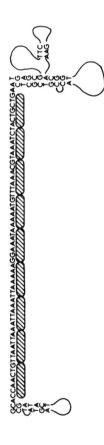

Figure 3 Sequence and conformational stability of the putative gene-32 mRNA operator site and vicinity. (*Top*) The DNA sequence (5'−3', noncoding strand only; the sequence as written corresponds to mRNA when T is replaced by U), with the beginning of the gene-32 protein sequence written above. The lowercase letters below the DNA sequence correspond to possible base-pairing interactions, i.e., the bases marked aaa can pair with the subsequent aaa sequence to form the stem of a hairpin structure, and so on. The values of ΔG°_{conf} (below) correspond to the calculated stability of the indicated hairpin structures. The lines (labeled *B−F*) correspond to the segments tested as potential operator sites. (*Bottom*) This structure is the preferred operator sequence, drawn in a gene-32 protein-saturated conformation showing the proposed flanking hairpin termini. (Reprinted, with permission, from von Hippel et al. 1982.)

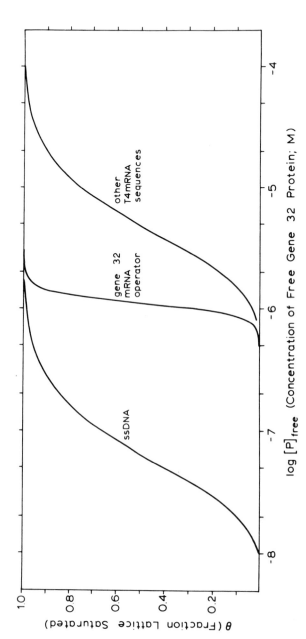

Figure 4 Binding curves summarizing the gene-32 protein autoregulatory system. The ssDNA curve is calculated using real T4 DNA sequences with an N=50 residue lattice length replication window and the infinite lattice calculation mode. The gene-32 mRNA operator curve is calculated for the putative operator structure shown at the bottom of Fig. 3 (line D). The other mRNA curve is calculated using real T4 sequences with an N=50 residue lattice length and the finite lattice calculation mode. The other T4 ribosome-binding sites studied (Gold et al. 1981; Stormo et al. 1982; G. Stormo, unpubl.) have more typical structures, and so would be repressed only at higher concentrations of gene-32 protein, as is observed. (Reprinted, with permission, from von Hippel et al. 1982.)

Nelson et al., in prep.) will provide altered repressor proteins for such a study. Physical characterization of the operator site to show that it is indeed unstructured would also be useful, although the sequence is, by inspection, remarkable. Genetic engineering and gene-fusion technologies will certainly be used to manipulate the operator site, better defining its properties (Krisch and Allet 1982).

Probably all systems that replicate DNA require a single-stranded DNA-binding protein. The gene-*32* solution to the problem of regulating such a protein is so elegant that we would have expected it to be used elsewhere. Several single-stranded DNA-binding proteins are known, but so far none seems to be regulated like gene *32*. The filamentous single-stranded DNA coliphages (fd, f1, M13) encode a single-stranded DNA-binding protein in their gene *5*. The synthesis of the binding protein itself is unregulated, yet it translationally represses the production of gene-*2* protein, a site-specific nicking enzyme required for DNA synthesis (Model et al. 1982; Yen and Webster 1982). The mechanism for this translational regulation seems not to use the specificity of an unstructured target, because gene-*32* protein does not repress gene-*2* expression, nor does gene-*5* protein repress gene-*32* expression in vitro (P. Model, pers. comm.). Although the gene-*5* protein binds single-stranded DNA without regard for sequence, the mode of binding to short oligonucleotides is somewhat different than for long polynucleotides (Alma et al. 1982), allowing the possibility of a sequence-specific nucleation event. The site of initial binding of the gene-*5* protein to viral DNA is probably near the beginning of gene *2* (Ray 1978), so that the operator for gene-*2* repression may be the mRNA site corresponding to the single-stranded DNA nucleation site.

Ribosomal Protein Operons of *E. coli*

The synthesis of ribosomes in *E. coli* is closely regulated with respect to the growth rate of the cells (Maaløe and Kjeldgaard 1966). The synthetic rates of the three rRNA components and the approximately 52 ribosomal proteins are balanced so as to respond coordinately to environmental changes (for review, see Gausing 1980). Approximately 16 transcriptional units, each having 1−11 genes, comprise the set of ribosomal protein operons (Lindahl and Zengel 1982). The first evidence that posttranslational controls operated on the ribosomal protein operons came from gene-dosage experiments (Geyl and Bock 1977; Dennis and Fiil 1979; Fallon et al. 1979a,b; Olsson and Gausing 1980). In *E. coli* strains merodiploid for various sets of the ribosomal protein genes, the synthesis of ribosomal protein mRNA (in RNA obtained after short pulses with radioactive precursors) was proportional to the gene dosage, yet the synthesis of the proteins themselves was not. It seemed that these genes were under posttranslational control such that excess mRNA was not translated into protein.

Results from experiments with gene-fusion plasmids provide in vivo evidence that one ribosomal protein from each operon acts as a translational repressor for many (but not necessarily all) proteins expressed from that operon. By placing various ribosomal protein genes under the control of an inducible promoter, such as *lac*, the regulatory effects of overproducing each protein were examined. In this way, in vivo regulatory roles for the following ribosomal proteins have been demonstrated: S4 (Dean and Nomura 1980), S7 (Dean et al. 1981a), S8 (Dean et al. 1981b), L1 (Dean and Nomura 1980), L4 (Lindahl and Zengel 1979; Zengel et al. 1980), and L10 (Yates et al. 1981). The identification of L1 as an in vivo repressor was further confirmed by the overproduction of protein L11 (the other protein under L1 control) in an *E. coli* mutant lacking protein L1 (Jinks-Robertson and Nomura 1981). The regulation of the operons containing genes controlled by these repressors illustrates the point that not all genes in a specific operon are necessarily controlled (Fig. 5). In two cases, the most promoter-proximal genes are not controlled by the ribosomal protein repressors, consistent with repression at the level of translation, and not transcription.

In vitro characterization of translational feedback repression by ribosomal proteins has come from coupled transcription-translation systems using cloned ribosomal protein genes and various purified ribosomal proteins. Addition of purified ribosomal repressor proteins to such reactions inhibited the synthesis of various proteins of the controlled operons. This approach confirms the in vivo identification of repressors: S4 (Yates et al. 1980), S7 (Dean et al. 1981a), S8 (Yates et al. 1980; Dean et al. 1981b), L1 (Yates et al. 1980; Yates and Nomura 1981), L4 (Yates and Nomura 1980), and L10 (Brot et al. 1980; Fukuda 1980; Robakis et al. 1981; Yates et al. 1981; Johnsen et al. 1982). In addition, repressor activity has been assigned to protein S20 by in vitro procedures only (Wirth and Bock 1980; Wirth et al. 1981). After uncoupling transcription from translation, several repressors have been shown to exert their inhibitory effects at the level of translation (Brot et al. 1980; Yates et al. 1980; Yates and Nomura 1980; Wirth et al. 1981).

Many of the repressor ribosomal proteins bind specifically to either 16S rRNA or 23S rRNA during ribosome assembly (Nomura and Held 1974; Nierhaus 1980). Added rRNA prevents the inhibitory action of the ribosomal protein repressors, presumably by competing with the mRNA for binding of the protein (Wirth et al. 1981; Yates and Nomura 1981; Johnsen et al. 1982). This is directly comparable to induction, in vivo and in vitro, of gene-*32* translation by single-stranded DNA (Gold et al. 1976; Russel et al. 1976; Lemaire et al. 1978). The in vitro studies have also been useful in defining the mRNA targets of the ribosomal protein repressors. DNA templates lacking various parts of an operon have been tested for "repressibility"; such experiments show that the repessors L1 and L10 act at single sites, overlapping the translational initiation codons (Yates and Nomura 1981; Yates et al. 1981; Johnsen et al. 1982). From a single target site, a repressor can affect the

L11 OPERON

	L1	L11	P	
	[L1] ←	L11	P	L11 OPERON
	+	+		in vitro
	(+)	+		in vivo

β OPERON

	β'	β	L7/12	L10	P	
←	-	-	+	[L10] +	P	in vitro
	-	-	+	(+)		in vivo

str OPERON

	EF-Tu	EF-G	S7	S12	P	
←	-	+	[S7] +	S12 -	P	in vitro
	-	+	(+)	-		in vivo

S10 OPERON

	S17	L29	L16	S3	(S19, L22)	L2	L23	L4	L3	S10	P	
←	-	-	-	-	-	±	+	[L4] +	+	+	P	in vitro
	+	+	+	+	+	+	+	+	(+)	+		in vivo

spc OPERON

	L15	L30	S5	L18	L6	S8	S14	L5	L24	L14	P	
	ND	ND	-	-	-	[S8] -	+	+	-	-	P	in vitro
	+	+	+	+	+	(+)	+	+	-	-		in vivo

α OPERON

	L17	α	S4	S11	S13	P	
←	-	-	[S4] +	+	+	P	in vitro
	+	±	(+)	+	+		in vivo

Figure 5 Organization and regulation of genes contained within the *str-spc* and *rif* regions of the bacterial chromosome. Genes are represented by the protein product. (→) For each operon, the direction of transcription from the promoter (P). It has recently been shown that the L11 and β operons are probably a single operon. That is, the L11 promoter functions as the major promoter for all genes contained within the L11 and β operons in exponentially growing cells (Bruckner and Matzura 1981; Nomura et al. 1982b). It appears that the β operon promoter previously identified is used only when transcription from the major upstream promoter ceases (Post et al. 1979). Regulatory ribosomal proteins are indicated by boxes. Effects of the boxed proteins on the in vitro and in vivo synthesis of proteins from the same operon are shown. L10 can function as a repressor in a complex with L12 (Fukuda 1980; Yates et al. 1981; Johnsen et al. 1982). + indicates specific inhibition of synthesis; −, no significant effect on synthesis; ±, weak inhibition of synthesis; (+), inhibition presumed to occur in vivo: ND, not determined. It has not been established how the regulation of the synthesis of ribosomal proteins S12, L14, and L24 is achieved. (Reprinted, with permission, from Nomura et al. 1982b.)

translation of all the sensitive downstream cistrons (see Fig. 5). This could be accomplished through "sequential translation," where the translation of downstream cistrons is dependent on translation (and perhaps termination) of the upstream target cistron in the regulatory unit (Nomura et al. 1982a,b; Nomura and Dean 1983; and see the related work of Oppenheim and Yanofsky 1980; Napoli et al. 1981; Schumperli et al. 1982). In most cases, the repression observed in vitro fails to extend as far downstream as that seen in vivo (Fig. 5). If the sequential translation hypothesis is correct, the larger effects seen on distal cistrons in vivo could be due to tight coupling of translation to that of the target cistron. In this view, translation of the downstream genes becomes "uncoupled" in vitro, freeing them from repression (Nomura et al. 1982a,b; Nomura and Dean 1983). This uncoupling could occur, e.g., by nucleolytic cleavage of the in vitro messages.

Because of the derepression by rRNA in vitro, the regions near the mRNA targets were searched for potential homologies with the protein-binding sites on the rRNA. Homologies involving structural and sequence features were proposed for S4 (Nomura et al. 1980), S7 (Nomura et al. 1980), S8 (Olins and Nomura 1981b), L4 (Olins and Nomura 1981a), L1 (Branlant et al. 1981; Gourse et al. 1981), and L10 (Johnsen et al. 1982). (For examples of these homologies, see Figs. 6 and 7.)

The above information has been integrated into a model for the autogenous control of ribosomal protein synthesis (for reviews, see Nomura et al. 1982a,b; Nomura and Dean 1983). The translation of a group of ribosomal protein genes encoded on a polycistronic mRNA is inhibited by one of the proteins encoded within that operon. Because of sequential translation the repressor can affect synthesis of all of the sensitive genes. The affinity of the repressor protein for its primary binding site on the rRNA is predicted to be greater than for the mRNA, thus ensuring that expression of these genes is turned off only when levels of free ribosomal proteins exceed rRNA levels. One of the factors contributing to the higher affinity for the primary ligand may be cooperative interactions between the different proteins binding to the rRNA (Nomura et al. 1982b; von Hippel et al. 1982; von Hippel and Fairfield 1983). According to this model, the cell need only control the rate of transcription of the rRNA genes in concert with the growth rate to control the rate of ribosomal biosynthesis via feedback repression. It is as yet unknown how the rRNA operons are regulated.

Various aspects of the model require experimental verification. Only one postulated mRNA target structure has been shown to bind repressor (L10-L12; Johnsen et al. 1982). The mechanism of sequential translation (and translational reinitiation) is not clearly understood. The ribosome-binding sites of downstream cistrons are not obviously impaired for direct loading by inspection within the context of *E. coli* ribosome-binding site determinants (Gold et al. 1981). The presence of a high-copy number plasmid carrying a ribosomal protein repressor gene is generally detrimental to the cell (presumably because its gene product inhibits the translation of ribosomal protein genes on the chromosome); one might select target-site mutations on the chromosome that

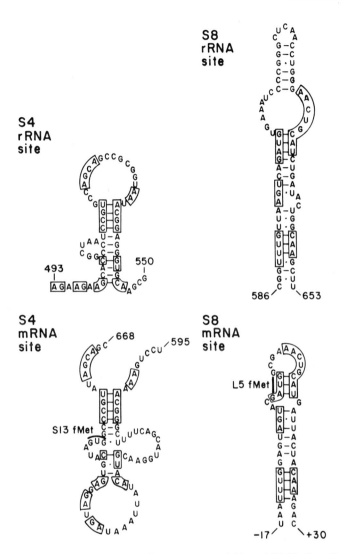

Figure 6   Predicted secondary structures of S4- and S8-binding sites on 16S rRNA and their respective mRNAs. The binding site of S4 on the mRNA overlaps the S13 initiation region; the mRNA site for S8 overlaps the L5 initiation region. Boxed sequences indicate homologies. (Adapted from Nomura et al. [1980] and Olins and Nomura [1981].)

are now insensitive to the action of the repressor (i.e., constitutives). So far this kind of selection has yielded largely mutants that depress the translation of the repressor from the plasmid, in addition to a few that behave as operator

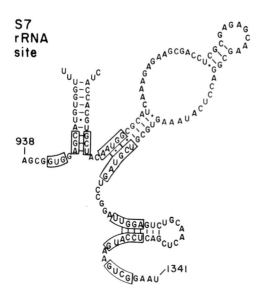

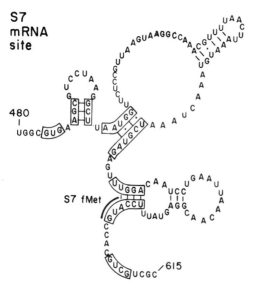

Figure 7 Predicted secondary structure of S7-binding sites on 16S rRNA and on its own mRNA. The S7-mRNA-binding site overlaps the S7 initiation region. Boxed sequences indicate homologies. (Adapted from Nomura et al. 1980.)

mutants (Fiil et al. 1980; J.D. Friesen, pers. comm.). Further work along these lines could provide an elegant genetic profile of the target sites for repressor action.

T4 regA Protein

The *regA* gene of bacteriophage T4 encodes a small ($\sim$12K [Cardillo et al. 1979]) protein synthesized in the early prereplicative phase of T4 development. Mutants in this gene have been isolated in several ways: as phage abnormally sensitive to the drug 5-fluorodeoxyuridine (Wiberg et al. 1973); as phage with a "white halo" plating phenotype under special plating conditions (Wiberg et al. 1977); and as pseudorevertants of leaky amber mutations in two T4 genes involved in DNA replication, *44* and *62* (Karam and Bowles 1974; Karam et al. 1977; Wiberg et al. 1977). The first two methods for isolating *regA* mutants cannot easily be explained in light of what else is known about the gene. The *regA* gene is located on the T4 chromosome between genes *43* and *62* (Karam and Bowles 1974; Wiberg et al. 1977). The clearest phenotype of a *regA* mutant is the relative overproduction of a subset of early gene products (Wiberg et al. 1973; Karam and Bowles 1974; Trimble and Maley 1976; Karam et al. 1977). The list of T4 genes controlled by *regA* includes genes involved in deoxyribonucleotide metabolism (*1*, *cd*, *56*, *42*; Wiberg et al. 1973; Trimble and Maley 1976) and in DNA replication (*r*IIA, *r*IIB, *44*, *62*, *45*; Karam and Bowles 1974; Karam et al. 1977), as well as the *regA* gene itself. Notice that genes *44* and *62* are *regA*-controlled, providing an explanation for the selection of *regA* mutants as suppressors of leaky ambers in these genes.

regA is autogenously controlled; it is itself included in the list of *regA*-controlled genes (Cardillo et al. 1979). Different alleles of *regA* differ in the degree to which they affect the synthesis of various proteins (Cardillo et al. 1979; Karam et al. 1982).

The overproduction of some early gene products seen in *regA*$^-$ infections suggested that the *regA*-gene product was a negative regulator of expression of these genes. Several lines of evidence indicate that *regA*-mediated repression occurs posttranscriptionally. The overall rate of T4 RNA synthesis is unaffected by the *regA* mutation (Karam and Bowles 1974). Comparison of patterns of proteins synthesized after T4 infection in the presence of the transcriptional inhibitor rifampicin revealed the typical pattern of overproduction in the *regA* mutant infection (Karam and Bowles 1974). In other words, *regA*$^-$ infections do not require continued transcription to yield proper derepression. In addition, one can isolate translatable mRNA (for several of the *regA*-controlled genes) from infected cells at times long after in vivo synthesis of those gene products has ceased (Trimble and Maley 1976). This implies that repression, in vivo, is at the translational level and that *regA* action does not permanently destroy the target mRNAs (Trimble and Maley 1976).

The effects of the *regA* mutation are dependent on the genetic background of the infecting phage. Differences between proteins synthesized in *regA*⁺ and *regA*⁻ infections are best observed when phage DNA synthesis is prevented (Wiberg et al. 1973; Karam and Bowles 1974) or when late transcription does not occur (Karam and Bowles 1974). Since late T4 transcription is dependent on DNA replication (Rabussay and Geiduschek 1977), these two conditions have in common the lack of late gene products. The failure to see the effects of a *regA* mutation when late genes are turned on may be due to the competition for ribosomes from late mRNAs, or it could be a direct result of the action of a late T4 protein.

The T4 *r*IIB gene has been studied as a representative *regA*-controlled gene because of the extensive collection of classical *r*IIB mutants (Benzer 1961; Barnett et al. 1967), as well as a more recent collection of translational mutants (Singer et al. 1981). The determination of the sequence of several mutations in the *r*IIB initiation region (D. Pribnow et al., in prep.) and the characterization of mutants mapping here (Singer et al. 1981) have allowed the study of *regA*-mediated repression of *r*IIB. The site on the *r*IIB mRNA recognized by *regA* has been defined by the screening of *r*IIB mutants for *regA* sensitivity (Karam et al. 1981; Campbell and Gold 1982). A mutation is defined as being in the *regA* recognition site if that mutation diminishes the sensitivity of *r*IIB expression to the *regA* protein. In these constitutive mutants, there is slight or no repression of *r*IIB expression in a *regA*⁺ infection, relative to a *regA*⁻ infection. Determination of the nucleotide sequence of the *regA* target site has come from sequencing known constitutive mutations (Fig. 8). All the known constitutive mutations are located in the region defined by deletion 326, in a stretch of 9 nucleotides starting approximately with the AUG and extending 3′ to it. This, and sequence homology with the initiation region of another *regA*-controlled gene, *45* (Spicer et al. 1982), suggests that *r*IIB's *regA* target is AUGUACAAU. Since this sequence is probably not involved in any stable structures within the *r*IIB message (D. Pribnow et al., in prep.), the *regA* protein probably recognizes mRNA sequence.

Although the *regA* protein recognizes this site on the *r*IIB mRNA, the mechanism of repression is not yet known. The protein probably binds and prevents ribosomal access to the initiator, in direct analogy to gene *32* and ribosomal proteins. However, *regA* could be a site-specific ribonuclease. Decreased functional stability of most (but not all) *regA*-sensitive messages in *regA*⁺ infections has been observed (Trimble and Maley 1976), but the magnitude of this effect seems too small to account for the amount of repression. The decreased stability may result from translational repression, as is the case for repressed ribosomal protein messengers (Fallon et al. 1979a). The *regA* protein is unique among known translational repressors in that it represses a large number of unlinked transcripts (Wood and Revel 1976).

Much work remains to be done before our understanding of the *regA* gene reaches that of gene *32* or the ribosomal protein genes. The repression of

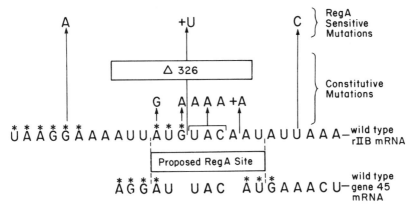

Figure 8   The proposed *regA* recognition site. (*) The initiator AUGs and the Shine and Dalgarno sequences. All but one of the assignments of mutants to the *regA*-sensitive or -constitutive class were previously made (Karam et al. 1981). We have recently found that the A-to-G change in the AUG (zAP10) is constitutive (C. Avery and K. Campbell, unpubl.). Two of the previously deduced positions of mutations (Karam et al. 1981) were slightly incorrect and are corrected here (D. Pribnow et al., in prep.); these are the substitution of G for A in the AUG (zAP10) and the insertion of the U after the AUG (FC302). The inclusion of the last U in the proposed site is due to (1) homology with the gene-*45* sequence and (2) the constitutive nature of the A insertion immediately preceding the U. (The only effect of the mutation on the sequence is to move the U 1 base rightward, which destroys the site.) The fact that the U insertion in the middle of the *regA* site leaves *r*IIB expression *regA*-sensitive suggests some flexibility in what the *regA* protein may inspect. Note, however, that the A insertion is not tolerated. We imagine that when all *regA* recognition sequences are known, they will cluster around a consensus sequence, exhibiting variations from gene to gene. (Reprinted, with permission, from Campbell and Gold 1982.) The sequence of gene *45* is in accordance with the corrected sequence of Spicer et al. (1982).

translation of the sensitive genes needs to be demonstrated in vitro using purified *regA* protein; the exact mechanism of action of *regA*-mediated inhibition of translation could then be determined. Work similar to the genetics of the *regA* target site within the *r*IIB mRNA should be started on some other *regA*-sensitive genes. Finally, in search of the overall significance to T4 of the *regA* gene, the relationship between *regA* function and phage yield must be examined. We have previously speculated about the role of *regA* in T4 DNA replication (Campbell and Gold 1982).

CONCLUDING REMARKS

In the following discussion about the evolution of translational control, we confront a topic considered at length by Francois Jacob in his book "The Possible and the Actual" (1982) and in his article "Evolution and Tinkering" (1977). Jacob notes that "for many biologists, every organism, every cell, every molecule has been refined to the last detail by an adaptation process that has been incessantly in action over millions of years and millions of generations.'' He further develops the counterintuitive idea that, in the face of adaptive pressure, a variety of constraints lead to molecular and organismic solutions that are not the best possible (such as one might achieve by facing each selective environment with limitless design opportunities). The major constraint is history. The organisms now extant have faced countless adaptive pressures sequentially; each survival selects for altered materials that become the starting point for future adaptations. Jacob warns us, as molecular biologists, not to consider the ''living world to be nearly as perfect as the one formerly attributed to divine creation.''

Jacob goes on to state that biologists invariably ask ''How does it work''? and ''How did it come about''? In the preceding sections, we reviewed some answers to the first question with respect to translational control. However, Jacob notes that the second question is ''deeper''; our best answers will usually be our best guesses. The systems under study are ''creations of history,'' and the critical details of that history cannot be known. Jacob imagines the process of natural selection as analogous to ''tinkering,'' rather than ''engineering,'' i.e., the design opportunities are limited by the materials at hand. Beautiful solutions and beautiful molecules do exist in biology; we are certain that some molecules have been tuned to perfection by ''incessant'' selection. In fact, we mean only that our collective imaginations have not yet suggested improvements that might be possible were we to engineer those species from scratch. The Jacob essays encourage us to propose small ideas concerning the evolution of translational regulation, both for the specific cases we have described and in general. Our general ideas must deal more with the ''possible'' than the ''actual.''

Evolutionary Pathways

Specific Cases

The RNA phages, of course, do not have the luxury of considering transcriptional regulation. However, both the coat and replicase proteins most probably served, at an earlier time, only their essential (nonregulatory) functions. The replicase must have once recognized only the 3' ends of both the plus and minus strands of the genome, thereby allowing progeny synthesis. Bursts were certainly low, since the unregulated coat cistron would allow translational initiation after plus-strand replication had begun. Given present dissociation rates for elongating ribosomes and replicase, the ensuing collisions would be lethal (Kolakofsky and Weissmann 1971; Taniguchi and Weissmann 1979). Survival of the RNA phages, which depended on recovering progeny in higher numbers than the parents, might have been facilitated by primitive ribosomes with slow initiation rates. As the protein synthetic apparatus evolved to initiate translation more quickly, the primitive RNA phages were put under adaptive pressure. One solution was the evolution of a target on the viral plus strand which, when complexed with replicase, did not allow ribosomes to start toward a lethal collision. The present RNA phages utilize this site as the primary ligand for replicase binding.

The replicase initiation sites on MS2 and $Q\beta$ are very similar, and each site can be complexed with its homologous coat protein to prevent translation. We believe that the primitive coat proteins might have bound viral RNA nonspecifically and with no preferred nucleation site. The evolution of a more efficient packaging mechanism may have yielded a nucleation site that overlapped the replicase initiation sites. We cannot imagine what selective advantage accrued to the phage by coupling nucleation of RNA packaging to repression of replicase translation.

Autogenous regulation of gene *32* at the translational level has seemed, from the earliest moments of our research, a straightforward solution to a set of adaptive pressures. The protein is essential for T4 DNA replication. Large phage bursts require sufficient gene-*32* protein to saturate all DNA replication complexes. However, excess free gene-*32* protein might inhibit a variety of essential functions (Gold et al. 1976; Lemaire et al. 1978). The primitive gene-*32* protein, which we imagine to have been unregulated, certainly contained a binding site for single-stranded nucleic acids. The present protein, and probably its ancestors, has no strong capacity to select ligands as a function of base sequence. Autogenous, translational regulation could have evolved by base changes around the translational initiation codon that diminished the potential secondary structure of that domain. This process is molecular mimicry; the gene-*32* ribosome-binding site began to look like single-stranded DNA. The nucleotides within the proposed operator include coding nucleotides, so that protein alterations may have been tested as the operator developed (we call this coevolution). This domain of the protein, the amino terminus, is involved in cooperative protein-protein interactions, and the value of the cooperativity constant is crucial in setting the free concentration of gene-*32* protein required for translational repression (von Hippel et al. 1982). Thus, we suggest that "tinkering" with a small region surrounding the translational initiation codon occurred as the most recent steps in the elaboration of this sophisticated regulatory circuit.

Regulation of the ribosomal protein operons in *E. coli* shows many similarities to that of gene *32*. We imagine that the components of early ribosomes were expressed constitutively; autogenous feedback regulation evolved as a means of reducing the free pools of individual ribosomal proteins. The different operons probably came under feedback regulation independently. Mechanistically, the targets for evolution toward regulation would include promoter-proximal mRNA and the ribosome-binding sites for downstream cistrons. The promoter-proximal target evolved to mimic the rRNA (Figs. 6

and 7); the downstream ribosome-binding sites evolved to require translational reinitiation rather than independent ribosome loading. Either step could have occurred first.

The last specific system reviewed was regulation of and by the T4 *regA*-gene product. Because translational feedback repression has evolved to control component proteins of macromolecular machines (the T4 replisome or the *E. coli* ribosome), we proposed that the *regA* protein is a member of a complex in which it fulfills its major function through binding to its primary ligand (Campbell and Gold 1982). We then reasoned, by analogy to the gene-*32* and ribosomal protein systems, that the mRNAs repressed by the *regA* protein are actually its secondary ligands (Campbell and Gold 1982). At this time, relevant data concerning the existence or identity of a primary ligand for *regA* protein do not exist. Likewise, no other function for *regA* protein, other than translational repression, has been established.

Adaptive Pressures in General

The major disadvantage of translational repression lies in the waste of energy. For *E. coli* growing aerobically on glucose, nearly 10^8 molecules of ATP are generated and used each minute (Lehninger 1965; Racker 1976); a single constitutive transcript of average gene length, synthesized at one complete molecule per minute, expends less than 0.002% of the available energy. This percentage would be higher under poor growth conditions. One should note that translation of an mRNA costs far more than the transcription of its gene; i.e., translational regulation of a transcriptionally constitutive gene saves *most* of the energy. The small energy cost of translational regulation might usefully be paid under either of two conditions: if a selective advantage is gained, or if a regulatory loop is established and fixed (by subsequent selection), such that the opportunity, e.g., for transcriptional regulation, is lost.

Some advantages are obvious for translational regulation. Repression can be fast, since no time is spent awaiting mRNA decay. The quickness of repression might save energy (that would be used during translation that occurs after *transcriptional* repression), which could compensate for the low-level energy drain from continued transcription. The speed of repression might be essential if the regulated gene product is toxic above some critical concentration. Translational regulation also provides faster induction, although this is more subtle. When an operon is transcriptionally induced, the first protein products appear at the same time as they would if a preformed RNA were translated upon induction (Schleif et al. 1973; Galas and Branscomb 1976). However, a pool of repressed mRNAs can provide a burst of new proteins after induction. If a regulated protein is needed in larger quantities than one can quickly derive from transcriptional induction, the regulation should ideally be translational.

Where then should we search for proteins whose synthesis is regulated translationally? First, any protein that binds to nucleic acids for its primary function has the evolutionary potential of becoming a feedback repressor of its own synthesis or the synthesis of any other product. In general, the nucleic-acid-binding domain of the protein could be recruited into serving a repressor function; the target for repression will evolve, by molecular mimicry, to look like the primary ligand. Note that a single-stranded DNA-binding protein could easily acquire the ability to recognize an mRNA, as could an rRNA-binding protein. Because RNA can form quite stable secondary structures, we would not be surprised if some double-stranded DNA-binding proteins (such as histones) were found to regulate their expression translationally.

A second prediction concerns abundant proteins (Lemaire et al. 1978). A primeval gene, expressed at a high level, may have had both a strong promoter and a relatively stable transcript. The subsequent evolution of regulation at the transcriptional level would require concomitant destabilization of the transcript if rapid repression were important for that gene. Conversely, evolution of translational regulation could occur without transcript destabilization and, hence, might confer a kinetic advantage. Furthermore, if the abundant product must be regulated around a narrow concentration range centered at a high level, translational regulation (in particular, rapid synthesis of large numbers of proteins after induction) may be favored. We note the appealing discussion of eukaryotic microtubules and actin filaments by Kirschner (1980); these very abundant proteins are shown to function best at a narrowly regulated concentration of free monomers. We would be delighted if actin or tubulin biosynthesis were translationally controlled.

Lastly, we note the extent of the study of the molecular biology of regulatory mechanisms in *E. coli* and other prokaryotes. The details of many systems are known; we are now awaiting cocrystallization data for repressors and their targets. However, even for *E. coli*, only a small number of the regulated genes have been studied. The study of eukaryotic gene expression lags far behind; the technology is only now suitable for studies that accurately describe a regulatory mechanism (except in yeast, where the tools have existed a bit longer). We hope that our discussion of some translationally regulated systems in *E. coli* will stimulate other scientists to make educated guesses about their systems and thereby shorten the path to experimental demonstrations of interesting regulatory mechanisms.

ACKNOWLEDGMENTS

We thank those who allowed us to cite unpublished work. We are grateful to Peter von Hippel for his valuable conversations about nucleic-acid-binding proteins. Thanks also go to those who allowed the use of previously published artwork. We extend special thanks to T. Walker, K. Piekarski, and A. Hill for invaluable help in preparing the manuscript.

REFERENCES

Alberts, B. and L. Frey. 1970. T4 bacteriophage gene 32: A structural protein in the replication and recombination of DNA. *Nature* **227:** 1313.

Alma, N.C.M., B.J.M. Harmsen, J.H. van Boom, G. van der Marel, and C.W. Hilbers. 1982. ^{1}H NMR studies of the binding of bacteriophage M13-encoded gene 5 protein to oligo(deoxyadenylic acid)s of varying length. *Eur. J. Biochem.* **122:** 319.

Altman, S., ed. 1978. *Transfer RNA.* MIT Press, Cambridge, Massachusetts.

Barnett, L., S. Brenner, F. Crick, R. Shulman, and R. Watts-Tobin. 1967. Phase-shift and other mutants in the first part of the rIIB cistron of bacteriophage T4. *Philos. Trans. R. Soc. Lond. B* **252:** 487.

Benzer, S. 1961. On the topography of the genetic fine structure. *Proc. Natl. Acad. Sci.* **47:** 403.

Beremand, M. and T. Blumenthal. 1979. Overlapping genes in RNA phage: A new protein implicated in lysis. *Cell* **18:** 257.

Bernardi, A. and P.F. Spahr. 1972. Nucleotide sequence at the binding site for coat protein on RNA of bacteriophage R17. *Proc. Natl. Acad. Sci.* **69:** 3033.

Blumenthal, T. and G.G. Carmichael. 1979. RNA Replication: Function and structure of $Q\beta$-replicase. *Annu. Rev. Biochem.* **48:** 525.

Branlant, C., A. Krol, A. Machatt, and J.-P. Ebel. 1981. The secondary structure of the protein L1 binding region of 23S RNA. Homologies with putative secondary structures of the L11 mRNA and a region of mitochondrial 16S rRNA. *Nucleic Acids Res.* **9:** 293.

Brot, N., P. Caldwell, and H. Weissbach. 1980. Autogenous control of *Escherichia coli* ribosomal protein L10 synthesis *in vitro*. *Proc. Natl. Acad. Sci.* **77:** 2592.

Bruckner, R. and H. Matzura. 1981. *In vivo* synthesis of a polycistronic messenger RNA for the ribosomal proteins L11, L1, L10 and L7/12 in *Escherichia coli*. *Mol. Gen. Genet.* **183:** 277.

Campbell, K. and L. Gold. 1982. Construction of bacteriophage T4 replication machine: Regulation of synthesis of component proteins. In *Interaction of translational and transcriptional control in the regulation of gene expression* (ed. M. Grunberg-Manago and B. Safer), p. 69. Elsevier, New York.

Cardillo, T.S., E.F. Landry, and J.S. Wiberg. 1979. regA protein of bacteriophage T4D: Identification, schedule of synthesis and autogenous regulation. *J. Virol.* **32:** 905.

Dean, D. and M. Nomura. 1980. Feedback regulation of ribosomal protein gene expression in *Escherichia coli*. *Proc. Natl. Acad. Sci.* **77:** 3590.

Dean, D., J.L. Yates, and M. Nomura. 1981a. Identification of ribosomal protein S7 as a repressor of translation within the *str* operon of *E. coli*. *Cell* **24:** 413.

——————— . 1981b. *Escherichia coli* ribosomal protein S8 feedback regulates a part of the *spc* operon. *Nature* **289:** 89.

Dennis, P.P. and N.P. Fiil. 1979. Transcriptional and post-transcriptional control of RNA polymerase and ribosomal protein genes cloned on composite ColE1 plasmids in the bacterium *Escherichia coli*. *J. Biol. Chem.* **254:** 7540.

Doherty, D.H., P. Gauss, and L. Gold. 1982. On the role of the single-stranded DNA binding protein of bacteriophage T4 in DNA metabolism. I. Isolation and genetic characterization of new mutations in gene 32 of bacteriophage T4. *Mol. Gen. Genet.* **188:** 77.

——————— . 1983. The single-stranded DNA binding protein of bacteriophage T4. In *Multifunctional proteins: Regulatory and catalytic/structural* (ed. J.F. Kane), p. 45. CRC Press, Cleveland, Ohio.

Fallon, A.M., C.S. Jinks, G.D. Strycharz, and M. Nomura. 1979a. Regulation of ribosomal protein synthesis in *Escherichia coli* by selective mRNA inactivation. *Proc. Natl. Acad. Sci.* **76:** 3411.

Fallon, A.M., C.S. Jinks, M. Yamamoto, and M. Nomura. 1979b. Expression of ribosomal protein genes cloned in a hybrid plasmid in *Escherichia coli*: Gene dosage effects on synthesis of ribosomal proteins and ribosomal protein messenger ribonucleic acid. *J. Bacteriol.* **138:** 383.

Fiil, N.P., J.D. Friesen, W.L. Downing, and P.P. Dennis. 1980. Post-transcriptional regulatory mutants in a ribosomal protein-RNA polymerase operon of *E. coli. Cell* **19**: 837.

Fukuda, R. 1980. Autogenous regulation of the synthesis of ribosomal proteins, L10 and L7/12, in *Escherichia coli. Mol. Gen. Genet.* **178**: 483.

Galas, D.J. and E.W. Branscomb. 1976. Ribosome slowed by mutation to streptomycin resistance. *Nature* **262**: 617.

Gausing, K. 1980. Regulation of ribosome biosynthesis in *E. coli.* In *Ribosomes: Structure, function and genetics* (ed. G. Chambliss et al.), p. 693. University Park Press, Baltimore, Maryland.

Geyl, D. and A. Bock. 1977. Synthesis of ribosomal proteins in merodiploid strains and in minicells of *Escherichia coli. Mol. Gen. Genet.* **154**: 327.

Gold, L., P.Z. O'Farrell, and M. Russel. 1976. Regulation of gene 32 expression during bacteriophage T4 infection of *Escherichia coli. J. Biol. Chem.* **251**: 7251.

Gold, L.M., P.Z. O'Farrell, B. Singer, and G. Stormo. 1973. Bacteriophage T4 gene expression. In *Virus research* (ed. C.F. Fox and W.S. Robinson), p. 205. Academic Press, New York.

Gold, L., D. Pribnow, T. Schneider, S. Shinedling, B.S. Singer, and G. Stormo. 1981. Translational initiation in prokaryotes. *Annu. Rev. Microbiol.* **35**: 365.

Gold, L., G. Lemaire, C. Martin, H. Morrissett, P. O'Conner, P.Z. O'Farrell, M. Russel, and R. Shapiro. 1977. Molecular aspects of gene 32 expression in *Escherichia coli* infected with the bacteriophage T4. In *Nucleic acid-protein recognition* (ed. H.J. Vogel), p. 91. Academic Press, New York.

Gourse, R.L., D.L. Thurlow, S.A. Gerbi, and R.A. Zimmerman. 1981. Specific binding of a prokaryotic ribosomal protein to a eukaryotic ribosomal RNA: Implications for evolution and autoregulation. *Proc. Natl. Acad. Sci.* **78**: 2722.

Horiuchi, K. 1975. Genetic studies of RNA phages. In *RNA phages* (ed. N.D. Zinder), p. 29. Cold Spring Harbor Laboratory, Cold Spring Harbor, New York.

Jacob, F. 1977. Evolution and tinkering. *Science* **196**: 1161.

——————— . 1982. *The possible and the actual.* Pantheon, New York.

Jinks-Robertson, S. and M. Nomura. 1981. Regulation of ribosomal protein synthesis in an *Escherichia coli* mutant missing ribosomal protein L1. *J. Bacteriol.* **145**: 1455.

Johnsen, M., T. Christensen, P.P. Dennis, and N.P. Fiil. 1982. Autogenous control: Ribosomal protein L10-L12 complex binds to the leader sequence of its mRNA. *EMBO J.* **1**: 999.

Karam, J.D. and M.G. Bowles. 1974. Mutation to overproduction of bacteriophage T4 gene products. *J. Virol.* **13**: 428.

Karam, J., C. McCulley, and M. Leach. 1977. Genetic control of mRNA decay in T4 phage-infected *Escherichia coli. Virology* **76**: 685.

Karam, J., L. Gold, B. Singer, and M. Dawson. 1981. Translational regulation: Identification of the site on bacteriophage T4 rIIB mRNA recognized by the regA gene function. *Proc. Natl. Acad. Sci.* **78**: 4669.

Karam, J., M. Dawson, W. Gerald, M. Trojanowska, and C. Alford. 1982. Control of translation by the *regA* gene of T4 bacteriophage. In *Interaction of translational and transcriptional control in the regulation of gene expression* (ed. M. Grunberg-Manago and B. Safer), p. 83. Elsevier, New York.

Kirschner, M.W. 1980. Implications of treadmilling for the stability and polarity of actin and tubulin polymers *in vivo. J. Cell Biol.* **86**: 330.

Kolakofsky, D. and C. Weissmann. 1971. Qβ replicase as a repressor of Qβ RNA-directed protein synthesis. *Biochim. Biophys. Acta* **246**: 596.

Krisch, H.M. and B. Allet. 1982. Nucleotide sequences involved in bacteriophage T4 gene 32 translational self-regulation. *Proc. Natl. Acad. Sci.* **79**: 4937.

Krisch, H., A. Bolle, and R.H. Epstein. 1974. Regulation of the synthesis of bacteriophage T4 gene 32 protein. *J. Mol. Biol.* **88**: 89.

Kruger, K., P.J. Grabowski, A.J. Zaug, J. Sands, D.E. Gottschling, and T.R. Cech. 1982. Self-splicing RNA...Auto-excision and autocyclization of the ribosomal RNA intervening sequence of *Tetrahymena*. *Cell* **31:** 147.

Lehninger, A.L. 1965. *Bioenergetics*. W.A. Benjamin, New York.

Lemaire, G., L. Gold, and M. Yarus. 1978. Autogenous translational repression of bacteriophage T4 gene 32 expression *in vitro*. *J. Mol. Biol.* **126:** 73.

Lindahl, L. and J. Zengel. 1979. Operon-specific regulation of ribosomal protein synthesis in *Escherichia coli*. *Proc. Natl. Acad. Sci.* **76:** 6542.

———. 1982. Expression of ribosomal genes in bacteria. *Adv. Genet.* **21:** 53.

Lodish, H.F. and H.D. Zinder. 1966. Mutants of the bacteriophage fZ. VIII. Control mechanism for phage-specific synthesis. *J. Mol. Biol.* **19:** 333.

Maaløe, O. and N.O. Kjeldgaard. 1966. *Control of macromolecular synthesis*. Benjamin Press, New York.

McGhee, J.D. and P.H. von Hippel. 1974. Theoretical aspects of DNA-protein interactions: Cooperative and non-cooperative binding of large ligands to a one-dimensional lattice. *J. Mol. Biol.* **86:** 469.

Meyers, F., H. Weber, and C. Weissmann. 1981. Interactions of Qβ replicase with Qβ RNA. *J. Mol. Biol.* **153:** 631.

MinJou, W., G. Haegeman, M. Ysebaert, and W. Fiers. 1972. Nucleotide sequence of the gene coding for the bacteriophage MS2 coat protein. *Nature* **237:** 82.

Model, P., C. McGill, B. Mazur, and W.D. Fulford. 1982. The replication of bacteriophage f1: Gene V protein regulates the synthesis of gene II protein. *Cell* **29:** 329.

Napoli, C., L. Gold, and B.S. Singer. 1981. Translational reinitiation in the rIIB cistron of bacteriophage T4. *J. Mol. Biol.* **149:** 433.

Newport, J., N. Lonberg, S.C. Kowalczykowski, and P. von Hippel. 1981. Interactions of T4-coded gene 32 protein with nucleic acids. II. Specificity of binding to DNA and RNA. *J. Mol. Biol.* **145:** 105.

Nierhaus, K.H. 1980. Analysis of the assembly and function of the 50S subunit from *Escherichia coli* ribosomes by reconstitution. In *Ribosomes: Structure, function and genetics* (ed. G. Chambliss et al.), p. 267. University Park Press, Baltimore, Maryland.

Nomura, M. and D. Dean. 1983. A model for the coordinate regulation of ribosomal protein synthesis. In *Nucleic acid research: Future development* (ed. I. Watanabe et al.), p. 457. Academic Press, Tokyo.

Nomura, M. and W.A. Held. 1974. Reconstitution of ribosomes: Studies of ribosome structure, function and assembly. In *Ribosomes* (ed. M. Nomura et al.), p. 193. Cold Spring Harbor Laboratory, Cold Spring Harbor, New York.

Nomura, M., D. Dean, and J.L. Yates. 1982a. Feedback regulation of ribosomal protein synthesis in *Escherichia coli*. *Trends Biochem. Sci.* **7:** 92.

Nomura, M., S. Jinks-Robertson, and A. Miura. 1982b. Regulation of ribosome biosynthesis in *Escherichia coli*. In *Interaction of translational and transcriptional control in the regulation of gene expression* (ed. M. Grunberg-Manago and B. Safer), p. 92. Elsevier, New York.

Nomura, M., J.L. Yates, D. Dean, and L.E. Post. 1980. Feedback regulation of ribosomal protein gene expression in *Escherichia coli*: Structural homology of ribosomal RNA and ribosomal protein mRNA. *Proc. Natl. Acad. Sci.* **77:** 7084.

Olins, P.O. and M. Nomura. 1981a. Regulation of the S10 ribosomal protein operon in *E. coli*: Nucleotide sequence at the start of the operon. *Cell* **26:** 205.

———. 1981b. Translational regulation by ribosomal protein S8 in *Escherichia coli*: Structural homology between rRNA binding site and feedback target on mRNA. *Nucleic Acids Res.* **9:** 1757.

Olsson, M.O. and K. Gausing. 1980. Post-transcriptional control of coordinated ribosomal protein synthesis in *Escherichia coli*. *Nature* **283:** 599.

Oppenheim, D.S. and C. Yanofsky. 1980. Translational coupling during expression of the tryptophan operon of *Escherichia coli*. *Genetics* **95:** 785.

Post, L.E., G.D. Strychartz, M. Nomura, H. Lewis, and P.P. Dennis. 1979. Nucleotide sequence of the ribosomal protein gene cluster adjacent to the gene for RNA polymerase subunit β in *Escherichia coli*. *Proc. Natl. Acad. Sci.* **76:** 1697.

Rabussay, D. and E.P. Geiduschek. 1977. Regulation of gene action in the development of lytic bacteriophages. In *Comprehensive virology* (ed. H. Fraenkel-Conrat and R.R. Wagner), vol. 8, p. 1. Plenum Press New York.

Racker, E. 1976. *A new look at mechanisms in bioenergetics*. Academic Press, New York.

Ray, D.S. 1978. *In vivo* replication of filamentous DNA. In *The single-stranded DNA phages* (ed. D.T. Denhardt et al.), p. 325. Cold Spring Harbor Laboratory, Cold Spring Harbor, New York.

Reed, R.E., M.F. Baer, C. Guerrier-Takada, H. Donis-Keller, and S. Altman. 1982. Nucleotide sequence of the gene encoding the RNA subunit (M1 RNA) of ribonuclease P from *Escherichia coli*. *Cell* **30:** 629.

Robakis, N., L. Meza-Basso, N. Brot, and H. Weissbach. 1981. Translational control of ribosomal protein L10 synthesis occurs prior to formation of the first peptide bond. *Proc. Natl. Acad. Sci.* **78:** 4261.

Russel, M.L., L. Gold, H. Morrissett, and P.Z. O'Farrell. 1976. Translational autogenous regulation of gene 32 during bacteriophage T4 infection. *J. Biol. Chem.* **251:** 7263.

Schleif, R., W. Hess, S. Finkelstein, and D. Ellis. 1973. Induction kinetics of the L-arabinose operon of *Escherichia coli*. *J. Bacteriol.* **115:** 9.

Schumperli, D., K. McKenney, D.A. Sobieski, and M. Rosenberg. 1982. Translational coupling at an intercistronic boundary of the *Escherichia coli* galactose operon. *Cell* **30:** 865.

Singer, B., L. Gold, S. Shinedling, M. Colkitt, L. Hunter, D. Pribnow, and M. Nelson. 1981. Analysis *in vivo* of translational mutants of the rIIB cistron of bacteriophage T4. *J. Mol. Biol.* **149:** 405.

Spicer, E.K., J.A. Noble, N.G. Nossal, W.K. Konigsberg, and K.R. Williams. 1982. Bacteriophage T4 gene 45 sequences of the structural gene and its protein product. *J. Biol. Chem.* **257:** 8972.

Steitz, J.A. 1979. RNA·RNA interactions during polypeptide chain initiation. In *Ribosomes* (ed. G. Chambliss et al.), p. 479. University Park Press, Baltimore, Maryland.

Stormo, G.D., T.D. Schneider, and L. Gold. 1982. Characterization of translational initiation sites in *E. coli*. *Nucleic Acids Res.* **10:** 2971.

Taniguchi, T. and C. Weissmann. 1979. *Escherichia coli* ribosomes bind to non-initiator sites of Qβ RNA in the absence of formylmethionyl t-RNA. *J. Mol. Biol.* **128:** 481.

Tinoco, I., P.N. Borer, B. Dengler, M.D. Levine, O.C. Uhlenbeck, D.M. Crothers, and J. Gralla. 1973. Improved estimation of secondary structure in ribonucleic acids. *Nat. New Biol.* **246:** 40.

Trimble, R.B. and F. Maley. 1976. Level of specific prereplicative mRNA's during bacteriophage T4 regA⁻, 43⁻ and T4 43⁻ infection of *Escherichia coli* B. *J. Virol.* **17:** 538.

von Hippel, P.H. and F.R. Fairfield. 1983. Quantitative approaches to the autogenous regulation of gene expression. In *Mobility and Recognition in Cell Biology* (ed. H. Sund and C. Veegeer), p. 213. de Gruyter, New York.

von Hippel, P.H., S.C. Kowalczykowski, N. Lonberg, J.W. Newport, L.S. Paul, G.D. Stormo, and L. Gold. 1982. Autoregulation of gene expression: Quantitative evaluation of the expression and function of the bacteriophage T4 gene 32 (single-stranded DNA binding) protein system. *J. Mol. Biol.* **162:** 775.

Weber, H., M.A. Billeter, S. Kahane, C. Weissmann, J. Hindley, and A. Porter. 1972. Molecular basis for repressor activity of Qβ replicase. *Nat. New Biol.* **237:** 166.

Wiberg, J.S., S.L. Mendelsohn, V. Warner, C. Aldrich, and T.S. Cardillo. 1977. Genetic mapping of regA mutants of bacteriophage T4D. *J. Virol.* **22:** 742.

Wiberg, J.S., S. Mendelsohn, V. Warner, K. Hercules, C. Aldrich, and J.L. Munro. 1973. SP62, a viable mutant of bacteriophage T4D defective in regulation of phage enzyme synthesis. *J. Virol.* **12:** 775.

Wirth, R. and A. Bock. 1980. Regulation of synthesis of ribosomal protein S20 *in vitro*. *Mol. Gen. Genet.* **178:** 479.

Wirth, R., V. Kohler, and A. Bock. 1981. Factors modulating transcription and translation *in vitro* of ribosomal protein S20 and isoleucyl-tRNA synthetase from *Escherichia coli*. *Eur. J. Biochem.* **114:** 429.

Wood, W.B. and H.R. Revel. 1976. The genome of bacteriophage T4. *Bacteriol. Rev.* **40:** 847.

Yates, J.L. and M. Nomura. 1980. *Escherichia coli* ribosomal protein L4 is a feedback regulatory protein. *Cell* **21:** 517.

————. 1981. Feedback regulation of ribosomal protein synthesis in *E. coli*: Localization of the mRNA target sites for repressor action of ribosomal protein L1. *Cell* **240:** 243.

Yates, J.L., A.E. Arfsten, and M. Nomura. 1980. *In vitro* expression of *Escherichia coli* ribosomal protein genes: Autogenous inhibition of translation. *Proc. Natl. Acad. Sci.* **77:** 1837.

Yates, J.L., D. Dean, W.A. Strycharz, and M. Nomura. 1981. *E. coli* ribosomal protein L10 inhibits translation of L10 and L7/L12 mRNAs by acting at a single site. *Nature* **294:** 190.

Yen, T.S.B. and R.E. Webster. 1982. Translational control of bacteriophage f1 gene II and gene X proteins by gene V protein. *Cell* **29:** 337.

Zengel, J.M., D. Muekl, and L. Lindahl. 1980. Protein L4 of the *E. coli* ribosome regulates an eleven gene r protein operon. *Cell* **21:** 523.

Recombinational Regulation of Gene Expression in Bacteria

Melvin I. Simon
Division of Biology
California Institute of Technology
Pasadena, California 91125
Michael Silverman
Agouron Institute
La Jolla, California 92037

INTRODUCTION

It is generally assumed that in a population derived from a single individual, e.g., a bacterial colony, the members of the population are almost all genetically identical. However, on closer inspection this assumption is found to be incorrect. Thus, e.g., Anderson and Roth (1977) have reviewed the evidence indicating that tandem duplication and amplification of regions of the *Escherichia coli* and *Salmonella* genomes arise at relatively high frequency (10^{-3}/cell/generation). Therefore, in a clonal population derived after $20-30$ generations of growth under selective conditions, individual cells may differ with respect to the number of copies of specific genes. Other more apparent heterogeneity is observed when bacteria are isolated and cloned from natural sources. The colonies are frequently found to be polymorphic with respect to a specific characteristic. One manifestation of this polymorphism is seen as sectored colonies. For example, when marine bacteria such as *Vibrio harveyi* are isolated from seawater and streaked on nutrient plates, the colonies, when observed in the dark, show sectors that are bioluminescent (see Fig. 1). If a small portion of the bioluminescent sector is picked and restreaked, each of the individual colonies is still found to be sectored, with part of the colony being bioluminescent and the other part dark (Keynan and Hastings 1961). Thus, these organisms show alternation of expression of bioluminescence, and the colony is said to be polymorphic with respect to bioluminescence.

Table 1 is a partial list of some observations of clonal dimorphism and polymorphism that have been reported. Some of these variations are complex. For example, the surface antigen of *Borrelia*, a spirochete that is associated with a disease called relapsing fever, has been shown to undergo repeated changes in its major surface antigen (Cunningham 1925). These changes are associated with relapses in the host and proliferation of the spirochete. Apparently, the switches in surface antigen allow the spirochete to survive the immune response in the host. *Borrelia hermsii* was cloned by infecting mice with single organisms, and the surface antigens of populations derived from

211

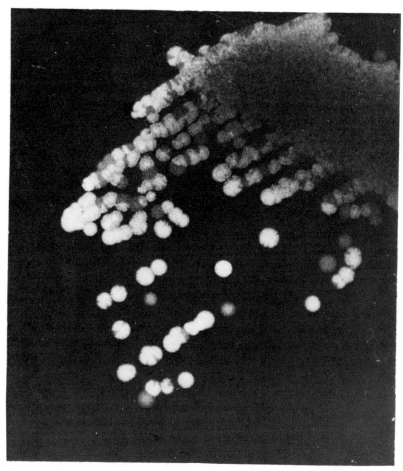

Figure 1   Colonies of bioluminescent *V. harveyi* photographed by their own light. An overnight culture was streaked on nutrient agar (Keynan and Hastings 1961) and grown at room temperature for 36 hr.

the clones were studied (Stoenner et al. 1982). Twenty-four distinguishable antigenic types were found without exhausting the repertoire of the population. The frequency of switching from one type to the other was estimated at $10^{-3}-10^{-4}$ per cell per generation. Furthermore, changes in serotype were shown to be correlated with changes in the electrophoretic properties of cell-surface proteins. The variable proteins were identified using monoclonal antibodies prepared against organisms with different serotypes (Barbour et al. 1982).

Although the variation found in other prokaryotic systems may not be as elaborate as that of the *Borrelia* surface antigen, there are clearly other examples of complex switching. The colony opacity of gonococci varies at high frequency. Changes in the electrophoretic behavior of the PII surface

Table 1   Examples of clonal polymorphism

Phenotype	Organism	Reference
Surface antigen	*Borrelia hermsii*	Stoenner et al. (1982)
H antigen	*Salmonella*	Lederberg and Iino (1956)
O antigen	*Salmonella*	Edwards and Ewing (1955)
Vi antigen	*Citrobacter freundii*	Baron et al. (1982)
Colony morphology (Flu)	*Escherichia coli*	Diderichsen (1980)
Colony morphology (Op)	*Neisseria gonococcus*	James and Swanson (1978)
Pilin variation	*Neisseria gonococcus*	Buchanan (1975)
Pigment variation	*Myxococcus xanthus*	Burchard and Dworkin (1966)
Pigment variation	*Rhodopseudomonas spheroides*	Griffiths and Stanier (1956)
Bioluminescence	*Vibrio harveyi*	Keynan and Hastings (1961)
(bright-dim)		
Pilin variation	*Escherichia coli*	Eisenstein (1981)
Pigment variation	*Serratia marcescens*	Bunting (1940)
Host range	Bacteriophage Mu	Howe (1980)
Host range	Bacteriophage P1	Iida et al. (1982)
Pigment variation	*Streptomyces reticuli*	Schrempf (1983)

protein were found to correlate with changes in opacity, and at least five different PII variants were found in one strain (Swanson 1982). The pilus antigen in *Neisseria gonorrhoeae* can also switch between a number of different forms (Buchanan 1975; Lambden et al. 1980). On the other hand, the systems that have been studied most extensively thus far appear to involve relatively simple dimorphisms. They also involve traits that are readily visible, e.g., colony morphology, surface antigens, and cell pigments. There may be many other functions that are variable in this way but affect properties of the cells or the colony that are not as easily observed.

Some metastable phenotypes in bacteria can be attributed to the segregation of an extra replicon. Thus, an unstable plasmid, a virus that is carried and replicates slowly, or the induction of a lysogen that carries genes determining a particular trait could all lead to apparent clonal polymorphism. However, a large number of observations exist where the explanation that the instability results from the presence of a gene on a segregating replicon has been ruled out. In many of these cases (see Table 1), the following characteristics are found: (1) Cells in the population can alternate between the expression of one phenotype or the other. The frequencies with which switching occurs from one form to the other range from one event in 50 cells per generation, e.g., Vi antigen variation (Baron et al. 1982), to 10^{-5} per cell per generation, e.g., flagellar phase variation in *Salmonella* (Stocker 1949). In general, the frequency is found to be genetically determined, and specific mutations that occur at low frequencies can inactivate the switching mechanism, stabilizing the cell in one form or the other. (2) The switching event occurs at the gene level and results in a specific heritable change in the genetic material. It is possible, therefore, to isolate the genes responsible for the dimorphism in either state and to study the changes in DNA structure that are correlated with changes in the state of the gene.

A variety of hypotheses have been suggested to explain the nature of the heritable change that determines the "state" of the gene. Some of these ideas include (1) covalent modification of the DNA corresponding to specific genes that affect the expression of the dimorphic trait, or (2) gene rearrangements, including duplications, insertions, inversions, deletions, and translocations, which could modify the gene itself and the regulatory elements that normally are required for gene expression. In this paper we have reviewed some examples of clonal polymorphism involving relatively simple gene rearrangements that affect gene expression. We have focused particularly on those cases that have been analyzed at the molecular level.

REGULATION BY INVERSION

Salmonella Flagellar Phase Variation

Flagellar antigen phase variation in *Salmonella* is one example of a large number of cell-surface antigenic variations that are found in clonal populations of the enterobacteria (Edwards and Ewing 1955). *Salmonella* were found to undergo transitions in the nature of the polypeptide that is the major structural component of bacterial flagella. Two genes, *H1* and *H2*, encode different forms of the flagellin protein, and only one of these genes is generally expressed in any cell. The organisms can switch from the expression of one gene to the other at frequencies that range from 10^{-3} to 10^{-5} per cell per generation (Stocker 1949). The *H1* and *H2* genes are widely separated from each other on the *Salmonella* genome. In some of the earliest gene-transfer experiments in bacteria, Lederberg and Iino (1956) used viral transduction to analyze phase variation. They showed that phase transition was regulated by a locus adjacent to the *H2* gene. This genetic element could exist in the "on" state or in the "off" state. In transduction experiments, both the *H2* gene and the state of regulation were transferred, suggesting that the gene was regulated by a heritable change at the DNA level. Lederberg and Iino proposed that the regulatory system involved in the control of phase variation was analogous to the genetic rearrangements that regulated gene expression (discovered by McClintock 1956) in maize. Subsequent molecular analysis of the elements involved in phase transition does in fact indicate that the genes required for switching are related to genes that are found associated with mobile genetic elements, e.g., transposons and viruses.

To determine the nature of the molecular events involved in switching, the DNA sequence corresponding to the *H2* region was cloned into *E. coli*. The cloned DNA was denatured and renatured and examined by electron microscopy for abnormal structures. A portion of the molecules were found to be heteroduplexes with a region composed of single-stranded DNA sequences corresponding to 995 bp (Zieg et al. 1977). These structures formed as a result of an inversion region adjacent to the *H2* gene. Phase transition was found to correlate with the inversion; i.e., when the region was in one orientation the *H2* gene was in the on state, and both the *H2*-gene product and the product of an adjacent gene, *rH1*, were formed. The *rH1* product corresponded to a 16,000-kD polypeptide that acted as a repressor of the *H1* gene (Silverman et al. 1979). When the inversion region was in the opposite orientation, the *H2* gene was in the off state, and the *rH1* repressor was not formed. Thus, the *H1* gene could be expressed.

Genetic analysis and subsequent DNA sequence determination of the region revealed the mechanism involved in phase transition (Silverman and Simon 1980). The results of this work are summarized schematically in Figure 2. The inversion region, including 995 bp, was found to be bounded on either side by a 14-bp inverted repeat sequence. Site-specific homologous recombination within the inverted repeated sequences resulted in inversion. The coding sequence of the *H2* structural gene begins 16 bp from one of the inverted repeat sequences (IRR). Within the inversion region a sequence of approximately 100 bp adjacent to one of the repeats (IRR) includes the promoter for the *H2* gene. In addition there is a large open reading frame that encodes a polypeptide called Hin. This polypeptide mediates the inversion. Mutations within the structural gene for the Hin function eliminate inversion and stabilize the organism in one phase or the other (Silverman and Simon 1980; Szekely and Simon 1981). Thus, it is the coupling and uncoupling of the promoter from the structural gene that regulates phase transition. The switching is the

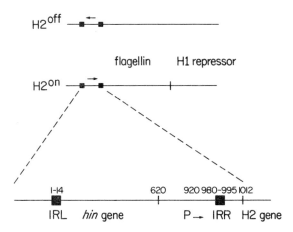

Figure 2   The components of the flagellar phase variation switch. (*Bottom line*) An enlargement of the invertible region. Numbers refer to the nucleotide numbering scheme described previously (Zieg and Simon 1980).

result of site-specific recombination mediated by the *hin*-gene product at the 14-bp inverted repeated sequence. We would expect, therefore, that in the presence of this gene product the nature of the reaction would depend on the configuration of repeated sites. Thus, if the sites were in direct repeat configuration, the *hin*-gene product should mediate site-specific recombination leading to deletion. Furthermore, if the recombinational sites were on different replicons, the presence of the *hin*-gene product should lead to the formation of cointegrates resulting from site-specific homologous recombination. All of these reactions have been demonstrated (Scott and Simon 1982). The relative efficiency with which each of these reactions occurs is different, and the inversion reaction is mediated most efficiently. However, inversion, deletion, and fusion have all been found when the appropriate substrates were built and inserted into a cell that contained an intact, expressed *hin* gene.

Further evidence suggesting that the *hin*-gene product itself is a recombinase comes from experiments that were done to vary the Hin product concentration in vivo and to measure the subsequent frequency of inversion. To increase the level of *hin*-gene product, the DNA fragment carrying the leftward promoter of λ was inserted adjacent to the *hin* gene (see Fig. 3). This plasmid was maintained in a cell that could produce temperature-sensitive (λ*c*I857) repressor. Thus, at low temperatures the λ repressor repressed *hin* synthesis. However, when the temperature was raised, the *hin*-gene product was synthesized rapidly, and it could be readily detected on acrylamide gels. By raising the temperature for different amounts of time, the intracellular level of the *hin*-gene product was varied (M. Bruist and M. Simon, in prep.). These cells were then infected with bacteriophage λ (immunity 434, so that λ could grow in the presence of the *c*I repressor) that carried an insert that allowed the number of switching events to be measured. Figure 3 illustrates the switching substrate. The structural gene for β-galactosidase was fused to the *H2* operon immediately downstream from the intact *H2* gene. This operon fusion was regulated by the inversion region with the flagellin promoter in the off configuration and contained a deletion in the *hin* gene. In the absence of *hin*-gene product, phage propagated in *E. coli* were found to be stable in the off configuration. Upon infection these phage did not lead to the formation of β-galactosidase. If, however, the λ was propagated in cells that carried the *hin*-gene function, then inversion occurred and the flagellin promoter became associated with the *H2-lac* fusion. The subsequent λ phage that emerged from this infection had the ability to form plaques, direct the formation of β-galactosidase on indicator plates, and restore motility in nonmotile flagellin-defective (Hag⁻) cells. Each phage that carried an inversion event formed a blue plaque in the presence of X-gal indicator, whereas colorless plaques resulted from phage that remained in the initial configuration and did not switch. Under normal growth conditions, in the absence of Hin and in strains that lacked basal switching activity, populations of λ were prepared that gave less than one blue plaque per 10^5

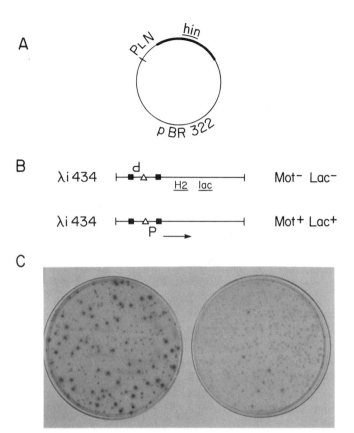

A

hin

PLN

pBR322

B

λi434 ⊢━■△━■━━━━━━━━┤ Mot⁻ Lac⁻
          d   H2 lac

λi434 ⊢━■━△━■━━━━━━━┤ Mot⁺ Lac⁺
                P ⟶

C

Figure 3 Assay system for *hin*-mediated inversion. (*A*) Overproducing plasmid. A DNA fragment produced by a restriction endonuclease carrying the leftward promoter of λ was inserted adjacent to the *hin* gene in a pBR322-derived plasmid. The plasmid was carried in an *E. coli* strain that has the *c*I857 repressor gene. At 30°C the repressor blocked p_L-mediated *hin* formation, whereas at 42°C the repressor was inactivated and *hin* was transcribed from p_L (M. Bruist and M. Simon, in prep.). (*B*) The λ phage used for scoring inversion events. λi434 carried an insert with a deletion in the *hin* gene and the invertible region in the off state (*top line*). The entire structural gene encoding β-galactosidase, but missing the promoter region, was inserted by the appropriate λ crosses adjacent to the *H2* gene. When inversion occurred (*bottom line*), the flagellar promoter activated transcription of both *H2* and *lac*. (*C*) Assay for inversion. (*Left*) Bacteriophage λ carrying the assay fragment were grown for one lytic cycle on cells that harbored the overproducing plasmid. The plaques on the *right* and *left* plates come from phage grown on uninduced (30°C) and induced (42°C) cells, respectively. The darker and apparently larger plaques on these X-gal plates are the ones that contain high levels of β-galactosidase. Each of these was picked and stabbed into motility plates, and the lysogens were found to be motile. The smaller light-colored plaques did not contain motile lysogens (M. Bruist and M. Simon, in prep.).

phage. The phage were then grown on cells that had been preinduced to yield increased levels of Hin, and the number of blue plaques increased dramatically (a factor of 10^4) even after only 10 minutes of full induction of Hin synthesis. Thus far (M. Bruist and M. Simon, in prep.), up to 15% of the total λ emerging from a burst have switched from the off to the on configuration or from the on to the off configuration. The simplest hypothesis consistent with all of these observations is that the *hin*-gene product is a recombinase that is ordinarily synthesized at very low levels in the cell. This results in very low frequencies of inversion. If the level of recombinase is increased, the rate of inversion increases.

Inversion in Other Contexts

Flagellar phase variation is only one example of the regulation of gene expression by inversion. Other examples of invertible segments exist, and the mechanism of inversion is similar to that found in the phase variation system. In bacteriophage Mu, there is a 3000-bp region, the G loop, which is found to invert (Kamp et al. 1979). This region contains parts of the structural genes encoding the tail fibers that determine the Mu phage host range (Howe 1980; Van de Putte et al. 1980). Figure 4 shows that the G loop exists in two configurations: $G(+)$, where it expresses the S and U genes, and $G(-)$, where the S' and U' genes are expressed (Giphart-Gassler et al. 1982). In Mu, the 3000-bp region is bounded by 34-bp inverted repeats (Kamp and Kahmann 1981; Plasterk et al. 1983a). On one side, outside of the inversion region, there is a gene that is required for inversion. This gene encodes a product, Gin, which mediates the site-specific inversion event. The promoter and part of the S gene are outside of the invertible region. Thus, the result of inversion is to disconnect or reconnect two sets of tail fiber genes to a stable promoter, so that one or the other set of host-determining functions is transcribed. Furthermore, the frequency of G-loop inversion was shown to depend upon the amount of Gin expressed (Plasterk et al. 1983b). The level of the polypeptide was increased by fusing the *gin* gene to a strong λ promoter. A switching substrate fixed in the off position of the G loop was constructed, and inversion resulted in the activation of a β-galactosidase gene (Plasterk et al. 1983b). The frequency of Gin-mediated switching was increased by more than two orders of magnitude in the overproducing strain.

A similar 3000-bp segment was found in bacteriophage P1 (Chow and Bukhari 1976). It was called the C loop, and it shows a great deal of homology with the Mu invertible region. A gene that maps outside of the invertible region is required for inversion. This sequence has been called *cin*, and it encodes a site-specific recombination function (Iida et al. 1982). The precise orientation of the inverted repeat segments is different in P1 from that found in Mu since, in addition to short inverted repeats (at which recombination occurs), there are 600-bp inverted repeat segments that flank the 3000-bp

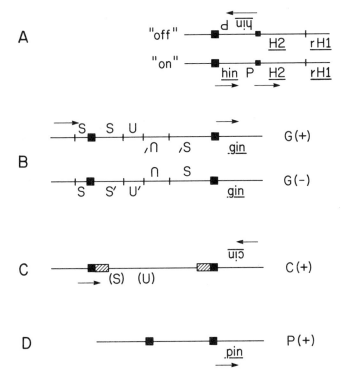

Figure 4 Schematic representation of the genes involved in invertible switching systems. (*A*) Flagellar phase variation. (*Top line*) The region in the off state when *H2* and *rH1* are disconnected from the flagellin promoter. (*Bottom line*) The system in the on state where both the *H2* and *rH1* genes are transcribed (Simon et al. 1980). (*B*) The G region of bacteriophage Mu. The dark boxes indicate the 34-bp repeated switch sites. (→) The direction of transcription (Van de Putte et al. 1980). (*C*) The C region of bacteriophage P1. Hatched boxes represent the 600-bp repeats that flank the C region. (S) and (U) represent the genes that probably encode host-range specificity (Kennedy et al. 1983). (*D*) The Pin region in *E. coli*. The *pin* gene has been sequenced (Plasterk et al. 1983a).

C loop. Both *gin* and *cin* functions have been shown to mediate site-specific deletion and replicon fusion, as well as inversion (Kennedy et al. 1983; Plasterk et al. 1983b). Finally, these three switching systems complement one another. Thus, if the *gin* gene is inactivated, Mu phage can still invert their G segment if they are grown in the presence of an active *hin* or *cin* gene (Kutsukake and Iino 1980; Kamp and Kahmann 1981). Conversely, *gin* and *cin* have been shown to be able to replace *hin* function. Furthermore, in DNA-DNA hybridization experiments, clear signals have been obtained when the *hin* gene was hybridized with bacteriophage Mu or bacteriophage P1. On the basis of the heat stability of these heteroduplexes, it was estimated that *hin* and *gin* are approximately 70–80% identical, whereas *cin* and *hin* are approximately 60–70% identical (Szekely and Simon 1981).

In addition to the *gin* and *cin* systems, a number of other examples of DNA inversions that regulate gene functions have been found. Enomoto and Stocker (1975) transferred the genes for the *Salmonella* phase variation function into *E. coli*. They showed that some *E. coli* strains included an endogenous function that was able to mediate switching at the *H2* locus. Subsequent work confirmed the existence of a cryptic recombinase function that could mediate flagellar phase variation in certain strains of *E. coli*. This gene, called *pin*, mapped at 26.5 minutes on the *E. coli* chromosome (Enomoto et al. 1983). More recently, bacteriophage Mu defective in *gin* function was grown in *E. coli* deficient in *pin*, but carrying a variety of plasmids from a *pin*-proficient strain. The plasmids were tested to see if they carried a function that would complement the defective *gin* and allow the Mu G region to invert (Plasterk et al. 1983a). The *pin* gene was found to complement Mu *gin* mutations, and the *pin* function mediated inversion of an adjacent 1600-bp sequence on the plasmid. Inversion of this region regulated the synthesis of two polypeptides: one, 65 kD, and the other, 42 kD. When the region was in the P(+) configuration, these polypeptides were found; when the region was inverted in the P(−) configuration, they were not synthesized (Plasterk et al. 1983a). *pin* appears to be more closely related to *gin* than it is to *hin*. The *pin* function is also able to complement both *hin* and *gin*. The nucleic acid sequences of the *pin*, *gin* (Plasterk et al. 1983a), and *hin* (Zieg and Simon 1980) genes have been determined, and the derived amino acid sequences, compared. There are extensive regions of identity between all three of these polypeptides.

Another system that includes a function that is complemented by *hin* has been found to be involved in surface antigen phase variation. In *Citrobacter freundii*, variation between Vi^+ and Vi^- forms is found. The transition involves the formation of a cell-surface antigen, and the variation can be identified by differences in colony morphology of the two forms (Snellings et al. 1981). Transitions occur at frequencies as high as 10^{-2} per cell per generation. Vi antigen synthesis involves two clusters of genes, one called ViaA, and the other, ViaB (Johnson et al. 1965). When ViaB genes were transferred from *Citrobacter* to *Salmonella* or to *E. coli*, it was found that *E. coli* could make the Vi antigen and was also able to alternate between the two different forms. This suggested that the transfer of the ViaB genes included a factor required for variation at this locus (Baron et al. 1982). To determine whether this new function belonged to the same complementation group as *hin*, *gin*, *cin*, and *pin*, strains carrying the ViaB region were used as hosts for λ phage that carried the *Salmonella H2* region in the off configuration and were defective for *hin*. The presence of the ViaB genes in these strains led to complementation of the defect in the *hin* gene and allowed inversion to occur adjacent to the *H2* gene, resulting in *H2*-gene expression (E.M. Johnson and D.J. Kopecko, in prep.). Thus, there may be a gene (*vin*) related to the other recombinase genes that could be involved in mediating Vi antigen switching. Figure 4 summarizes the properties of the systems mediated by inversion.

Recombinases

All of the metastable systems that we have described thus far—flagellar phase variation, bacteriophage host range variation, and Vi variation—are driven by a site-specific recombination event that occurs within a short inverted repeat sequence. Table 2 lists the sequences that have been shown to be involved in the recombination event. In both the *gin* and *pin* system, there are 34-bp inverted repeats within which recombination occurs. In the *hin* system, the two repeats, IRR and IRL, are different. IRL retains homology with the 34-bp repeats found adjacent to *gin* in Mu and adjacent to *pin* in *E. coli*. On the other hand, IRR retains only 14 bp of the inverted repeat, and the recombination event occurs within this sequence. The differences in the arrangement of the repeats have led to the suggestion that flagellar phase variation may have evolved from a system in which a mobile element that included one 34-bp sequence adjacent to the recombinase gene was inserted near the flagellar gene promoter (Kamp and Kahmann 1981; Silverman and Simon 1983; Szekely and Simon 1983). Partial homology between sequences in the region that regulates flagellin transcription and the 34-bp repeat could have resulted in the initial inversion event and the formation of the 14-bp IRR site. There is indeed evidence that these kinds of low-frequency events occur and result in the formation of new gene arrangements. Iida and his co-workers (Kennedy et al. 1983) found that partial homology of a plasmid sequence with one of the *cin* sites (the recombination site involved in C inversion) led to the formation, at low frequencies, of stable cointegrate forms. Thus, there may be a range of sequences related to the consensus sequence (Table 2) recognized by the recombinases that act as inefficient substrates for inversion, deletion, and

Table 2 Inverted repeat DNA sequences where site-specific recombination occurs

	[a]Hin IRR	AAAATTTTCCTTTTGGAAGGTTTTTGATAACCAA
	Hin IRL	TTGGTTCTTGAAAACCAAGGTTTTTGATAAAGCA
G(+)	[b]Gin IRR	CCGTTTCCTGTAAACCGAGGTTTTGGATAATGGT
	[c]Gin IRL	CTGTTTATCCAAAACCTCGGTTTACAGGAAACGG
P(−)	[b]Pin IRR	CTCCTTCTCCCAAACCAACGTTTATGAAAATGAA
	Consensus	TT--TC---AAACC----GGTTT---GA--AA

The appropriate DNA strand was chosen to align and compare the sequences.
[a]From Zieg and Simon (1980).
[b]From Plasterk et al. (1983a).
[c]From Kamp and Kahmann (1981).

plasmid fusion or integration. These events could form the basis for the evolution of pathways of recombinational regulation.

There are other recombinase systems, some of them related in amino acid sequence to the recombinases involved in inversion, that recognize totally different subsets of nucleotide sequence. Thus, the TnA transposons have been found to encode a resolvase gene that mediates site-specific recombination between 19-bp directly repeated sequences (Sherratt et al. 1981; Grindley et al. 1982). Resolvases encoded by members of the TnA transposon family have been found to have 30−40% amino acid identity with the Hin recombinase (Simon et al. 1980), and the conserved amino acids are almost all included in the amino acid sequences that are found to be identical in the Hin, Gin, and Pin sequences (Plasterk et al. 1983a; Silverman and Simon 1983). However, there is no apparent relationship between the nucleotide sequence to which the resolvase gene binds and at which it mediates recombination and the sequences recognized by the Hin or Gin recombinase. Furthermore, resolvase appears to preferentially mediate recombination between directly repeated sequences, whereas the Hin recombinases appear to prefer inverted repeat substrates (Scott and Simon 1982; Kennedy et al. 1983; Plasterk et al. 1983b). Thus, these two families of recombinases may have been derived from a common ancestor but have evolved and diverged considerably. The conserved sequences may indicate regions of the proteins that are involved in mediating the strand exchange reaction, whereas the divergent sequences could be involved in forming the domain of the proteins required for the recognition of specific DNA sequences. Other systems that mediate site-specific recombination, e.g., the bacteriophage λ integrase (*int-att*; Mizuuchi and Mizuuchi 1980) and the bacteriophage P1 recombinase (*lox-cre*; Abremski et al. 1983), have no distinctive amino acid sequence or nucleic acid sequence homology with either resolvase or the Hin family of recombinases. They may have evolved from different sources, or they may no longer have features that are recognizably homologous to Hin or resolvase at the amino acid sequence level.

REGULATION BY REARRANGEMENTS OTHER THAN INVERSION

There are a number of examples of rearrangements that occur at relatively high frequency and affect gene expression. Thus, gene amplification occurs by the formation of tandem duplications at frequencies of 10^{-3} in both phage and bacteria and can be stabilized and studied under appropriate selective conditions (Anderson and Roth 1977). In addition to duplication, there are examples of transposition events that affect gene expression. A number of transposons have been shown to carry promoters that are able to direct transcription from sequences that are adjacent to the transposon (Iida et al. 1983; Kleckner 1983). Thus, the insertion of a transposon within or adjacent to a cryptic gene can result in expression initiated by promoters carried on the transposable element. The insertion of a transposon may also act to generate a new promoter. An interesting variation of this kind of gene activation was found to occur with the cryptic *bgl* genes in *E. coli* (Reynolds et al. 1981). The *bgl* operon contains three structural genes required for the catabolism of β-glucosides. The genes are not normally expressed or induced in wild-type *E. coli* K12. However, mutants occurred at frequencies of 10^{-5} that were able to induce the expression of the genes in the *bgl* operon. A number of these mutants were characterized, and some of them where shown to be the result of the insertion of IS*1* or IS*5* elements into a small target region adjacent to the *bgl* genes. Activation of these genes, however, was not a result of juxtaposition of a promoter on the transposon to the structural genes since the *bgl* operon, although activated by the insertion element, still required the presence of a suitable inducer. Furthermore, point mutations in the target region also activated the *bgl* operon. These findings suggested the hypothesis that the cryptic nature of the *bgl* genes was the result of the inability of their normal promoter to initiate transcription. Changes in the DNA structure at the promoter activated it so that it could be transcribed. This hypothesis was supported by the subsequent finding that the *bgl* operon can be activated by mutations that map in the *gyrB* or *gyrA* genes (DiNardo et al. 1982). These genes are required for the synthesis of gyrase, which is necessary for maintaining the superhelical structure of the DNA. Thus, changes in superhelicity may be sufficient to activate the *bgl* operon under certain specific conditions. Therefore, in addition to the direct mechanisms of activation or inactivation of genes by insertion into or adjacent to the gene, transposons may act by changing the local structure of the DNA in the regulatory region of the gene.

In addition to the relatively simple rearrangements that regulate gene activity, there is a variety of examples of more complex changes. One of these is the regulation of pilin phase variation in *N. gonorrhoeae*. Recent work has shown that changes in the expression of the pilin genes are correlated with rearrangements in chromosomal structure. However, the precise mechanism that operates in this system is not known. Meyer et al. (1982) have isolated DNA sequences that correspond to one of the pilin genes. They were able to hybridize this specific probe to DNA derived from cloned populations of cells that had been passaged for a number of generations. They showed that there were changes in the endonuclease restriction fragment distribution in these regions, and the apparent rearrangement accompanied changes in colony morphology. Although it is clear that those sequences detected with the pilin gene probes underwent rearrangements, the precise nature of the rearrangement is unclear. There may be multiple copies of the pilin genes, and differential expression of these genes may depend on duplication and transposition in a manner similar to the expression of different cell-surface antigens in trypanosomes (Borst 1983).

CONCLUSIONS

It is clear from this brief survey that there are a variety of systems in bacteria where gene rearrangements may accompany changes in gene expression. It is difficult, however, to establish the precise nature of the rearrangement that underlines specific examples of variation. To establish a mechanism, we must have sophisticated understandin , of both the genetics and the molecular biology of the system. In most cases there is a paucity of data in either area or in both areas. Thus, e.g., specific inversions have been found in staphylococcal plasmids (Murphy and Novick 1980), but they have not been correlated with biological functions. On the other hand, there are many phenotypic switches for which there are no molecular or genetic probes (Table 1), and it is therefore difficult to look for rearrangements. As recombinant DNA techniques are applied to more diverse bacterial systems, the appropriate probes will become available.

Variation within clonal populations exists in prokaryotes. Some of the variation is introduced by apparently random events, e.g., recombination leading to duplication and eventual amplification of regions of the genome. Other kinds of variation appear to be programmed, i.e., built into the genome together with genetic regulation of the frequency of variation. The degree to which these mechanisms are involved in the propagation and survival of natural populations of bacteria is not clear. The requirement that bacteria, such as pathogens, adapt to rapidly changing environments and surfaces to establish an infection could be met by using recombinational switches to generate diversity in populations with respect to specific phenotypes. As we understand more about the molecular basis of complex bacterial behavior, we may find that these kinds of regulatory mechanisms are ubiquitous.

Current work suggests that programmed variation could have evolved as a result of the interaction of mobile elements capable of inducing gene rearrangement with resident, stable genes. The resulting regulatory circuits would involve specific recombinational events that regulate gene expression. They would also interact with regulatory mechanisms that control the behavior of genes on mobile elements. This notion was first introduced by McClintock (1956) to explain her observations in maize, and it provides a useful paradigm for studying the role of gene rearrangement in regulating gene expression in prokaryotes.

ACKNOWLEDGMENTS

Research was supported by a grant from the National Science Foundation. We thank M. Druist, D. Kopecko, and R. Plasterk for allowing us to quote their unpublished results.

REFERENCES

Abremski, K., R. Hoess, and N. Sternberg. 1983. Studies on the properties of P1 site-specific recombination. *Cell* **32:** 1301.

Anderson, R. and J.R. Roth. 1977. Tandem genetic duplication in phage and bacteria. *Annu. Rev. Microbiol.* **34:** 473.

Barbour, A.G., S.L. Tessier, and H.G. Stoenner. 1982. Variable major proteins of *Borrelia hermsii. J. Exp. Med.* **156:** 1312.

Baron, L.S., D.J. Kopecko, S.M. McCowen, N.J. Snellings, E.M. Johnson, W.C. Reid, and C.A. Life. 1982. Genetic and molecular studies of the regulation of atypical citrate utilization and variable Vi antigen expression in enteric bacteria. In *Genetic engineering of microoganisms for chemicals* (ed. A. Hollander), p. 175. Plenum Press, New York.

Borst, P. 1983. Antigenic variation in trypanosomes. In *Mobile genetic elements* (ed. J.A. Shapiro), p. 619. Academic Press, New York.

Buchanan, T.M. 1975. Antigenic heterogeneity of gonococcal pili. *J. Exp. Med.* **141:** 1470.

Bunting, M.I. 1940. The production of stable populations of color variants of *Serratia marcescens* no. 274 in rapidly growing cultures. *J. Bacteriol.* **40:** 69.

Burchard, R.P. and M. Dworkin. 1966. Light-induced lysis and carotenogenesis in *Myxococcus xanthus. J. Bacteriol.* **91:** 535.

Chow, L.T. and A. Bukhari. 1976. The invertible DNA segments of coliphage Mu and P1 are identical. *Virology* **74:** 242.

Cunningham, J. 1925. Serological observations on relapsing fever in Madras. *Trans. R. Soc. Trop. Med. Hyg.* **19:** 11.

Diderichsen, B. 1980. *flu,* a metastable gene controlling surface properties of *E. coli. J. Bacteriol.* **141:** 859.

DiNardo, S., K.A. Voelkel, R. Sternglanz, A. Reynolds, and A. Wright. 1982. *E. coli* DNA topoisomerase I mutants have compensatory mutations in DNA gyrase genes. *Cell* **31:** 43.

Edwards, P.R. and W.H. Ewing. 1955. Identification of *Enterobacteriaceae.* Burgess, Minneapolis, Minnesota.

Eisenstein, B. 1981. Phase variation of type 1 fimbriae in *E. coli* is under transcription control. *Nature* **214:** 337.

Enomoto, M. and B.A.D. Stocker. 1975. Integration at *hag* or elsewhere of *H2* genes transduced from *Salmonella* to *Escherichia coli. Genetics* **81:** 595.

Enomoto, M., H. Oosawa, and H. Momotu. 1983. Mapping of the *pin* locus for a site-specific recombinase that causes flagellar phase variation in *E. coli* K-12. *J. Bacteriol.* **156:** 663.

Giphart-Gassler, M., R.H.A. Plasterk, and P. Van de Putte. 1982. G inversion in bacteriophage Mu. A novel way of gene splicing. *Nature* **297:** 339.

Griffiths, M. and R. Stanier. 1956. Some mutational changes in the photosynthetic pigment system in *Rhodopseudomonas spheroides. J. Gen. Microbiol.* **14:** 698.

Grindley, N.D.F., M.R. Lauth, R.G. Wells, R.J. Wityk, J.J. Salvo, and R.R. Reed. 1982. Transposon-mediated site-specific recombination. *Cell* **30:** 19.

Howe, M. 1980. The invertible G segment of phage Mu. *Cell* **21:** 605.

Iida, S., J. Meyer, and W. Arber. 1983. Prokaryotic IS elements. In *Mobile genetic elements* (ed. J.A. Shapiro), p. 159. Academic Press, New York.

Iida, S., J. Meyer, K. Kennedy, and W. Arber. 1982. A site-specific, conservative recombination system carried by bacteriophage P1. *EMBO J.* **1:** 1445.

James, J.F. and J. Swanson. 1978. Studies on gonococcus infection. XIII. Occurrence of color/opacity colonial variants in clinical cultures. *Infect. Immun.* **19:** 332.

Johnson, E.M., B. Krauskoph, and L.S. Baron. 1965. Genetic mapping of Vi and somatic antigenic determinants in *Salmonella. J. Bacteriol.* **19:** 302.

Kamp, P. and R. Kahmann. 1981. The relationship of two invertible segments in bacteriophage Mu and *Salmonella typhimurium* DNA. *Mol. Gen. Genet.* **184:** 564.

Kamp, P., L.T. Chow, T.R. Broker, D. Kwoh, D. Zipser, and R. Kahmann. 1979. Site-specific recombination in phage Mu. *Cold Spring Harbor Symp. Quant. Biol.* **43:** 1159.

Kennedy, K., S. Iida, J. Meyer, M. Stalhammer-Carlmalm, R. Hiestand-Nauer, and W. Arber. 1983. Genome fusion mediated by the site-specific DNA inversion system of bacteriophage P1. *Mol. Gen. Genet.* **189:** 413.

Keynan, A. and J.W. Hastings. 1961. Bioluminescence in *V. harveyi. Biol. Bull.* **121:** 375.

Kleckner, N. 1983. Transposon Tn10. In *Mobile genetic elements* (ed. J.A. Shapiro), p. 261. Academic Press, New York.

Kutsukake, K. and T. Iino. 1980. Inversion of specific DNA segments in flagellar phase variation of *Salmonella* and inversion systems of bacteriophage P1 and Mu. *Proc. Natl. Acad. Sci.* **77:** 7338.

Lambden, P.R., J.N. Robertson, and P.J. Watt. 1980. Biological properties of two distinct pilus types produced by isogenic variants of *Neisseria gonorrhoeae. J. Bacteriol.* **141:** 393.

Lederberg, J. and T. Iino. 1956. Phase variation in *Salmonella. Genetics* **41:** 743.

McClintock, B. 1956. Intranuclear systems controlling gene action and mutation. *Brookhaven Symp. Biol.* **8:** 58.

Meyer, T.F., N. Mlawer, and M. So. 1982. Pilus expression in *Neisseria gonorrhoeae* involves chromosomal rearrangement. *Cell* **30:** 45.

Mizuuchi, M. and K. Mizuuchi. 1980. Integrative recombination of bacteriophage. *Proc. Natl. Acad. Sci.* **77:** 3220.

Murphy, E. and R.P. Novick. 1980. Site-specific recombination between plasmids of *Staphylococcus aureus. J. Bacteriol.* **141:** 316.

Plasterk, R.H.A., A. Brinkman, and P. Van de Putte. 1983a. DNA inversion in the chromosome of *E. coli* and in bacteriophage Mu. *Proc. Natl. Acad. Sci.* **80:** 5355.

Plasterk, R.H.A., T.A.M. Ilmer, and P. Van de Putte. 1983b. Site-specific recombination by Gin of bacteriophage Mu: Inversions and deletions. *Virology* **127:** 24.

Reynolds, A.E., J. Felton, and A. Wright. 1981. Insertion of DNA activates the cryptic *bgl* operon in *E. coli* K12. *Nature* **293:** 625.

Scott, T.N. and M.I. Simon. 1982. Genetic analysis of the mechanism of the *Salmonella* phase variation site-specific recombination system. *Mol. Gen. Genet.* **188:** 313.

Schrempf, H. 1983. Reiterated sequences within the genome of *Streptomyces*. In *Proceedings of the Fifth John Innes Symposium* (ed. K. Chater, et al.), p. 130. Croom Helm, London.

Sherratt, D.J., A. Arthur, and M. Burke. 1981. Transposon-specified site-specific recombination systems. *Cold Spring Harbor Symp. Quant. Biol.* **45:** 275.

Silverman, M. and M. Simon. 1980. Phase variation: Genetic analyses of switching mutants. *Cell* **19:** 845.

————— . 1983. Phase variation and related systems. In *Mobile genetic elements* (ed. J.A. Shapiro), p. 537. Academic Press, New York.

Silverman, M., J. Zieg, and M. Simon. 1979. Flagellar phase variation: Isolation of the *rhl* gene. *J. Bacteriol.* **137:** 517.

Simon, M., J. Zieg, M. Silverman, G. Mandel, and R. Doolittle. 1980. Phase variation: Evolution of a controlling element. *Science* **209:** 1370.

Snellings, N.J., E.M. Johnson, D.J. Kopecko, H.H. Collins, and L.S. Baron. 1981. Genetic regulation of variable Vi antigen expression in a strain of *Citrobacter freundii. J. Bacteriol.* **145:** 1010.

Stocker, B.A.D. 1949. Measurement of the rate of mutation of flagellar antigenic phase in *Salmonella typhimurium. J. Hyg.* **47:** 398.

Stoenner, H.G., T. Dodd, and C. Larsen. 1982. Antigenic variation of *Borrelia hermsii. J. Exp. Med.* **156:** 1297.

Swanson, J. 1982. Colony opacity and protein II composition of gonococci. *Infect. Immun.* **37:** 359.

Szekely, L. and M. Simon. 1981. Homology with the invertible DNA sequence that controls flagellar-phase variation. *J. Bacteriol.* **148:** 829.

———— . 1983. The DNA sequence adjacent to flagellar genes and the evolution of flagellar phase variation. *J. Bacteriol.* **155:** 74.

Van de Putte, P., S. Cramer, and M. Giphart-Gassler. 1980. Invertible DNA determines host specificity of bacteriophage Mu. *Nature* **286:** 218.

Zieg, J. and M. Simon. 1980. Analysis of the nucleotide sequence of an invertible controlling element. *Proc. Natl. Acad. Sci.* **77:** 4196.

Zieg, J., M. Silverman, M Hilmen, and M. Simon. 1977. Recombinational switch for gene expression. *Science* **196:** 170.

Some Bacterial Transposable Elements: Their Organization, Mechanisms of Transposition, and Roles in Genome Evolution

William S. Reznikoff
Department of Biochemistry
University of Wisconsin
Madison, Wisconsin 53706

Transposable elements are DNA sequences that encode catalytic functions for and participate in the genetic process termed transposition. This process (which is pictorially described in Fig. 1) involves the translocation of the transposable element from one site to a second site (termed a target site and indicated in Fig. 1). This target site is duplicated during transposition with one copy of it being found on either side of the transposed element. In many cases (perhaps all), the transposition process involves replication *and* recombination events—a copy of the transposable element remains in its original site, and a copy appears at the new site (a possible exception to this is discussed later).

Transposition events generate insertion mutations, which disrupt the integrity of the target DNAs. In addition, since transposable elements can carry transcription initiation and/or termination signals, they can alter downstream gene expression. Transposable elements also catalyze other genetic events, including (1) deletion formation, (2) inversion formation, and (3) replicon fusion (cointegrate formation). The deletion and inversion events can be explained as being products of intrareplicon transposition events (Shapiro 1979).

Cointegrates are intermediates in the transposition pathway for some transposable elements, and they may be the product of an alternative pathway for other transposable elements. Cointegrate formation involves the fusion of the replicon carrying a transposable element with a target site on a second replicon, with the concomitant duplication of the transposable element. This process (shown schematically in Fig. 2) has been elucidated by the experiments of Muster and Shapiro (1981), in which a monomeric λ::Tn*3* molecule was found to mediate a replicon fusion event with the predicted duplication of Tn*3*. It should be noted that cointegratelike structures can also be formed by "transposition" of a transposable element—donor replicon—transposable element unit from an oligomeric donor molecule. For the purposes of this review, the use of the word cointegrate is restricted to structures resulting from replication/recombination replicon fusion events from "monomer" donor mol-

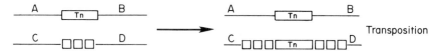

Figure 1   Transposition. A schematic presentation of replicative transposition in which a transposable element (Tn) translocates to a new site (□□□) with concomitant duplication of the site. (This diagram is similar to those presented by J. Shapiro in a number of articles.)

ecules and not to those formed by transposition from oligomeric donor molecules. The term cointegrate has not been used to describe cointegratelike structures generated by two-step events involving transposition from one replicon to another (Fig. 1) followed by homologous or site-specific recombination fusing the two replicons (Fig. 10).

The fact that transposable elements participate in these many types of genetic events suggests that they could have played a profound role in genome evolution. Since similar transposable elements exist in eukaryotes, the principles discovered for the transposable elements to be considered in this review could have general applicability.

There are three general classes of transposable elements in *Escherichia coli*:

1. Insertion sequences (ISs). These transposable elements encode no function except those related to the transposition process. A simple hypothetical type of insertion sequence is pictured in Figure 3. The critical components of this insertion sequence are specific sites at its ends (usually identical sequences in inverted relationship to each other) and a gene encoding a "transposase"—a protein involved in the recognition of the terminal repeats. Those insertion-sequence elements that have been studied in detail (including IS*50* and γδ) have been shown to be more complex than this simple picture.

2. Transposons (Tns). These are transposable elements in which the terminal sites bracket the transposase gene(s) plus additional genetic material encoding some property (or properties) functionally unrelated to the transposition process. Some transposons are composite structures in which two entire insertion-sequence elements bracket the additional genetic function(s).

3. Bacteriophage Mu. This is a lysogenic bacteriophage (Taylor 1963) that is a special class of transposon. Mu transposition functions are carried on the

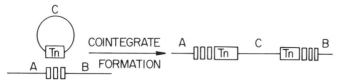

Figure 2   Cointegrate formation. Replicon fusion mediated by a transposable element.

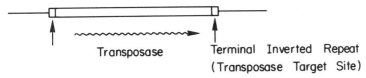

Transposase   Terminal   Inverted   Repeat
                          (Transposase   Target   Site)

Figure 3   A transposable element. The minimal components of a transposable element include a gene encoding a transposase and terminal target sites for the transposase. These sites are typically the same sequence in inverted orientation.

same genetic unit as other genes, in this case, the genes involved in the synthesis of progeny phage.

In this particular paper, the genetic organization of transposable elements has been described primarily using two model systems. These are γδ-Tn3 and IS50. γδ (sometimes called IS1000) is an insertion sequence that is related to Tn3 (a transposon encoding ampicillin resistance), Tn501 (encoding Hg++ resistance), and Tn1721 (encoding tetracycline resistance). Information derived from studies on γδ and Tn3 has been treated as if it dealt with a single type of transposable element. The reader should keep in mind that although Tn3 and γδ are very similar, they are not identical. Later in the text, it is shown how Tn3 may have evolved from a γδ-type element. IS50 is the insertion sequence associated with Tn5, the transposon encoding neomycin/ kanamycin resistance. These two examples were chosen because they are well-studied representatives of two classes of transposable elements. γδ and IS50 differ as follows: (1) γδ transposes through two distinct steps, cointegrate formation followed by resolution, whereas IS50 has not been shown to involve a cointegrate as an intermediate in the transposition process. (2) γδ encodes two distinct proteins whose sequential activities catalyze the two steps mentioned above, whereas IS50 probably synthesizes only one protein required for transposition. (3) γδ has two types of sites required for transposition (terminal inverted repeat sites and an internal resolution site), whereas IS50 appears to have only terminal inverted repeat sites. (4) γδ and IS50 regulate their transposition by entirely different mechanisms.

These differences are discussed in more detail subsequently. Many of the transposable elements that resemble IS50 in general properties may be quite different in detail and, in some cases (such as the transposition regulatory properties of Tn10 and its insertion sequence IS10), these are specifically described.

Having established a picture of the organization of these two types of transposable elements, I have described possible mechanisms of transposition and how these two types of transposable elements may be different in this regard. Finally, I have discussed how transposable elements have been proposed to play a role in bacterial genome evolution specifically in regard to the formation of Hfrs and F's.

Clearly, this paper is a very selective treatment of transposable elements. For a broader coverage of this field, the reader should also see a variety of recent books ("DNA Insertion Elements, Plasmids, and Episomes," ed. A. Bukhari et al. [1977]; "Movable Genetic Elements," *Cold Spring Harbor Symp. Quant. Biol.*, vol. 45 [1981]; "Mobile Genetic Elements," ed. J. Shapiro [1983]) and review articles (Calos and Miller 1980; Starlinger 1980; Kleckner 1981).

ORGANIZATION OF TRANSPOSABLE ELEMENTS—
GENES ENCODING PROTEINS REQUIRED FOR TRANSPOSITION

IS*1000* Transposition-specific Proteins—The Transposase

As mentioned previously, the transposition of $\gamma\delta$, Tn*3*, and related transposable elements proceeds by a two-step process: cointegrate formation followed by resolution of the cointegrate. The first step, which must involve the recognition of the ends of the element, is catalyzed in part by a 120K protein called the transposase. This has been determined from the observation that mutations preventing the synthesis of this protein fail to transpose and form cointegrates. These mutations (which define the Tn*3* *tnpA* gene pictured in Fig. 4) can be complemented in *trans* (Chou et al. 1979b; Gill et al. 1979). The *tnpA* protein from Tn*3* has been purified (Fennewald et al. 1981), but the only interesting property detected so far is that it binds nonspecifically to single-stranded DNA.

$\gamma\delta$ Transposition-specific Proteins—The Resolvase

Resolution of $\gamma\delta$ cointegrates can follow one of two routes: homologous recombination catalyzed by the host *recA* system or site-specific recombination catalyzed by the $\gamma\delta$-encoded resolvase protein. As pictured in Figure 4, this

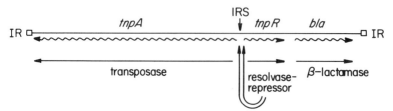

Figure 4 Tn*3*. Tn*3* encodes the synthesis of two enzymes active in transposition: a transposase that catalyzes cointegrate formation and a resolvase that resolves the cointegrate through a site-specific recombination event at a region termed IRS (internal resolution site). The resolvase also functions as a repressor, regulating the transcription of its own gene (*tnpR*) and the gene for the transposase (*trpA*). Tn*3* also carries the *bla* gene, which codes for β-lactamase. The insertion sequence $\gamma\delta$ (IS*1000*) is similar to Tn*3*, except that it does not carry the *bla* gene.

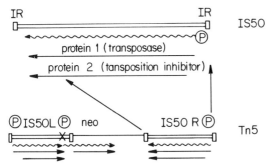

Figure 5   IS50 and Tn5. IS50 is an insertion sequence that is associated with the neomycin/kanamycin-resistance-encoding transposon Tn5. IS50 encodes two proteins in the same reading frame. Protein 1 synthesis is required in *cis* for transposition to occur, and it is presumed to be the transposase. It is unknown whether or not protein 2 is also part of the transposase. Protein 2 is a *trans*-active inhibitor of transposition. Tn5 is a composite transposon in which one of the IS50 elements (IS50L) has a single base-pair difference (×) from the transposition-proficient element IS50R. This single base-pair difference generates a nonsense codon prematurely terminating protein 1 and 2 synthesis. This results in both proteins being inactive. It also generates a promoterlike sequence responsible for transcription of the *neo* gene.

trans-active 21K protein is encoded by the *tnpR* gene (Arthur and Sherratt 1979; Kostriken et al. 1981; Reed 1981a). This protein has been purified and found to catalyze the proposed site-specific recombination in vitro using negatively supercoiled substrate DNA with the only required cofactor being Mg^{++} (Reed 1981b).

IS50 Transposition-specific Proteins — The Transposase

IS50 encodes two proteins: protein 1 (58K) and protein 2 (which is encoded by the same reading frame as protein 1 but lacks the aminoterminal 40 amino acids [see Fig. 5]). All mutations (including an ochre mutation) that fail to make protein 1 fail to transpose (Rothstein et al. 1980; Rothstein and Reznikoff 1981; Isberg et al. 1982). Thus, protein 1 is a required component of the IS50 transposase. Some of the analyzed mutants still make protein 2. Since these are transposition-defective, we conclude that protein 2 by itself is not the transposase, although theoretically it could be a component of the transposase. For reasons that are discussed subsequently, I favor the model in which protein 1 is the transposase and its amino terminus is a critical component of the specific DNA-binding activity or of the catalytic site. This protein has not yet been purified. IS50 mutations that fail to synthesize protein 1 are not complemented efficiently in *trans* (Isberg and Syvanen 1981; Johnson et al. 1982). Thus, the IS50 transposase is *cis*-active in contrast to the Tn3 transposase (see above) but similar to the Tn10/IS10 transposase (Morisato et al. 1983). The IS*50* and IS*10* transposase proteins also appear to share two other related properties: Transposase-defective mutations can be complemented in *cis* (Rothstein and Reznikoff 1981; K. Kendrick and R. Johnson, unpubl.; D. Morisato et al., in prep.), but the efficiency of this complementation decreases as the distance between the gene encoding the functional transposase and the defective insertion-sequence element increases (D. Morisato et al., in prep.; R. Johnson, unpubl.). These observations suggest that newly made transposase preferentially binds to DNA adjoining the gene that encodes it and that transposase stays associated with that DNA molecule as it acts. The probability that it will stay bound must somehow be a function of its migration distance along the DNA.

Many other transposable elements have had potential transposase-coding sequences identified (e.g., the existence of extended open reading frames which, when interrupted by mutation, give rise to transposition-defective phenotypes), but in most cases, the proteins have not yet been detected.

IS*50* Transposition-specific Proteins — A Resolvase?

It is believed that IS*50* does not encode a resolvase. This is because (1) no mutations similar to those that define *tnpR* in Tn*3* have been identified. That is, no transposition-defective mutations have been isolated that allow cointegrate formation (nobody has looked seriously for them either). (2) There are no coding sequences present in IS*50* other than those already committed to the synthesis of protein 1 (and its shortened version, protein 2; Fig. 5). (3) IS*50* cointegrate-type structures are relatively stable when propagated in *recA*⁻ hosts; thus, IS*50* does not normally manifest any resolvase activity in vivo. (Isberg and Syvanen 1981; Hirschel et al. 1982a,b). Reasons 1 and 2 above might be expected if the resolvase were one of the enzymatic activities of protein 1 (and/or protein 2) and not contained on a separate peptide. Reason 3 could occur if the resolvase activity were regulated.

ORGANIZATION OF TRANSPOSABLE ELEMENTS — *cis*-ACTIVE SITES

Terminal Sites — IS*50* and γδ

Since the terminal sites of transposable elements are conserved during transposition and define the ends of transposable-element-catalyzed deletion and inversion events, they are thought to be important recognition sites for the enzyme(s) involved in the recombination/replication events. Indeed, there are two types of observations that confirm this concept: (1) Deletion mutations that remove one entire end of a transposable element are transposition-defective (e.g., see Heffron et al. 1977; Rothstein et al. 1980). (2) For all transposable elements studied, the sequences at (or near) the two ends are related to each other as inverted repeats. IS50 has hyphenated 9-bp inverted repeats at its ends (CTG$_I^A$CTCTT; Berg et al. 1982), whereas γδ has 35-bp inverted repeats (Broker et al. 1977; Reed et al. 1979). Important aspects of the IS50 terminal sites are just now being determined. The IS50 inverted repeats differ by a single base pair (CTGACTCTT is the "outside" [O] repeat and CTGTCTCTT is the "inside" [I] repeat). This single base-pair difference may be important because transposition occurs more frequently with the following structure, O - - - - O (the normal Tn5 structure), rather than with I - - - - I or O - - - - I (Berg et al. 1982). An alternative interpretation is that the neighboring sequences (inside the inverted repeats) are critical to the transposition process. This is now known to be the case. A deletion analysis of the ends required for the O - - - - O transposition process has indicated that an additional sequence 10−14 bp long inside of the outside inverted repeat is required at both ends to achieve efficient transposition (Johnson and Reznikoff 1983). Thus, for Tn5, the end targets are not just the small inverted repeats. It is interesting to note that the sequence from 7−16 bp of the outside end target shows a remarkable similarity to a homologous 9-bp sequence that is found in repeated form in gram-negative bacterial origins of replication. Thus, the ends of Tn5 (and one end of IS50) may be recognized not only by the IS50 transposase but also by the host replication system (Johnson and Reznikoff 1983).

γδ Internal Resolution Site

Resolution of γδ cointegrates (the final step in the transposition process) requires the protein product of the *tnpR* gene (the resolvase, as described before) and a target site for the resolvase, termed the internal resolution site (IRS). The IRS has been localized to a region between the *tnpA* and *tnpR* genes (as shown in Fig. 4) by virtue of the fact that deletions in Tn3 that cut into or remove this region fail to undergo the resolving reaction in vivo in the presence of the resolvase. The crossover point within the IRS has been located by sequence analyses of the products of resolution events between γδ and Tn3 (Heffron et al. 1981; Kostriken et al. 1981; Reed 1981a). The two transposable elements are similar enough to undergo a heterologous resolution reaction, but they are not identical. Therefore, the crossover point can be located by the position of the switch from Tn3 sequences to γδ sequences in the product molecules. By this analysis, the crossover site within the IRS has been localized to a 19-bp sequence that is identical in Tn3 and γδ. A more precise definition of the IRS has come from in vitro studies of the resolvase-IRS interaction, which have defined the exact DNA cleavage point involved in the recombination event (Reed and Grindley 1981), as well as probable binding sites required for the resolvase activity (Grindley et al. 1982; one binding site is identical to the conserved 19 bp).

IS50 and other similar transposable elements are not known to have an IRS.

If such a site does exist in IS*50*, and if it plays an essential role in transposition, it must be located within 23 bp of the outside ends of IS*50*. Tn*5* mutants, in which all but the outside 23 bp of both IS*50* elements have been removed, still can transpose when transposase is supplied by another IS*50* on the same replicon (Johnson and Reznikoff 1983).

REGULATION OF TRANSPOSITION

The transposition of the γδ-Tn*3* transposable element family and of IS*50* (and of other similar transposable elements) is an infrequent event. This is not surprising since genome degeneration would occur if it were not. One explanation for this low frequency is that the transposition process is regulated. Although transposition regulation may be a common feature of many, if not all, transposable elements, the mechanisms for achieving this regulation are quite different for different transposable elements. For instance, γδ regulates the *expression* of the transposase and resolvase genes (*tnpA* and *tnpR*). IS*50*, on the other hand, appears to accomplish its regulation by synthesizing low amounts of the transposase and regulating its *activity*. The regulation of IS*10* transposition is proposed to be at the level of regulating the synthesis of the transposase but by a totally different mechanism from that used by γδ.

Regulation of the γδ-Tn3 Transposase and Resolvase Genes

Mutations have been isolated in Tn*3* that overproduce the transposase (*tnpA* product) and the resolvase (*tnpR* product). These mutations are in the *tnpR* gene (see Fig. 4; Chou et al. 1979a, Gill et al. 1979). From this result comes the conclusion that the *tnpR* gene encodes a protein with a repressor function, as well as the previously mentioned resolvase function. Since the 160 bp between the *tnpA* and *tnpR* genes encode the promoters for both genes and the IRS, it seems likely that the *tnpR* protein recognizes the same sequences for both of its functions: resolution of cointegrates and repression of *tnpR* and *tnpA* expression. Presumably, one or more of the three resolvase-binding sites defined by Grindley et al. (1982) is the operator(s).

Regulation of IS50 Transposition

Biek and Roth (1981) first described experiments that indicated that Tn*5* (IS*50*) transposition was regulated. Introduction of Tn*5* into a ''naive'' cell (a cell with no resident Tn*5* sequences) resulted in a much higher frequency of transposition than introduction into an ''experienced'' cell. Subsequent experiments with mutants of Tn*5* in the recipient cell have led to the conclusion that the transposition inhibition property of experienced cells was due to the synthesis of protein 2 (the shorter of two proteins encoded by IS*50*R; see Fig. 5) (Isberg et al. 1982; Johnson et al. 1982). The activity of protein 2 has also been studied in steady-state assays in which the effect of protein-2 production on the following was examined: (1) the transposition frequency of resident Tn*5*, (2) the transcription of β-galactosidase mRNA from IS*50-lac* fusions, and (3) the synthesis of protein 1−β-galactosidase fusion proteins (Isberg et al. 1982; Johnson et al. 1982; R. Johnson, unpubl.). The results of these experiments indicated that protein 2 inhibits transposition in the steady-state circumstance as well as in the infection assay, that the critical factor determining the transposition frequency appears to be the ratio of protein 2 to protein 1, and that protein 2 does not affect the level of transcription or translation of protein 1. One model to explain these results is that protein 2 (which is identical to protein 1 except that it lacks the aminoterminal 40 amino acids) interacts with protein 1 such that it inactivates it (much as the *lacI*−d repressor monomers interact with the wild-type *lac* repressor monomers to inactivate them [Miller 1978]). Other possible models include the competition of protein 2 with protein 1 for the target site or some host function specific for IS*50* transposition.

Although both protein 1 and protein 2 are made in low amounts, presumably because their translation initiation signals are not optimal, protein 2 is produced in significantly greater quantities than protein 1 (Isberg et al. 1982; Johnson et al. 1982). Thus, one would expect that there would be an excess of protein 2 present in most cells to inactivate protein 1. In some newly infected cells, however, protein 1 might be synthesized before any protein 2 was present, thus leading to the higher transposition frequency in naive cells.

Regulation of IS*10* Transposase Synthesis

The frequency of Tn*10* transposition is also regulated (Kleckner 1983; Simons and Kleckner 1983). Multiple copies of one of the Tn*10* insertion-sequence elements (IS*10*) depress in *trans* the frequency of Tn*10* transposition. This inhibition appears to be correlated with decreased translation of the transposase (as measured using transposase-*lacZ* protein fusions) but not a decrease in transposase mRNA synthesis (as measured using transposase-*lacZ* operon fusions). In fact, the entire IS*10* element is not necessary for contributing the inhibitory function, merely the outer segment (~170 bp in length; see Fig. 6). This portion of IS*10* encodes the following series of genetic signals: (1) p_{IN}, the promoter that programs the synthesis of the transposase mRNA with a transcription start site at 81 bp; (2) the transposase gene ribosome-binding site and AUG codon (the ATG occurs at 108−110 bp); and (3) p_{OUT}, a promoter that programs the synthesis of a regulatory mRNA. This transcript has start sites at 115 bp and 116 bp, which means that it overlaps and is complementary to the first 35−36 bases of the transposase mRNA. Since p_{OUT} is a "stronger" promoter than p_{IN}, an excess of p_{OUT} transcripts will normally be present. The current model is that these transcripts inhibit translation of the transposase mRNA by an RNA-RNA interaction.

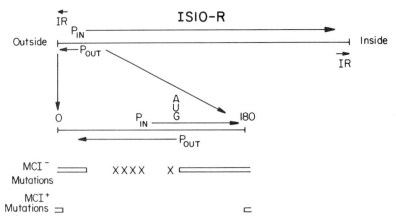

Figure 6   IS*10* and its regulatory elements. This diagram is an adaptation of one presented by Kleckner (1983). It describes the complex organization of the outside 180 bp of IS*10*, which contains a promoter facing inward (p_{IN}) that programs the synthesis of the transposase and a promoter facing outward (p_{OUT}), whose RNA product is believed to inhibit translation of the transposase. MCI⁻ mutations ($\supset$, deletions; or ×, point mutations) decrease the transactive inhibition function. MCI⁺ deletion mutations do not.

HOST FUNCTIONS IN TRANSPOSITION

The host is known to play a role in the transposition process in three different ways. First, it provides the target sites in the recipient DNA. Second, it might provide some of the enzymatic machinery involved in the transpostion process. Third, it presumably provides the enzymes that generate the proper topology for the donor and recipient DNAs.

Site Specificity

Transposable elements show two levels of site specificity. They manifest a tendency to pick sites within sequences characterized by some general property. For instance, Tn*3* and IS*1* (Tn*9*) preferentially select targets in AT-rich regions (Galas et al. 1980; Casadaban et al. 1981). The data on Tn*5*, in this regard, are somewhat conflicting. Berg et al. (1980) found that Tn*5* transposition into the *lacY* gene occurs more frequently than into the *lacZ* gene. This implies a bias toward AT-rich sequences, but Bossi and Ciampi (1981), on the basis of a sequence analysis of a small numer of Tn*5* insertions, found that Tn*5* prefers GC-rich sequences in the region surrounding its insertion sites. Some transposable elements also have a high probability of transposing into particular localized regions termed "hotspots." For Tn*3*, one target hotspot on plasmid pTU4 has been shown to have a region of extensive homology with the Tn*3* terminus (15 of 17 bp of a sequence within the hotspot are identical to 15 of 18 nucleotides within the Tn*3* terminus) (Tu and Cohen 1980). The data of Bossi and Ciampi (1981) imply that a similar, though weaker, homology exists between the ends of Tn5 and its insertion sites. Tn10 transposition hotspots are known to resemble the heptamer sequence GCTNAGC, and this sequence is typically found within the 9-bp duplicated target site (Halling and Kleckner 1981). This heptamer bears no obvious relationship to the ends of Tn10.

Host Enzymes

Isberg and Syvanen (1982) have found that Tn5 transposition is inhibited by coumermycin (an inhibitor of gyrB protein) and, at restrictive temperatures, in a gyrA temperature-sensitive mutant. Since gyrA and gyrB encode the two subunits of DNA gyrase, these results clearly implicate DNA gyrase as an important host factor in Tn5 (and IS50) transposition. An obvious role for gyrase is the generation of the proper topology in the donor and/or target DNA (negative supercoils). In fact, the data in Isberg and Syvanen (1982) are consistent with the "target" DNA's topology being most critical. Similar experiments have not been performed for $\gamma\delta$, but $\gamma\delta$ cointegrate resolution (the last step in its transposition) in vitro requires negatively supercoiled substrate DNA (Reed 1981b).

A second host enzyme that has been implicated in Tn5 transposition is DNA polymerase I. Some host transposition-defective mutations map in polA, and some polA mutations lower the frequency of Tn5 (IS50) transposition (Clements and Syvanen 1981; Sasakawa et al. 1981). DNA polymerase I actually has several associated enzymatic activities (the polymerase, the $3'-5'$ exonuclease, and the $5'-3'$ exonuclease), and it is not clear which is the critical activity for Tn5 transposition. However, since the target sequence is duplicated as a result of the transposition event, it would not be surprising if the polymerase activity is essential.

Other important host functions may exist, but their discovery awaits the isolation and characterization of additional host mutants that fail to support the normal transposition frequency.

MODELS TO EXPLAIN TRANSPOSITION

The molecular mechanism of transposition is unknown. A variety of models have been proposed. Three common models are discussed here, and their features are related to the known properties of both the $\gamma\delta$-Tn3 transposable element family and the transposable elements similar to IS50. These models are illutrated in three figures chosen from the literature. Figure 7 (from Muster and Shapiro 1981) is similar to that first described by Shapiro (1979). The salient features of this replication-recombination model are that both transposable element ends are selected at the initiation step and that cointegrates are a required intermediate. Figure 8 (from Galas and Chandler [1981] and similar

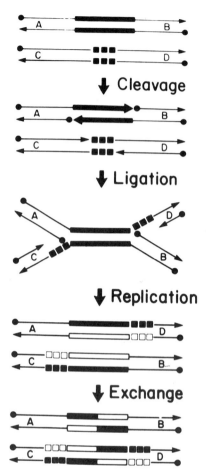

Figure 7 An obligatory cointegrate formation model for transposition. (Reprinted, with permission, from Muster and Shapiro 1981.)

to that of Harshey and Bukhari [1981]), presents a replication-recombination model in which one end of the transposable element is selected at the initiation of transposition (hence, transposition would be a polarized phenomenon), and the product of the reaction is either a cointegrate or a true transposition. The choice of the product is dictated by the nature of the final reaction or its efficiency. Figure 9 (from Berg 1977) raises the formal possibility of nonreplicative transposition. Note that for this model to have any validity, it requires either that cointegrate formation does not occur for the transposable element in question or that it occurs by a different pathway than transposition, since cointegrate formation is, by definition, a product of a recombination-replication reaction.

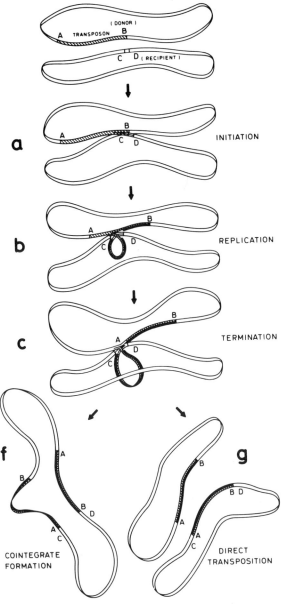

Figure 8 A polarized transposition model. (Reprinted, with permission, from Galas and Chandler 1981.)

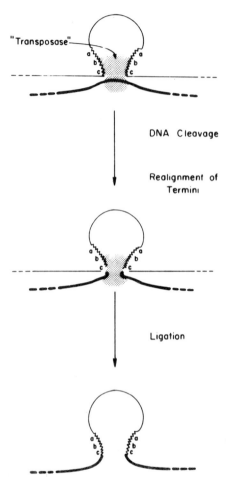

DNA Cleavage

Realignment of
Termini

Ligation

Figure 9 A nonreplicative transposition model. (Reprinted, with permission, from Berg 1977.)

Obligatory Cointegrate Model

The model pictured in Figure 7 is clearly compatibile with everything known about γδ and Tn*3*, namely that their transposition does occur via the formation and resolution of cointegrates. Superficially, this model does not appear to be compatible with the properties of IS*50* transposition. The relevant observations are, first, IS*50* cointegrates are stable when propagated in a *recA* background (e.g., there is no apparent resolvase activity) (Hirschel et al. 1982a,b), and second, no transposition-defective mutations have been isolated that allow cointegrate formation. These are not rigorous criticisms of the applicability of this model, because (1) IS*50* transposition may involve cointegrate formation as a coupled or concerted step in the reaction, such that dissociated cointegrates are dead ends; and (2) the transposase might have a resolvase function (thus, transposition-defective mutations that would form cointegrates should be difficult to isolate) and resolution, like transposition, would be regulated (thus, resolution should be rare in experienced cells).

There is, however, another set of observations specific to IS50 and IS10 (and not other superficially similar transposable elements) that are not compatible with the model in Figure 7. Formation of IS50 cointegratelike structures is recA-stimulated (which is not true of transposition) and is extremely rare in recA⁻ backgrounds (Hirschel et al. 1982a; N. Kleckner, pers. comm.). These observations raise the question as to whether IS50 cointegratelike structures are in fact formed by a true cointegrate process. This possibility is discussed in a later section.

Polarized Replicative Transposition

The model pictured in Figure 8 is particularly attractive because it is easily compatible with the properties of all known transposable elements. There is even some evidence from experiments with Tn602 that transposition is directional, i.e., prefers to transpose sequences on one side of an insertion-sequence element rather than on the other (S. Stibitz, unpubl.). This is supportive of a polarized transposition model.

Nonreplicative Transposition

The nonreplicative transposition model (Fig. 9) is clearly not compatible with the properties of the γδ transposable element family. What about transposable elements similar to IS50? For many of these (such as IS1), cointegrate formation in recA⁻ cells occurs at a frequency not dissimilar to the transposition frequency (Shapiro and MacHattie 1979; D. Berg and M. Chandler, pers. comm.). Thus, these elements clearly catalyze recombination-replication reactions (cointegrate formation); to postulate that nonreplicative transposition occurs for them is to say that they encode two quite different recombination pathways. For IS50, this model cannot be excluded, although there is no evidence that supports it either. The particular property of IS50 of interest is that cointegrate formation is rare in recA⁻ cells. One interpretation of this observation is that IS50 cointegrates are really the product of transposition events originating from rare oligomeric donor molecules (D. Berg, pers. comm.). This could be ruled out if the frequency of oligomeric molecules was found to be too low to account for the observed frequency of apparent cointegrates. Alternatively, IS50 cointegrates could be rare, because in the transposition process, they would be efficiently resolved. For instance, cointegrate formation for Tn3 is rare unless the resolvase or the IRS is defective (F. Heffron, pers. comm.). Finally, the nonreplicative transposition model does not readily explain how inversions and deletions occur.

TRANSPOSABLE ELEMENTS AND REPLICON EVOLUTION

Transposable elements appear to have been instrumental in facilitating many forms of bacterial replicon evolution. The known mechanisms of Hfr and F′ formation in *E. coli* and the evolution of R factors demonstrate the various potential roles that transposable elements can play in this process. These roles can be summarized as follows:

1. Transposable elements can serve as movable regions of genetic homology that provide substrates for *recA*-mediated recombination events.
2. Transposable elements that encode a resolvase/internal resolution site system can serve as movable regions facilitating *recA*-independent, site-specific recombination events.
3. Transposble elements can fuse two replicons via formation of cointegrate structures.
4. Transposable elements can catalyze the formation of smaller replicons (e.g., an F′ from an Hfr) via their catalysis of deletion formation.
5. Transposable elements can bracket a genetic marker, thereby creating a transposon that can subsequently transpose as a unit to a second replicon.

Processes 1, 2, and 3 are known to have been the mechanisms of Hfr formation, and processes 1, 2, and 4 have been instrumental in F′ (and R′) formation. (Hfr-F′ formation need not be linked by the same pathway. An Hfr formed by route 1 could give rise to F′ by routes 1, 2, or 4). Process 5 has been implicated in R-factor formation.

Movable Regions of Genetic Homology

The heteroduplex mapping studies of F and various F′s by Norman Davidson and his colleagues (Davidson et al. 1974; Hu et al. 1975) have clearly indicated that homologous reciprocal recombination events between insertion sequences (such as IS2) carried on both the chromosome and F are an important mechanism for the formation of Hfrs and F′. Schematic diagrams of these two processes are presented in Figures 10 and 11. Since Hfr formation has been shown to occur at a much higher frequency in *recA*⁺ cells (Cullum and Broda 1979), this is most likely the major route for Hfr formation. It should be noted that since insertion-sequence elements can ''move'' through the transposition process, the actual locations for F insertion into the chromosome, and the endpoints for chromosome incorporation onto F′ upon excision of F by this mechanism, are limited only by the frequency of insertion-sequence transposition into the various sites.

Site-specific IS-IS Interaction

It is clear that the γδ resolvase system can catalyze site-specific γδ-γδ recombination events which, depending upon the location of the participating

Figure 10 Hfr formation via IS-IS recombination. Integration of an F factor into the chromosome can occur by means of a recombination event between identical IS elements found on the F factor and the chromosome. This can occur via homologous recombination or, in cases in which the IS elements encode a site-specific resolvase, by site-specific recombination.

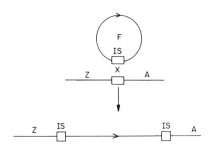

$\gamma\delta$ elements, can resemble the processes pictured in Figures 10 and 11. One example of such an event is the *recA*-independent fusion of plasmids F155 and KLF5 using their respective $\gamma\delta$ elements (Palchaudhuri et al. 1976).

Cointegrate Fusion of Replicons

A cointegrate formed between the chromosome and a fertile plasmid carrying a transposable element, as pictured in Figure 2, generated an Hfr that gave rise to F14 (Davidson et al. 1974). This process has also been implicated as a means by which Tn*3* mediates mobilization of nonconjugative plasmids (Crisona et al. 1980).

Replicon "Excision" by Deletion Formation

Transposable elements are known to catalyze deletion formations, presumably by means of intramolecular transposition events. Characteristically, these deletions initiate at one end of the transposable element and proceed to some other site, possibly related to a preferred target site for the transposable element. The evolution of shortened F' plasmids from ORF203 and F13 via $\gamma\delta$-mediated deletion formation has been studied by Deonier et al. (1983). Starting with an Hfr with an appropriately located transposable element, one can generate F' plasmids by this route (Fig. 12). This is the mechanism postulated to explain

Figure 11 Excision for F factor. The F factor can excise from the Hfr integrated state to reform the free F factor or an F' via IS-IS element recombination. The ability to form F' by this route and the type of F' formed (type I carries DNA from one side of the integrated element, and type II carries DNA from both sides) would be a function of the fortuitous presence of IS elements at different locations on the chromosome.

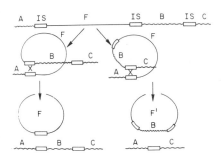

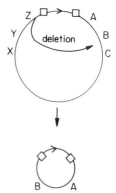

Figure 12   F′ formation by deletion formation.

R′ formation by RP4::mini-Mu (Van Gijsegen and Toussaint 1982). Alternatively, the F′ may arise from the "deleted" DNA.

Transposons

If two copies of a specific type of transposable element were located on a chromosome (as shown in Fig. 13), this would generate three different transposable elements: ISa, ISb, and ISa-×-ISb. The latter structure would, by definition, be a transposon. This type of transposon is termed a "composite" structure; it involves the "bracketing" of a gene (or group of genes) by two identical (or nearly identical) transposable elements. As outlined by Iida et al. (1981), the bracketing of a gene by a transposable element can occur by a variety of events known to be catalyzed by these elements, such as transposition (Fig. 14), inversion, and cointegrate formation coupled with a deletion.

   The transposition properties of transposons will lead to their acquisition by plasmids presumably at a frequency dictated by their transposition proficiency and the existence of appropriate target sites on the plasmids. Whether plasmids carrying a given transposon are discovered would be a function of the selective pressure for the presence of the particular transposon-encoded characteristic and the fertility of the particular plasmid. Thus, it is not surprising that genes encoding antibiotic resistance and heavy metal resistance have been found to be carried on transposons located on fertile plasmids. However, other genetic systems such as *his-gnd* (Wolf 1980), *lac* (Cornelis et al. 1978), and heat-stable toxin production (So et al. 1979) have also been found on transposons.

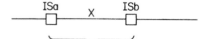

Figure 13   Composite transposon.

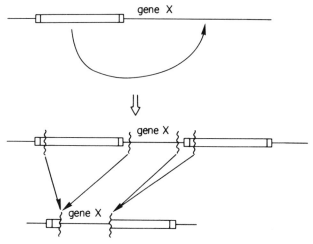

Figure 14   Formation and evolution of a transposon. A composite transposon can form by bracketing a gene with identical insertion sequences. This can occur by transposition (as shown) or by insertion-sequence-mediated inversion or cointegrate formation followed by a deletion (Iida et al. 1981). A composite transposon carries unnecessary genetic information, including the second copy of the transposase gene and the internal recognition sites, and thus might evolve to lose this information.

Presumably, any bacterial gene can be a component of a transposon.

Given the general requirements for transposable element structure (specific terminal sites and a transposase; see Fig. 3), we can make two predictions about composite transposons. First, in general, it should not matter whether the bracketing insertion sequences are in a direct repeat or inverted repeat arrangement. This is because it is the terminal repeats located at the ends of each insertion sequence that are recognized in transposition and not the total insertion sequence. An example of these two alternatives is shown by Tn602 and Tn903 (Fig. 15). Both carry the same gene, and the insertion sequences bracketing this gene are essentially the same (S. Stibitz, unpubl.). Tn602 and Tn903 have qualitatively the same transposition properties. However, the detailed properties of these two types of transposons can be nonequivalent if the two end target sites of the individual insertion sequences are nonidentical (as is the case for IS50; Berg et al. 1982) and/or if transposition is polarized preferentially picking an end near the amino terminus or the carboxyl terminus of the transposase gene (Galas and Chandler 1981). Second, the composite transposon carries redundant genetic information: the duplicated transposase gene and the internal insertion sequence end target sites. The transposon could evolve by deleting these without losing transposition functionality. An extreme example of this is postulated in Figure 14. Tn5 (Fig. 5) could be postulated to have taken the first step down this pathway, having acquired a single base-pair change, which generates a pseudotransposase/regulation gene in IS50L (Roth-

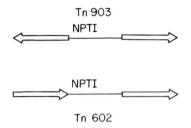

Figure 15   Tn*903* and Tn*602*. Two composite transposons structured from nearly identical insertion sequences, bracketing the same antibiotic-resistance function and manifesting similar transposition properties, yet their insertion sequences are oriented differently (E.S. Stibitz, pers. comm.).

stein and Reznikoff 1981). Tn*3* (Fig. 4) might be considered an extreme example of the evolutionary pattern pictured in Figure 14, although another possibility for Tn*3* evolution is described later.

A different implication of the composite model for the origin of transposons is that the insertion sequences (or their remnants) and the genetic system that they bracket should be distinct genetic units. There are three cases in which this distinction may have broken down. (1) The promoter for the NPTII gene in Tn*5* is within IS*50*L (the mRNA startpoint is ~84 bp within the inside edge of IS*50*L) (see Fig. 5; Auerswald et al. 1981; Rothstein and Reznikoff 1981; Rothstein et al. 1981). This "linkage" of the NPTII gene with IS*50*L could be viewed as resulting from the accidental exclusion of the original NPTII-gene promoter during the formation of Tn*5* and the subsequent selection for an "up" promoter mutation in an IS*50*L sequence that fortuitously resembled a promoter. (2) The transposition functions of the Hg^{++}-resistant transposon Tn*501* are induced by incubation of cells in the presence of Hg^{++} (Schmitt et al. 1981). Thus, the Hg^{++}-resistance regulatory functions also control the transposition functions. This is thought to be due to readthrough transcription and may merely represent the enhanced expression of the transposition functions above that programmed by the normal transposition system promoter. (3) Incubation of Tn*10* containing cells with tetracycline induces synthesis of a small RNA molecule that specifically hybridizes to the outer 400 bp of IS*10* (Schmidt et al. 1981). This RNA species may only be a stable fragment from a readthrough transcript originating in the *tet*-resistance-encoding region of Tn*10*.

A final note on the evolution of transposons: There are alternative models to explain the evolution of transposons such as Tn*3*. For instance, Machida et al. (1982) have discovered that IS*102* can evolve into a transposon carrying tetracycline-resistance determinants by recruiting a new target site on the distal side of the *tet* locus. The proposed mechanism is diagrammed in Figure 16. Note that it does not require the previous formation of a composite structure but, rather, is dependent on the placement of the insertion-sequence element near the locus in question and the fortuitous presence of a sequence resembling the terminal recognition site on the other side of the locus.

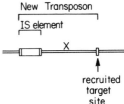

Figure 16   Evolution of a transposon by recruitment of a new target site. This model is similar to that proposed for the generation of a *tet*^R transposon by IS*102*, as described by Machida et al. (1982).

CONCLUSION

This paper has addressed three separate issues: (1) What is the genetic organization of transposable elements? (2) How does transposition occur? (3) What role might transposable elements have played in the evolution of genomes?

Although one can hypothesize a simple structure for transposable elements (see Fig. 2) I have used information about two well-studied cases (Tn*3*/γδ and IS*50*) to show that transposable elements can have more complex structures. For instance, Tn*3* encodes two separate enzyme activities (a transposase and a resolvase whose consecutive functioning is required for transposition). Tn*3* also carries an internal *cis*-active site (as well as terminal repeats) that is required for transposition. Both transposable elements have evolved mechanisms to regulate transposition. Tn*3* regulates the synthesis of its transposase and resolvase, whereas IS*50* encodes a second protein in the same reading frame as its transposase, which regulates the activity of the transposase. Additional information was presented about the IS*10* transposition regulation system, which suggested that still other transposition regulatory strategies may have evolved. Thus, although one could explain the properties of transposable elements with a very simple model, known systems suggest that the reality may be more complex.

Three models were described that have been presented in the literature to describe how transposition occurs. One model (nonreplicative transposition [Fig. 9]) clearly is not consistent with the properties of Tn*3*/γδ transposition and that of some other transposable elements. It may be applicable to IS*50* transposition. There exists no critical evidence that would lead me to favor one of the other two models (obligatory cointegrate formation [Fig. 7] and polarized transposition [Fig. 8]). One important area of study that would help distinguish these two models would be to examine the role of cointegrates in transposition by transposable elements such as IS*50*.

Retrospective structural studies and the knowledge of the genetic events mediated by transposable elements suggest that they have been important agents in the evolution of bacterial genomes. This was discussed by analyzing how they presumably played a role in Hfr, F′, and R-factor formation. Since similar elements have been discovered in the other biological kingdoms, transposable elements may have been important in the evolution of all genomes.

ACKNOWLEDGMENTS

I thank Drs. Berg, Chandler, Galas, and Shapiro for permission to reprint figures from their publications; Drs. Kleckner and Jaskunas for sending me preprints; and the members of my laboratory for agonizing over my presentation of the material. Previously unpublished experiments by R. Johnson and E.S. Stibitz were supported in part by grant GM-19670 from the National Institutes of Health.

REFERENCES

Arthur, A. and D.J. Sherratt. 1979. Dissection of the transposition process: A transposon-encoded site-specific recombination process. *Mol. Gen. Genet.* **175:** 267.

Auerswald, E.A., G. Ludwig, and H. Schaller. 1981. Structural analysis of Tn*5*. *Cold Spring Harbor Symp. Quant. Biol.* **45:** 107.

Berg, D.E. 1977. Insertion and excision of the transposable kanamycin resistance determinant Tn*5*. In *DNA insertion elements, plasmids, and episomes* (ed. A.I. Bukhari et al.), p. 205. Cold Spring Harbor Laboratory, Cold Spring Harbor, New York.

Berg, D.E., A. Weiss, and L. Crossland. 1980. Polarity of Tn*5* insertion mutations in *Escherichia coli*. *J. Bacteriol.* **142:** 439.

Berk, D.E., L. Johnsrud, L. McDivitt, R. Ramabhadran, and B.J. Hirschel. 1982. Inverted repeats of Tn5 are transposable elements. *Proc. Natl. Acad. Sci.* **79:** 2632.

Biek, D. and J.R. Roth. 1981. Regulation of Tn*5* transposition. *Cold Spring Harbor Symp. Quant. Biol.* **45:** 189.

Bossi, L. and M.S. Ciampi. 1981. DNA sequences at the sites of three insertions of the transposable element Tn5 in the histidine operon of *Salmonella*. *Mol. Gen. Genet.* **183:** 406.

Broker, T., L. Chow, and L. Soll. 1977. The *E. coli* gamma-delta recombination sequence is flanked by inverted duplications. In *DNA insertion elements, plasmids, and episomes* (ed. A.I. Bukhari et al.), p. 575. Cold Spring Harbor Laboratory, Cold Spring Harbor, New York.

Calos, M.P. and J.H. Miller. 1980. Transposable elements. *Cell* **20:** 579.

Casadaban, M.J., J. Chou, P. Lemaux, C.-P.D. Tu, and S.N. Cohen. 1981. Tn*3*: Transposition and control. *Cold Spring Harbor Symp. Quant. Biol.* **45:** 269.

Chou, J., M. Casadaban, P. Lemaux, and S.N. Cohen. 1979a. Identification and characterization of a self-regulated repressor of translocation of the Tn3 element. *Proc. Natl. Acad. Sci.* **76:** 4020.

Chou, J., P. Lemaux, M. Casadaban, and S.N. Cohen. 1979b. Transposition protein of Tn3: Identification and characterization of an essential repressor-controlled gene product. *Nature* **282:** 801.

Clements, M. and M. Syvanen. 1981. Isolation of a *polA* mutation that affects transposition of insertion sequences and transposons. *Cold Spring Harbor Symp. Quant. Biol.* **45:** 201.

Cornelis, G., D. Shosal, and H. Saedler. 1978. Tn 951: A new transposon carrying a lactose operon. *Mol. Gen. Genet.* **160:** 215.

Crisona, N.J., J.A. Nowak, H. Nagaishi, and A.J. Clark. 1980. Transposon-mediated conjugational transmission of nonconjugative plasmids. *J. Bacteriol.* **142:** 701.

Cullum, J. and P. Broda. 1979. Chromosome transfer of Hfr formation by F in rec^+ and $recA$ strains of *Escherichia coli* K12. *Plasmid* **2:** 358.

Davidson, N., R.C. Deonier, S. Hu, and E. Ohtsubo. 1974. Electron microscope heteroduplex studies of sequence relations among plasmids of *Escherichia coli*. X. Deoxyribonucleic acid sequence organization of F and F-primes, and the sequences involved in Hfr formation. In *Microbiology-1979* (ed. D. Schlessinger), p. 56. American Society for Microbiology, Washington, D.C.

Deonier, R.C., K. Yun, and M. Kupperman. 1983. γδ-mediated deletions of chromosomal segments on F-prime plasmids. *Mol. Gen. Genet.* **190**: 42.

Fennewald, M.A., S.P. Gerrard, J. Chou, M. Casadaban, and N.R. Cozzarelli. 1981. Purification of the Tn3 transposase and analysis of its binding to DNA. *J. Biol. Chem.* **256**: 4687.

Galas, D.J., and M. Chandler. 1981. On the molecular mechanisms of transposition. *Proc. Natl. Acad. Sci.* **78**: 4858.

Galas, D.J., M.P. Calos, and J.H. Miller. 1980. Sequence analysis of Tn9 insertions in the *lacZ* gene. *J. Mol. Biol.* **144**: 19.

Gill, R., F. Heffron, and S. Falkow. 1979. Identification of the protein encoded by the transposable element Tn3 which is required for its transposition. *Nature* **282**: 797.

Grindley, N.D.F., M.R. Lauth, R.G. Wells, R.J. Wityk, J.J. Salvo, and R.R. Reed. 1982. Transposon-mediated site-specific recombination: Identification of three binding sites for resolvase at the *res* sites of γδ and Tn3. *Cell* **30**: 19.

Halling, S.M. and N. Kleckner. 1981. A symmetrical six-basepair target site sequence determines Tn10 insertion specificity. *Cell* **28**: 155.

Harshey, R.M. and A.I. Bukhari. 1981. A mechanism of DNA transposition. *Proc. Natl. Acad. Sci.* **78**: 1090.

Heffron, F., P. Bedinger, J. Champoux, and S. Falkow. 1977. Deletions affecting the transposition of an antibiotic resistance gene. *Proc. Natl. Acad. Sci.* **74**: 702.

Heffron, F., R. Kostriken, C. Morita, and R. Parker. 1981. Tn3 encodes a site-specific recombination system: Identification of essential sequences, genes, and the actual site of recombination. *Cold Spring Harbor Symp. Quant. Biol.* **45**: 259.

Hirschel, B.J., D.J. Galas, and M. Chandler. 1982a. Cointegrate formation by Tn5, but not transposition, is dependent on *recA*. *Proc. Natl. Acad. Sci.* **79**: 4530.

Hirschel, B.J., D.J. Galas, D.E. Berg, and M. Chandler. 1982b. Structure and stability of transposon 5-mediated cointegrates. *J. Mol. Biol.* **159**: 557.

Hu, S., E. Ohtsubo, and N. Davidson. 1975. Electron microscope heteroduplex studies of sequence relations among plasmids of *E. coli*. XI. The structure of F13 and related F-primes. *J. Bacteriol.* **122**: 749.

Iida, S., J. Meyer, and W. Arber. 1981. Genesis and natural history of IS-mediated transposons. *Cold Spring Harbor Symp. Quant. Biol.* **45**: 27.

Isberg, R.R. and M. Syvanen. 1981. Replicon fusions promoted by the inserted repeats of Tn5: The right repeat is an insertion sequence. *J. Mol. Biol.* **150**: 15.

——————. 1982. DNA gyrase is a host factor required for transposition of Tn5. *Cell* **30**: 9.

Isberg, R.R., A.L. Lazaar, and M. Syvanen. 1982. Regulation of Tn5 by the right-repeat proteins: Control at the level of transposition reaction? *Cell* **30**: 883.

Johnson, R.C. and W.S. Reznikoff. 1983. DNA sequences at the ends of transposon Tn5 required for transposition. *Nature* **304**: 280.

Johnson, R.C., J.C.-P. Yin, and W.S. Reznikoff. 1982. Control of Tn5 transposition in *Escherichia coli* is mediated by protein from the right repeat. *Cell* **30**: 873.

Kleckner, N. 1981. Transposable elements in prokaryotes. *Annu. Rev. Genet.* **15**: 341.

——————. 1983. Transposon Tn10. 1983. In *Mobile genetic elements* (ed. J.A. Shapiro), p. 261. Academic Press, New York.

Kostriken, R., C. Morita, and F. Heffron. 1981. Transposon Tn3 encodes a site-specific recombination system. Identification of essential sequences, genes, and actual site of recombination. *Proc. Natl. Acad. Sci.* **78**: 4041.

Machida, Y., C. Machida, and E. Ohtsubo. 1982. A novel type of transposon generated by insertion element IS102 present in a pSC101 derivative. *Cell* **30**: 29.

Miller, J.H. 1978. The *lacI* gene. In *The operon* (ed. J.H. Miller and W.S. Reznikoff), p. 31. Cold Spring Harbor Laboratory, Cold Spring Harbor, New York.

Morisato, D., J.C. Way, H.-J. Kim, and N. Kleckner. 1983. Tn10 transposase acts preferentially on nearby transposon ends *in vivo*. *Cell* **32**: 799.

Muster, C.J. and J.A. Shapiro. 1981. Recombination involving transposable elements: Replicon fusion. *Cold Spring Harbor Symp. Quant. Biol.* **45:** 239.

Palchaudhuri, S., E. Ohtsubo, and W.K. Maas. 1976. Fusions of two F-prime factors in *Escherichia coli* studied by electron microscope heteroduplex analysis. *Mol. Gen. Genet.* **146:** 215.

Reed, R.R. 1981a. Resolution of cointegrates between transposons γδ and Tn3 defines the recombination site. *Proc. Natl. Acad. Sci.* **78:** 3428.

————. 1981b. Transposon-mediated site-specific recombination: A defined *in vitro* system. *Cell* **25:** 713.

Reed, R.R. and N.D.F. Grindley. 1981. Transposon-mediated site-specific recombination *in vitro*: DNA cleavage and protein-DNA linkage at the recombination site. *Cell* **25:** 721.

Reed, R.R., R.A. Young, J.A. Steitz, N.D.F. Grindley, and M.S. Guyer. 1979. Transposition of the *Escherichia coli* insertion element γδ generates a five base-pair repeat. *Proc. Natl. Acad. Sci.* **76:** 4882.

Rothstein, S.J. and W.S. Reznikoff. 1981. The functional differences in the inverted repeats of Tn5 are caused by a single base pair nonhomology. *Cell* **23:** 191.

Rothstein, S.J., R.A. Jorgensen, K. Postle, and W.S. Reznikoff. 1980. The inverted repeats of Tn5 are functionally different. *Cell* **19:** 795.

Rothstein, S.J., R.A. Jorgensen, J.C.-P. Yin, Z. Yong-di, R.D. Johnson, and W.S. Reznikoff. 1981. Genetic organization of Tn5. *Cold Spring Harbor Symp. Quant. Biol.* **45:** 99.

Sasakawa, C., Y. Uno, and M. Yoshikawa. 1981. The requirement for both DNA polymerase and 5' to 3' exonuclease activities of DNA polymerase I during Tn5 transposition. *Mol. Gen. Genet.* **182:** 19.

Schmidt, F.J., R.A. Jorgensen, M. DeWilde, and J.E. Davies. 1981. A specific tetracycline-induced, low-molecular-weight RNA encoded by the inverted repeat of Tn10 (IS10). *Plasmid* **6:** 148.

Schmitt, R., J. Attenbuchner, and J. Grinsted. 1981. Complementation of transposition functions by transposons Tn501 (Hg^R) and Tn1721 (Tet^R). In *Molecular biology, pathogenicity, and ecology of bacterial plasmids* (ed S.B. Levy et al.), p. 359. Plenum Press, New York.

Shapiro, J. 1979. Molecular model for the transposition and replication of bacteriophage Mu and other transposable elements. *Proc. Natl. Acad. Sci.* **76:** 1933.

Shapiro, J.A. and L.A. MacHattie. 1979. Integration and excision of prophage λ mediated by the IS1 element. *Cold Spring Harbor Symp. Quant. Biol.* **43:** 1135.

Simons, R.W. and N. Kleckner. 1983. Translational control of IS10 transposition. *Cell* **34:** 683.

So, M., F. Heffron, and B.J. McCarthy. 1979. The *E. coli* gene encoding heat stable toxin (ST) is a bacterial transposon flanked by inverted repeats of IS1. *Nature* **277:** 453.

Starlinger, P. 1980. IS elements and transposons. *Plasmid* **3:** 241.

Taylor, A.L. 1963. Bacteriophage-induced mutation in *Escherichia coli*. *Proc. Natl. Acad. Sci.* **50:** 1043.

Tu, C.-P.D. and S.N. Cohen. 1980. Translocation specificity of the Tn3 element: Characterization of sites of multiple insertions. *Cell* **19:** 151.

Van Gijsegen, F. and A. Toussaint. 1982. Chromosome transfer and R-prime formation by an RP4::mini-Mu derivative in *Escherichia coli*, *Salmonella typhimurium*, *Klebsiella pneumoniae*, and *Proteus mirabilis*. *Plasmid* **7:** 30.

Wolf, R.E. 1980. Integration of specialized transducing bacteriophage λcI857St68h80d*gnd his* by an unusual pathway promotes formation of deletions and generates a new translocatable element. *J. Bacteriol.* **142:** 588.

Genetic Analysis of Protein Localization

Spencer A. Benson* and Thomas J. Silhavy†
*Cancer Biology Program and †Genetic Engineering Laboratory,
LBI-Basic Research Program
National Cancer Institute-Cancer Research Facility
Frederick, Maryland 21701.

INTRODUCTION

Protein localization is the process by which a protein is exported from its site of synthesis in the cytoplasm to any one of a number of different cellular compartments. During the last decade, this aspect of cellular biogenesis has been the focus of considerable research. Despite obvious differences in the subcellular structure of prokaryotic and eukaryotic cells, all cells seem to use similar mechanisms of protein export. For example, experiments using recombinant DNA technology have shown that the intragenic information specifying export in a eukaryotic gene (ovalbumin, insulin) can be recognized by *Escherichia coli* and vice versa (alkaline phosphatase [PhoA]; β-lactamase [Bla]) (Fraser and Bruce 1978; Talmadge et al. 1980; Roggenkamp et al. 1981; Müller et al. 1982). This conservation of export mechanisms has fostered exchange of information among scientists working in areas as diverse as eukaryotic cell biology and prokaryotic molecular genetics.

In this paper we discuss the genetic techniques that have been applied successfully to the study of protein localization in *E. coli*. Because of its relative sophistication, genetic analysis in *E. coli* surpasses that currently possible with other systems. However, research in the field of protein localization is multidisciplinary, and other cells offer distinct advantages with respect to biochemical analysis. Therefore, to familiarize the reader with current concepts and unsolved problems, we begin by briefly reviewing the relevant biochemical data. For a more detailed account of these studies, the reader is referred to cited reviews.

PROTEIN LOCALIZATION IN EUKARYOTIC CELLS

The basic principles of protein export were established by the work of George Palade and co-workers (Palade 1975). They demonstrated that proteins destined for secretion, e.g., are synthesized by ribosomes that are tightly bound to the rough endoplasmic reticulum. Such proteins are never found in completed form in the cytoplasm; rather, they are segregated immediately into the lumen of the organelle. From this location they are transported through the Golgi to secretory vesicles where they remain stored until secretion. This intracellular

253

transport generally is assumed to occur via vesicular intermediates that bud from one organelle and fuse with another. Secretion occurs by a process that resembles exocytosis. This intracellular routing mechanism is often referred to as the Palade pathway.

Proteins destined for many cellular locations are routed through the Palade pathway. Besides secretory proteins, soluble proteins destined for lysosomes, membrane proteins destined for the plasmalemma, or the organelles of the pathway itself are also transported by the same route. The hallmark of proteins exported by the pathway is glycosylation. Protein glycosylation occurs inside of, and is mediated in discrete steps by enzymes that are localized to, the endoplasmic reticulum and the Golgi. Indeed, the intracellular location of a protein passing through the pathway can be ascertained by determining the degree of glycosylation (see Zilberstein et al. 1980).

Signal Hypothesis

The signal hypothesis, which offers a plausible explanation for the molecular events responsible for the cotranslational segregation of nascent polypeptide chains across the rough endoplasmic reticular membrane, was proposed by Blobel and Dobberstein (1975). According to this model, a protein destined to be secreted from cells is synthesized initially as a larger precursor with 15−30 additional amino acids at the aminoterminal end of the molecule. This peptide extension (the signal sequence) was proposed to initiate binding of the translation complex to the rough endoplasmic reticular membrane. This binding results in the formation of a transient pore through which the nascent peptide chain passes as synthesis proceeds. The net result is a vectorial transfer of the protein across the rough endoplasmic reticular membrane to the lumen of the organelle. The signal sequence is removed by a specific protease (signal peptidase), probably before synthesis of the secretory protein is completed.

The signal hypothesis made several experimental predictions that have been substantiated by subsequent findings. First, many exported proteins are indeed synthesized in larger precursor form with a typical signal sequence at the aminoterminal end of the molecule. Second, a specific export machinery associated with the rough endoplasmic reticulum does exist. Several components of this machinery have been purified and characterized. Walter and Blobel (1980) purified a membrane-associated protein complex (m.w. = 250,000) composed of six polypeptide chains. This complex is released from the rough endoplasmic reticular membrane by a high-salt wash. In similar experiments, Meyer and Dobberstein (1980) identified a peptide fragment that is released from microsomes by a high-salt wash and limited proteolysis. By raising antibodies against the purified fragment, Meyer et al. (1982) were able to demonstrate that the fragment corresponds to a 72K membrane protein. The 250K complex functions by binding to the signal sequence as it emerges from the ribosome and halting further translation.

When stripped, salt-washed microsomes are added, translation resumes, and the export process begins (Walter and Blobel 1981). This complex, called signal recognition protein (SRP), functions to couple synthesis and export. This is a critical function because export cannot occur if translation proceeds too far beyond the signal sequence (Rothman and Lodish 1977). The 72K membrane protein is the component that relieves the SRP-mediated translation block (Meyer et al. 1982). This component has been termed docking protein. A diagram describing the signal hypothesis and the function of SRP and docking protein is shown in Figure 1.

Signal Hypothesis and Membrane Proteins

Many membrane proteins behave as incompletely secreted proteins. Perhaps the most thoroughly studied example of such a protein is the glycoprotein of vesicular stomatitis virus (VSV) (Lodish and Rothman 1979). This protein is situated in the membrane with its glycosylated aminoterminal end facing out and a small portion of its carboxyterminal end facing the cytoplasm. The protein is synthesized on membrane-bound ribosomes in a manner analogous to secreted proteins, except that the completed protein is never released into the lumen of the organelle. It remains anchored in the membrane by a sequence of 20−25 hydrophobic amino acids located near the carboxyl terminus of the molecule. This stretch of hydrophobic amino acids has been termed the stop transfer sequence (Blobel 1980). Thus, the signal hypothesis can be adapted to explain the insertion of the glycoprotein into the membrane by the simple proposal that a second export signal is located downstream from the signal sequence in the primary structure of the protein (see Fig. 1).

Many questions concerning the signal hypothesis and membrane proteins remain unanswered. For example, all membrane proteins probably do not have a topology as simple as the VSV glycoprotein. Most probably span the bilayer more than once. The signal hypothesis can be adapted to explain complex topology, but additional intragenic export signals are required (Blobel 1980; Sabatini et al. 1982). At present, there is no experimental evidence that would define this additional information.

Posttranslational Export: A Second Mechanism

Not all proteins that are localized to intracellular organelles are routed through the Palade pathway. Protein specified by nuclear DNA, destined for localization in mitochondria or chloroplasts in plants, and proteins destined for peroxisomes are exported by a mechanism different from that seen with the rough endoplasmic reticulum. This was first demonstrated with the chloroplast protein ribulose-1,5-bisphosphate carboxylase (Dobberstein et al. 1977; Chua and Schmidt 1978; Highfield and Ellis 1978). These investigators showed that the protein is synthesized in precursor form by ribosomes that are not mem-

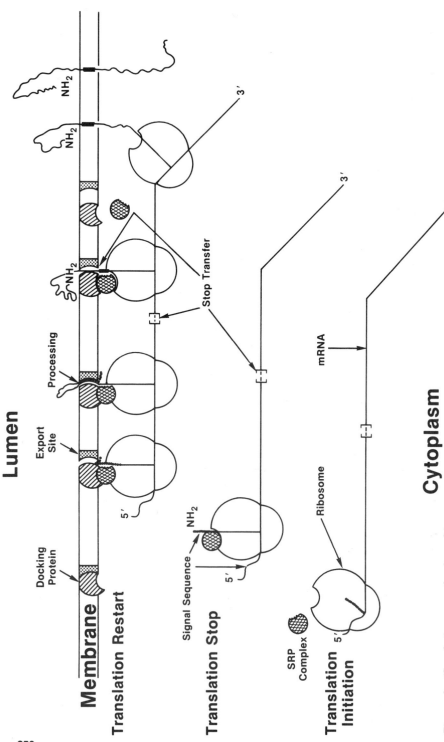

Lumen

Membrane

Cytoplasm

Docking
Protein

Export
Site

Processing

Stop Transfer

Translation Restart

Signal Sequence

Translation Stop

SRP
Complex

Ribosome

mRNA

Translation
Initiation

Figure 1 (*See facing page for legend.*)

256

brane-bound. It is posttranslationally taken up by chloroplasts, and the precursor is processed proteolytically during or immediately after import. The precursor contains an aminoterminal peptide extension of 44 amino acids, which presumably functions as the uptake signal (Schmidt et al. 1979). Similar results have now been obtained for various chloroplast and mitochondrial proteins (for reviews, see Kreil 1981; Neupert and Schatz 1981).

BIOCHEMICAL STUDIES OF PROTEIN EXPORT IN *E. coli*

Gram-negative bacteria, such as *E. coli*, contain four cellular compartments: the cytoplasm, an inner or cytoplasmic membrane, an outer membrane, and an aqueous space between the two membranes called the periplasm. Each of these compartments contains a specific set of proteins. For the most part, studies of protein localization in *E. coli* have focused on a relatively few key proteins. The relevant characteristics and properties of these proteins are summarized in Table 1. Many of these proteins are known by several names. To avoid confusion we use a similar abbreviation for both the protein and the structural gene, e.g., the *lamB* gene codes for the LamB protein.

Existing evidence suggests that most proteins destined to be localized to the periplasm or the outer membrane are exported by a cotranslational mechanism similar to that described for eukaryotic cells. (An apparent exception is Bla, which may be exported by a posttranslational mechanism; see Koshland and· Botstein 1982.) These proteins are synthesized in precursor form with a signal

Figure 1   Schematic illustration of cotranslational export. The export process begins at the *bottom*: The ribosome assembles and initiates translation of the mRNA at the 5′ end. The signal sequence (represented by the wavy line in the mRNA or by small open circles in the protein) emerges from the ribosome and is recognized by the SRP complex, which interacts with the ribosome and the nascent polypeptide chain-stopping translation (*middle*). This translation block is relieved when the complex (ribosome, SRP, and nascent polypeptide chain) interacts with the membrane docking protein at the export site (*top*). The signal sequence is comprised of two segments, an aminoterminal-charged segment and a hydrophobic segment. During the initial stages of polysome binding to the export sites, the positively charged segment interacts with the inner leaflet of the membrane bilayer or a component of the export site. The hydrophobic segment then "loops" into the bilayer (Inouye and Halegoua 1979), and a functional export site is formed. As translation proceeds, the nascent chain is transferred vectorially across the membrane bilayer. Proteolytic processing of the signal sequence from the polypeptide chain is achieved by a peptidase activity located at the outer face of the membrane. Such processing probably occurs before synthesis of the protein is complete. The model shows the existence of a second information signal (stop transfer) located within the protein. As this signal emerges from the ribosome, it results in a dissociation of the ribosome from the membrane and release of the SRP complex. Subsequent translation of the mRNA completes the carboxyterminal end of the protein in the cytoplasm, leaving the protein embedded in a *trans*-membrane fashion with the amino terminus facing the lumen and the carboxyl terminus facing the cytoplasm.

Table 1 Commonly studied exported proteins in E. coli

Protein	Structural gene	Regulation	Mutant phenotype	Cellular location	Topology
LamB (λ receptor)	lamB	maltose-inducible	λ^R, Dex$^-$	outer membrane	transmembrane carboxyl terminus facing out
OmpA (3a,II)	ompA	constitutive	K3^R, TuII*^{R} Col^R, Con$^-$ EDTA^S	outer membrane	transmembrane carboxyl terminus facing in
Lpp (lipoprotein)	lpp	constitutive	resistance to globomycin	outer membrane	integral carboxyl terminus facing in
MalE (maltose-binding protein)	malE	maltose-inducible	Mal$^-$	periplasm	—
PhoA (alkaline phosphatase)	phoA	phosphate-repressible	XP	periplasm	—
Bla (TEM-1 β-lactamase)	bla	constitutive	Amp^S	periplasm	—
f1 or M13 major coat	gene VIII	—	defective phage	inner membrane	transmembrane carboxyl terminus facing in
f1 or M13 minor coat	gene III	—	defective phage	inner membrane	transmembrane carboxyl terminus facing in

Phenotypes are abbreviated as follows: Dex, growth on maltodextrin; ColL, colicin L; Con, conjugation; Mal, growth on maltose; Amp, ampicillin; Amp^R, ampicillin-resistant; Amp^S, ampicillin-sensitive; XP, enzyme activity detectable with bromo-chloro-indolylphosphate. This compilation of data was obtained from many different sources (for recent reviews, see Di Rienzo et al. 1978; Osborn and Wu 1980; Hall and Silhavy 1981a; Hofnung 1982; Michaelis and Beckwith 1982).

sequence at the aminoterminal end of the molecule (Michaelis and Beckwith 1982). Ribosomes engaged in the synthesis of these proteins are bound to the inner face of the cytoplasmic membrane (Davis and Tai 1980). With respect to protein export, the cytoplasmic membrane of *E. coli* is functionally analogous to the rough endoplasmic reticulum of eukaryotic cells.

Inner membrane proteins present a more complicated picture. Some of these proteins are synthesized in precursor form (Ichihara et al. 1981; Pratt et al. 1981), whereas others are not (Ehring et al. 1980; Nielsen et al. 1981). The latter proteins may contain a signal sequence that is not removed. Alternatively, they may not be cotranslationally inserted into the membrane. These proteins are highly hydrophobic, and they may simply partition into the membrane after synthesis is complete.

The most thoroughly studied inner membrane protein is the f1 (M13) major coat protein. This protein exists transiently in the inner membrane before its incorporation into phage heads. When present in the inner membrane, it assumes the topology analogous to the VSV glycoprotein; it spans the membrane once and its amino terminus faces out (Wickner 1979). Indeed, this protein is, in almost all respects, the prokaryotic equivalent of a eukaryotic viral glycoprotein. The mechanism of coat protein export is controversial. Experimental evidence suggests both a cotranslational (Chang et al. 1978, 1979) and a posttranslational (Wickner 1980) mechanism. Recently, Watts et al. (1981) reconstituted a purified preparation of leader peptidase into vesicles composed of *E. coli* phospholipids. They demonstrated that these vesicles could bind, process, and correctly insert radiochemically pure precursor coat protein (Silver et al. 1981) into the bilayer. These results indicate a posttranslational export mechanism. Furthermore, they suggest that leader (signal) peptidase is the only cellular component required for coat protein localization.

To account for the apparent posttranslational export mechanism of coat protein, Wickner (1979) proposed a model for protein localization, called the membrane trigger hypothesis, which is distinctly different from the signal hypothesis. According to this model, the signal sequence promotes folding of the completed precursor into a soluble export-competent conformation. Upon exposure to a hydrophobic environment, the protein is "triggered" into a different conformation that allows spontaneous membrane insertion. Cleavage of the signal sequence would drive the reaction and make it irreversible. In a recent review, Wickner (1980) applied this model to the export of numerous other proteins.

Although strongly suggestive, the experiments of Watts et al. (1981) cannot be regarded as conclusive proof of the membrane trigger hypothesis. Neither the preparation of leader peptidase nor precursor coat protein was chemically pure. The reconstituted vesicles contained demonstrable amounts of detergent and, in addition, correct insertion of the protein into the bilayer was inefficient. Finally, it must be noted that the coat protein is not a typical membrane protein. It exists in the membrane transiently. The protein also functions in the phage coat, where it almost certainly plays a role in phage infection. Accordingly, like its viral glycoprotein counterparts, it may promote fusion of the virus with lipid bilayers (White and Helenius 1980). If the coat protein has such properties, in vitro studies may yield confusing results.

GENETIC ANALYSIS OF PROTEIN EXPORT

The most straightforward genetic approach to the study of protein localization is to isolate a number of export-defective mutants. Subsequent analysis of these mutations would reveal the nature and location of intragenic information specifying export, and it would define cellular components of the export machinery and shed light on their mechanistic roles. Although conceptually simple, this approach is technically difficult. The fundamental problem is that there is no phenotypic difference between a mutation that prevents export and a mutation that prevents synthesis. Both mutations will confer a null phenotype. In the absence of a selectable or scorable phenotype, one is faced with the necessity of a brute force biochemical screen. For example, one could screen a population of mutants that exhibit a null phenotype for those in which accumulation of precursor can be detected. Besides being labor-intensive and time-consuming, this approach, historically, has not been successful. With very few exceptions, none of the thousands of mutations that are known to cause a null phenotype for an exported protein affect the signal sequence or cause a block in export.

The most notable exception is a mutation in lipoprotein (Lpp). Wu and Lin (1976) used a "suicide" selection based on the exclusive biosynthesis of Lpp in the absence of three amino acids not found in the protein. One of the mutants contained a missense mutation that changed glycine at position 14 of the signal sequence to aspartic acid (Table 2). Despite this alteration, protein export to the outer membrane still occurs (Lin et al. 1978). Why this selection yielded this particular mutation is not clear. Nevertheless, it represents the first example of a signal sequence mutation. This result demonstrates that removal of the signal sequence is not required for protein export to the outer membrane.

In the absence of a method for direct selection of export-defective mutants, a second approach, aimed at defining intragenic export components, was taken. This approach used a series of nonsense mutations located throughout the gene in question. This series of chain-terminating mutations produces a collection of truncated polypeptides of varying lengths. By determining the cellular location of the truncated peptide using specific antisera, one can define the site of intragenic export information.

Two *malE* nonsense mutations have been studied in detail (Ito and Beckwith 1981). One produces a peptide that is 90% complete. This protein is processed and exported to the periplasm as evidenced by the fact that it is released by cold osmotic shock. The second mutation produces a peptide one-third the length of wild-type MalE. This peptide is processed but is not released by osmotic shock. It is, however, exported from the cytoplasm, since the fragment is sensitive to externally added trypsin. These results demonstrate that the carboxyterminal two-thirds of MalE is not required for either export or processing.

Koshland and Botstein (1980) have obtained similar results with Bla. They did not find any peptides that were released by osmotic shock, although all were processed. Originally, this result was interpreted to indicate that the peptides remained in the cytoplasm. However, subsequent experiments, using externally added trypsin, demonstrated that the peptides are outside of the cytoplasm (Koshland and Botstein 1982). Here again, the carboxyterminal portion of the protein is not required for export from the cytoplasm.

The advent of recombinant DNA technology permitted the application of a related experimental approach that also results in the production of truncated peptides. Once the gene for an exported protein has been cloned, portions of the gene coding for the carboxyterminal end of the protein can be removed using restriction enzymes. This approach has been extensively applied to the outer membrane protein, OmpA (Bremer et al. 1980, 1982). The mature protein contains 325 amino acids. Only the 193 amino acids at the amino terminus are required for export from the cytoplasm and for stable incorporation into the outer membrane. What is remarkable is the fact that these 193 amino acids are functional. This truncated peptide complements all of the phenotypes of *ompA* except sensitivity to EDTA (Table 1). In contrast, a peptide of 133 amino acids is nonfunctional and very unstable. A similar approach has also been used to study export of the f1 gene III protein to the inner membrane (Boeke and Model 1982). Here again, the aminoterminal portion of the protein was shown to be unnecessary for export from the cytoplasm. However, deletions were found that altered the cellular location of the truncated peptide. These results shed light on intragenic export information and are discussed further below.

The results with OmpA emphasize the problems associated with approaches that use nonsense fragments or aminoterminal deletions. Both suffer the same limitations. First, truncated peptides are often very unstable, and second, short fragments are likely to be nonfunctional and may not cross-react with antisera. Although the results that have been obtained demonstrate that export information must lie at the aminoterminal end of the protein, the inherent limitations of this approach preclude precise genetic analysis.

Application of Gene-fusion Technology

A solution to the inherent problems associated with truncated peptides came with the application of gene-fusion technology. This technique allows one to specifically label aminoterminal fragments of an envelope protein with a marker that is stable and simple to assay. Beckwith and co-workers exploited

Table 2 Mutational alterations of signal sequences

Protein	Charged segment	Hydrophobic segment
f1 or M13 major coat	MET LYS LYS SER LEU VAL LEU LYS|ALA	SER VAL ALA VAL ALA THR LEU VAL PRO MET LEU SER PHE ALA ALA↓
f1 or M13 minor coat	MET LYS LYS|LEU LEU PHE ALA ILE PRO	LEU VAL VAL PRO PHE TYR SER HIS SER ALA↓
PhoA	MET LYS|GLN SER THR ILE ALA LEU ALA LEU LEU PRO	LEU LEU PHE THR PRO VAL THR LYS ALA ARG↓
		↓GLN ↓ARG
MalE	MET LYS ILE LYS THR GLY ALA ARG ILE LEU ALA LEU SER ALA LEU	THR THR MET MET PHE SER ALA LYS↓
		↓PRO ↓GLU ↓ ↓ ↓
		LYS ARG ARG
		ARG
Bla	MET SER ILE GLN HIS PHE ARG VAL ALA LEU ILE PRO PHE PHE ALA ALA PHE CYS LEU PRO VAL PHE ALA HIS↓
		LEU/SER°
		ALA PHE LEU PHE↓
		°VAL↓
		LEU ARG HIS PHE ALA PHE LEU PHE
	PRO|CYS ARG	

262

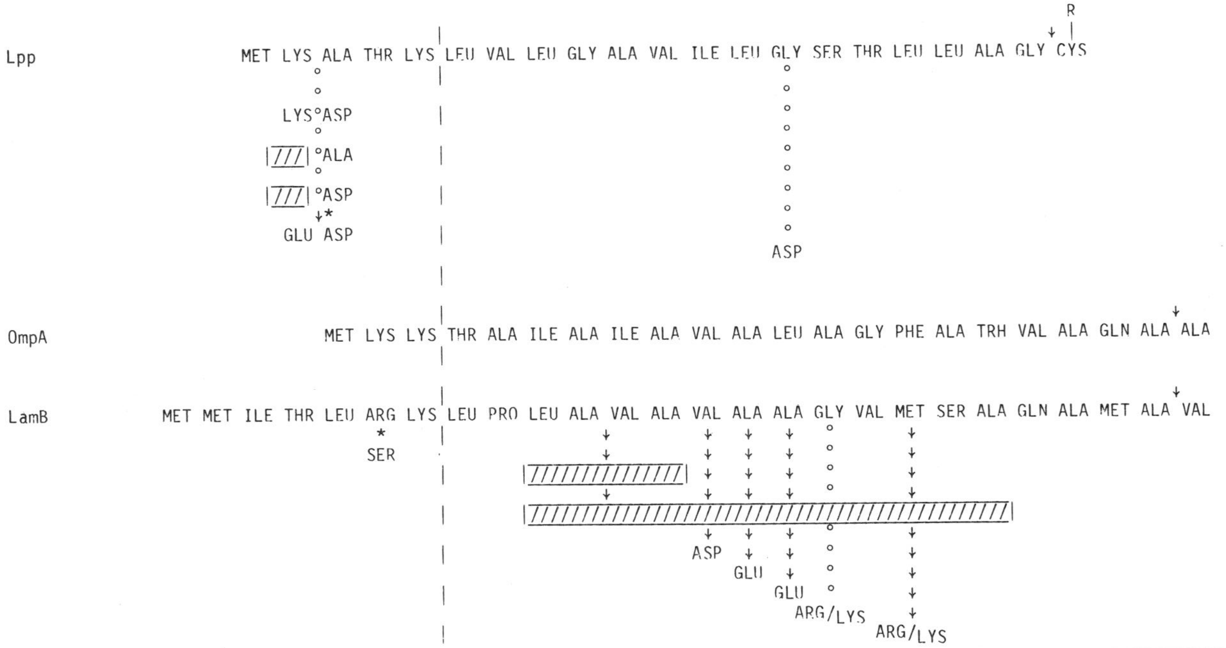

Lpp

$\overset{\text{R}}{\underset{\downarrow}{|}}$

MET LYS ALA THR LYS LEU VAL LEU GLY ALA VAL ILE LEU GLY SER THR LEU LEU ALA GLY CYS

LYS°ASP
°ALA
|///|°ALA
|///|°ASP
GLU ASP

OmpA

MET LYS LYS THR ALA ILE ALA ILE ALA VAL ALA LEU ALA GLY PHE ALA TRH VAL ALA GLN ALA ALA

LamB

MET MET ILE THR LEU ARG LYS LEU PRO LEU ALA VAL ALA VAL ALA ALA GLY VAL MET SER ALA GLN ALA MET ALA VAL

SER
ASP
GLU
GLU
ARG/LYS
ARG/LYS

The amino acids of the signal sequences for the proteins listed in Table 1 are shown. These include the major (Sugimoto et al. 1977) and minor (Schaller et al. 1978) coat proteins of f1 or M13 phage, which are located in the inner membrane; the periplasmic proteins, PhoA (H. Inouye et al. 1982), Bla (Sutcliffe 1978), and MBP (Bedouelle et al. 1980); and the outer membrane proteins, Lpp (Inouye et al. 1977), OmpA (Beck and Bremer 1980), and LamB (Hedgpeth et al. 1980). The arrow above each of the sequences indicates the processing site. The R group above the last cysteine residue of the Lpp sequence designates the thioetherdiglyceride that is present on the mature protein (Braun 1975). Specific mutations within the signal sequences are shown below each of the wild-type sequences. Mutations designated by an arrow block localization of the protein; mutations designated by a circle do not. Deletions are shown by a box, and mutations that appear to decrease translation are designated by an asterisk. See the text for a complete description of the phenotypes conferred by each of these mutations.

the sophistication of *lac* genetics and the unusual properties of the cytoplasmic enzyme β-galactosidase to develop techniques that allow fusion of *lacZ* (codes for β-galactosidase) to any gene in *E. coli* (Casadaban 1976; for review, see Silhavy and Beckwith 1983). Such fusions specify a hybrid protein composed of an aminoterminal sequence from the target gene product and a large, functional carboxyterminal portion of β-galactosidase (Fig. 2). By constructing a series of fusions differing only in the amount of target gene DNA contained in the hybrid gene and determining the cellular location of the hybrid protein that is produced, investigators have been able to more accurately define the location of intragenic export information.

Gene-fusion technology has been applied extensively to study the localization of two periplasmic proteins, MalE (Bassford et al. 1979) and PhoA (Michaelis and Beckwith 1982; Michaelis et al. 1983a,b), and the outer membrane protein LamB (Silhavy et al. 1977, 1979; Hall et al. 1982). To a lesser extent, the technique has also been used to study export of the inner membrane protein MalF (Silhavy et al. 1976, 1979; Shuman et al. 1980) and the outer membrane protein OmpF (Hall and Silhavy 1981a,b).

Results from studies using gene fusions show a clear pattern. Fusions constructed to contain a substantial portion of a gene specifying a noncytoplasmic protein produce a hybrid protein that is exported, at least to some degree, from the cytoplasm. Conversely, fusions that contain only a small portion of a gene specifying an exported protein produce a hybrid protein that remains in the cytoplasm. This demonstrates that export information is contained within the structural gene and indicates that the information must lie at a position corresponding to the aminoterminal end of the protein.

Using gene fusions, investigators have shown that β-galactosidase can be exported to both an inner and an outer membrane location. However, β-galactosidase has not been successfully exported into the periplasm. Even fusions that contain nearly all of the gene coding for a periplasmic protein produce a hybrid protein that remains stuck in the cytoplasmic membrane. Apparently, β-galactosidase contains amino acid sequences that are incompatible with passage through the membrane. Attempted export of β-galactosidase to the periplasm jams export sites, causing the intracellular accumulation of precursors of other exported proteins (Silhavy et al. 1979; Michaelis and Beckwith 1982).

The fact that β-galactosidase sequences can be exported to the inner or outer membrane but not to the periplasm is significant. It suggests that although export of periplasmic and membrane proteins may be similar in the early stages, the export pathways diverge before completion. Genes that specify membrane proteins may contain additional export information (i.e., stop transfer sequences) that halts vectorial transfer before synthesis is complete (see Fig. 1), thus preventing β-galactosidase sequences from entering the membrane. If true, then it should be possible to fuse the *lacZ* gene to a gene coding a membrane protein such that information specifying export initiation

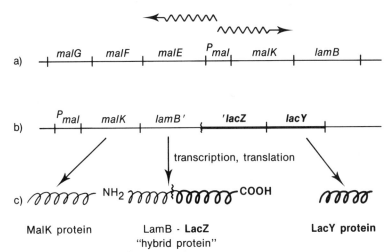

Malk protein   Lamb - **Lacz**   **LacY protein**
                "hybrid protein"

Figure 2 (*a*) The two divergent operons that comprise the *malB* locus in *E. coli* (Hofnung 1982). These two operons specify five proteins, which together make up the activity transport system for maltose and maltodextrins. Both operons, and thus the synthesis of all five proteins, are induced by maltose in the growth media. At least three of these proteins are destined to be localized to noncytoplasmic compartments. The *malE* gene specifies the periplasmic MalE; *malF*, *malG*, and *malK* specify inner membrane proteins, and *lamB* specifies the outer membrane protein LamB. Transcription of these two operons is initiated from a central promoter region (p_{mal}). Direction of transcription is shown by wavy arrows. (*b*) Genetic structure of a *lamB-lacZ* fusion. The fusion joint is designated by a short, vertical, wavy line. Transcription and subsequent translation of the operon results in the production of three proteins. (*c*) The MalK protein is required for maltose transport. The LacY protein is required for lactose transport. The structure of the LamB-LacZ hybrid protein is shown. The hybrid protein is comprised of LamB sequences at the amino terminus and a major function portion of β-galactosidase at the carboxyl terminus. All of the hybrid proteins discussed here have essentially identical amounts of β-galactosidase sequences at the amino terminus. For *E. coli* to express a Lac⁺ phenotype, strains must have both LacZ (β-galactosidase) and LacY (lactose transport) protein activities.

but not stop transfer is present. Such a fusion should produce a hybrid protein that is exported in a manner analogous to that described for periplasmic proteins. This is what is observed: *lamB-lacZ* fusions that contain approximately 60% (codes for the signal sequences plus 241 amino acids) of the *lamB* gene produce a hybrid protein that is exported efficiently to the outer membrane, whereas fusions that contain approximately 40% (codes for the signal sequence plus 173 amino acids) of *lamB* jam the export machinery and cause precursor accumulation (Silhavy et al. 1979; Emr et al. 1980a).

Finally, fusions have been constructed that specify a hybrid protein that contains the complete LamB signal sequences plus 15 amino acids of mature

LamB. Despite the presence of the signal sequence, the hybrid protein remains in the cytoplasm (Moreno et al. 1980). This suggests, at least for LamB, that the signal sequence is not sufficient to cause export from the cytoplasm. Other export information located downstream in the *lamB* gene appears to be required.

Mutant Selections Based on *lacZ* Gene Fusions

The information gained from the biochemical analysis of fusion strains has provided certain insights into the mechanism(s) of protein localization. However, the most significant contribution of gene-fusion technology stems from the unusual phenotypes often exhibited by fusion strains. These phenotypes are the consequence of the cell's attempt to export a hybrid protein containing sequences of β-galactosidase and can be exploited to isolate export-defective mutants. More importantly, the properties of *lacZ* fusions allow one to distinguish mutations that block export from those that prevent synthesis. Although the number of examples is not large, it appears that these phenotypes will be exhibited generally.

The first unusual phenotype is characteristic of fusion strains that produce a hybrid protein that is incorrectly or inefficiently exported. High-level synthesis of such hybrid proteins is lethal, probably because these proteins jam the export machinery preventing localization of other essential envelope proteins. In the case of *malE-lacZ* or *lamB-lacZ* fusions, this overproduction lethality is observed when maltose is added to the growth medium to induce high-level synthesis of the hybrid protein. Accordingly, such strains are sensitive to maltose (Mal^S). Mutations that relieve the Mal^S phenotype but do not prevent synthesis of the hybrid protein (Lac$^+$) are export-defective.

The second unusual phenotype is characteristic of fusion strains that produce a hybrid protein that is efficiently localized to a membrane. In such an environment, the β-galactosidase portion of the hybrid protein exhibits reduced enzymatic activity, probably because it cannot assume an active conformation. Accordingly, these fusion strains are unable to grow on lactose (Lac$^-$). By selecting for Lac$^+$, one can obtain mutations that prevent export of the hybrid protein. These selection procedures have permitted the isolation of numerous export-defective mutants. Such mutants provide the basis for a genetic analysis of protein export.

Signal Sequence Mutations

By using the selection procedures described in the previous section, numerous mutations have been obtained that alter the signal sequences of three genes (*lamB, malE, phoA*), which were then characterized at the level of the DNA sequence. These mutations shed some light on the functions performed by the various molecular components of the signal sequence (Table 2). Because our work deals primarily with the LamB signal sequence, we use it as a focus for discussion.

Like all signal sequences, the LamB signal sequence can be divided into two distinct domains: an aminoterminal hydrophilic segment and a central hydrophobic core that extends near to the site of processing. These two domains are generally separated by one or two basic amino acids, especially in prokaryotic sequences (Table 2). Sequence comparison of all known signal sequences reveals no other striking homologies except for the presence of an amino acid with a small side chain (Ala, Ser, Gly, Cys) at the processing site.

The aminoterminal domain, excluding the basic amino acid residues, does not appear to play a critical role in export. Several lines of evidence support this claim. First, no export-defective mutations are known that lie in this region. Second, this sequence varies in composition and varies in size from as small as 1 (methionine, specified by the initiation codon) to as large as 7. Indeed, Talmadge et al. (1981) have placed 18 extra amino acids in this region of the signal sequence of preproinsulin. Export from the cytoplasm of *E. coli* appears to occur normally.

The central hydrophobic core clearly plays an important role in the initiation of protein export. All of the export-defective *lamB* mutations alter this region of the signal sequence (Emr et al. 1980b). Similarly, all export-defective *malE* (Bedouelle et al. 1980), *phoA* (Michaelis 1982; Michaelis et al. 1983b), and *bla* (produced by in vitro mutagenesis; Koshland et al. 1982) mutations alter this region as well. Since no sequence homology can be recognized between various signal sequences and since all export-defective *lamB* mutations, and nearly all others as well, result in the presence of a charged amino acid in this region, this sequence may function simply because it is hydrophobic.

Although we do not doubt the obvious importance of hydrophobicity, we believe that certain amino acid residues in this region play a more critical role in export initiation. In the LamB signal sequence, the presence of a charged residue at positions 14, 15, 16, or 19 blocks export to more than 95%. However, a charged residue, either acidic or basic, at position 17 has essentially no effect on the export of LamB to the outer membrane (Emr and Silhavy 1982). Thus, it is the position of the alteration, not simply the presence of a charge, that determines the effect of the mutation.

We believe that the residues at positions 14, 15, 16, and 19 define an important recognition site. These 4 residues probably interact directly with a cellular component of the protein export machinery. If one of these residues is altered by mutation, the critical recognition cannot occur, and the export process does not initiate. The result is the accumulation of precursor in the cytoplasm. The data we have obtained by genetic analysis of the LamB signal sequences are consistent with this proposal. All of the export-defective *lamB* mutations (14 base substitutions and 13 deletion mutations) alter one or remove one or more of these critical 4 residues. The two point mutations that do not alter one of these residues do not block export.

An apparent exception to this statement that all export-defective *lamB* mutations alter at least one of these critical 4 residues is the small deletion mutation *lamB*S78. This 12-bp deletion removes amino acids 10, 11, 12, and 13 from the LamB signal sequence. It blocks export to more than 95%. Although this deletion certainly does not alter one of the four critical amino acids directly, evidence that we have obtained recently (Emr and Silhavy 1983) indicates that the mutation alters the recognition site indirectly by altering the secondary conformation of residues 14, 15, and 16.

Using rules to predict peptide secondary structure (Chou and Fasman 1978), Bedouelle and Hofnung (1981) determined that the hydrophobic core of the LamB signal sequence most probably exists in an α-helical conformation. Since two amino acids in this core region, proline at position 9 and glycine at position 17, destabilize helical structures, the helix is predicted to terminate in the region of these two residues. According to these rules, none of the point mutations that alter the LamB signal sequence would alter this secondary structure. However, the small deletion mutation *lamB*S78, which removes residues 10, 11, 12, and 13, would alter the secondary structure, because in the mutant signal sequence the helix-destabilizing residues proline and glycine are too close to each other (3 residues apart instead of 7, as in the wild-type sequence) to permit a helix to form between them. Consequently, the critical residues 14, 15, and 16 cannot form the helical conformation required for recognition.

The hypothesis that the *lamB*S78 deletion alters the secondary structure (as described above) has been tested genetically. Since the critical recognition site is still intact in the mutant signal sequence, we predicted that function would be restored by a second mutation that permits the critical region to assume an α-helical conformation. We found that secondary mutations that change the proline at position 9 to leucine or that change the glycine at position 17 to cysteine restore function to the mutant signal sequence. Both of these changes permit the recognition site to assume an α-helical conformation (Emr and Silhavy 1983).

The various molecular components of the LamB signal sequence and the function each appears to perform in initiating protein export can be summarized as follows: (1) The aminoterminal domain does not appear to be required. (2) The central hydrophobic core is essential. Furthermore, this core must be able to assume an α-helical conformation to allow recognition to occur. (3) A critical subset of four amino acids contained within the hydrophobic core comprises a recognition site that interacts directly with a component of the cellular export machinery.

The nature of the cellular component that interacts with the recognition site in the hydrophobic core is not known. We presume, however, that the components will be defined genetically by mutations like *prlA*. Such mutations alter a cellular component and restore recognition of mutationally altered signal sequences (see below). We do not mean to imply that the only function performed by the signal sequence is in the intiation of export. Protein localization is likely to be a multistep process. Conceivably, the signal sequence could function in several of these steps. The function of the basic amino acid residues that separate the two signal sequence domains remains unclear. None of the export-defective mutations that we have isolated alters one of these residues, suggesting that these residues do not function in export initiation. Recently, Schwartz et al. (1981) isolated a mutant in which the arginine at position 6 of the LamB signal sequence is changed to a serine. Results that were obtained with this mutant suggest that the mutation may interfere with the cellular mechanism that couples export and translation (Hall and Schwartz 1982; Hall et al. 1983). Analogous mutations have been constructed "in vitro" in the gene coding for Lpp, a major outer membrane protein of E. coli, and similar results were obtained (S. Inouye et al. 1982). These results are discussed in more detail below.

Other Intragenic Information Specifying Export and Membrane Insertion

The signal sequence functions in the initiation of protein export. Although this sequence is essential, studies with lamB-lacZ fusions, e.g., suggest that it is not sufficient to cause export from the cytoplasm; other information, located within lamB, is required in addition. Moreover, there must be information within lamB, and other genes that codes for membrane proteins causing membrane insertion and sorting signals that specify localization to the correct cellular membrane. To determine the nature and location of this additional intragenic export information within lamB, we devised techniques to permit the isolation of a series of in-frame deletion mutations internal to the structural gene. These techniques make use of a large lamB-lacZ fusion that encodes 241 amino acids of LamB. By cloning this fusion onto a small plasmid and then using a unique restriction site within the lamB sequence coupled with treatment by the endonuclease BAL 31, we were able to generate a series of deletions starting at this site. Since the hybrid gene encodes a functional β-galactosidase activity, selection for loss of the unique restriction site and screening for clones that retained a functional β-galactosidase activity allowed us to easily identify a number of in-frame deletions that were internal to the lamB sequence. By analyzing the effects of these deletions on the export of both a LamB-LacZ hybrid protein and an otherwise wild-type LamB, we were able to more precisely define the location of export information within the lamB gene (Benson and Silhavy, 1983).

The data obtained from the analysis of in-frame lamB deletions and from the various lamB-lacZ fusions are consistent with the hypothesis that discrete regions of the protein play critical roles in the localization process. The signal sequence defines the first discrete region and appears to function in the initiation of protein export. A second region is defined by the amino acids 15–70 of mature LamB. If the signal sequence plus 15 amino acids of mature

LamB are attached to a large functional carboxyterminal portion of β-galactosidase, the hybrid protein produced remains in the cytoplasm (Moreno et al. 1980). Conversely, an otherwise identical fusion containing approximately 175 amino acids of LamB is exported to the outer membrane with an efficiency of about 40% (Silhavy et al. 1977). Clearly, this segment of the *lamB* gene contains sufficient information to direct β-galactosidase to an outer membrane location. The in-frame deletions that we isolated demonstrated that amino acids 70−220 can be removed without affecting export of either LamB-LacZ hybrid proteins or LamB itself. Accordingly, we conclude that information required for export from the cytoplasm and incorporation into the outer membrane must be contained in a region of *lamB* corresponding to amino acids 15−70. The nature and function of this critical region is a major focus of current research.

A third region of *lamB*, although not required for localization, appears to increase the efficiency of the export process. This region has only been definable through the use of *lamB-lacZ* fusions. Deletion analysis indicates that this region lies after amino acid 235 of LamB. Whether this region functions in LamB export remains to be demonstrated.

Using techniques of recombinant DNA to produce truncated peptides, Boeke and Model (1982) have identified a functional stop transfer sequence in the f1 gene III protein. This inner membrane protein spans the bilayer once and has a topology resembling a eukaryotic viral glycoprotein. The amino terminus is hydrophobic and probably anchors the protein in the membrane. Deletions that produce truncated peptides lacking this carboxyterminal sequence cause export to the periplasm. These results are similar to those obtained with the membrane-bound and secreted forms of IgM μ chain (Rogers et al. 1980). Here again, removal of a functional stop transfer sequence prevents membrane insertion and allows complete passage of the peptide through the membrane. Further work is required to establish the precise nature and function of the hydrophobic carboxyterminal sequence.

Cellular Components of the Protein Export Machinery

Export-defective Mutants

An important goal of studies on protein export is to identify and characterize cellular components of the protein export machinery. One approach to this problem is to isolate mutants that exhibit a pleiotropic export-defective phenotype. Various schemes have been used to identify such mutants. Some of these are based on the assumption that mutations that prevent protein export will be lethal. For example, a collection of conditional lethal mutants isolated by Y. Hirota that are temperature-sensitive for growth have been screened for accumulation of precursors of exported proteins at the nonpermissive temperature without success (K. Ito, pers. comm.). The failure of such brute-force techniques demonstrates the need for specific selection or screens for export-defective mutants.

Wanner et al. (1979) have designed a procedure for the isolation of mutants unable to secrete periplasmic proteins. This procedure is based upon the assumption that some factor(s) exists that is necessary for the secretion of periplasmic proteins, e.g., alkaline phosphatase, but that is not necessary for the synthesis of cytoplasmic proteins, e.g., β-galactosidase. Mutants are detected on agar medium that includes indicators for both alkaline phosphatase and β-galactosidase synthesis. In addition, since mutations defective in the secretion of envelope proteins may be lethal to the cell, the general approach used by Schedl and Primakoff (1973) for the detection of conditional lethal mutants was adopted. The mutation *perA* was isolated in this manner. Although this mutation has pleiotropic effects, we do not know whether the mutation affects secretion or regulation, or both. The mutation appears to lie in a gene *envZ*, which is known to be involved in the transcriptional regulation of the genes specifying the major outer membrane proteins OmpF and OmpC (Wandersman et al. 1980; Hall and Silhavy 1981a).

Another method for obtaining export-defective mutants requires foreknowledge of a chromosomal locus known to be involved in protein localization. This technique uses local mutagenesis (Silhavy and Beckwith 1983) to increase the frequency of desired mutants. Using this procedure, Ito et al. (1983) isolated a temperature-sensitive mutant that affects protein export. The target for the local mutagenesis was the large ribosomal gene cluster, a locus known to contain the *prlA* mutation (see the following section). All cell manipulations were done at 30°C, and transductants were screened for the inability to grow at 42°C. Temperature-sensitive mutants were screened for an export defect by looking for accumulated precursor MalE after short exposure to the nonpermissive temperature. The precursor protein was identified by antibody precipitation.

A direct selection for mutants affecting the export machinery has been developed based on the Lac⁻ phenotype exhibited by certain *malE-lacZ* fusion strains. As described above, mutations that prevent export of this hybrid protein confer a Lac⁺ phenotype. The properties of this strain suggest that only a slight degree of internalization would provide sufficient β-galactosidase activity for growth on lactose. Mutations have been obtained that define two new genetic loci in *E. coli: secA* and *secB* (Oliver and Beckwith 1981; Oliver et al. 1982).

The *secA* mutation has been characterized in some detail. Strains containing this mutation grow normally at 30°C; however, at 42°C, growth stops and cells form filaments and accumulate precursor forms of various exported proteins. However, not all exported proteins are affected by *secA*. Some proteins are localized normally at the nonpermissive temperature. The mutation is not strictly conditional; some precursor can be detected at permissive temperature.

Subsequent studies have shown that the *secA* gene maps at 2.5 minutes on the *E. coli* chromosome. Using *secA*-transducing phages and ultraviolet-irradiated cells, Oliver and Beckwith (1982a) have shown that the gene product is a protein of 92,000 m.w. *secA-lacZ* fusions have been constructed, and the hybrid protein has been used to obtain antisera directed against SecA. In addition, a *secA* amber mutation has been isolated. Studies using SecA antibody and various *secA* mutant strains indicate that the SecA protein is essential for growth and is regulated in response to the secretion requirements of the cell; i.e., if the export machinery is jammed with a hybrid protein or if protein export is blocked by a *secA*[ts] mutation, expression of *secA* is derepressed at least tenfold (Oliver and Beckwith 1982b).

The *secB* mutants have been more difficult to characterize because no conditional lethal mutations are known. The mutants accumulate precursors of various exported proteins, and the mutation maps at 80 minutes on the *E. coli* chromosome (Oliver et al. 1982).

Suppressors of Export-defective Mutations

As already noted, one approach to identifying cellular components of the export machinery is to isolate mutants that are generally export-defective. Another approach is to devise a selection for mutants in which an internalized protein is exported. The mutants we have isolated in which the precursor of the *lamB*-gene product is found in the cytoplasm provide a selection for the export of an internalized protein. These mutant strains do not localize LamB properly because the export machinery cannot recognize the mutationally altered signal sequence. We reasoned, therefore, that it should be possible to alter the export machinery by mutation to restore recognition of the signal sequence. Such mutations would define components of the export machinery. The availability of mutants in which the export process is altered should allow the identification of important gene products, and this, in turn, should provide a means to analyze the export pathway biochemically.

In mutant strains that produce a LamB protein with a defective signal sequence, the protein is found in soluble form in the cytoplasm with the altered signal sequence still attached. Consequently, such strains exhibit a typical LamB⁻ phenotype, i.e., inability to use maltodextrins as a carbon source and resistance to bacteriophage λ. Using these strains, we isolated second-site pseudorevertants by selecting for the ability to grow on maltodextrins. The mutation responsible for pseudoreversion suppresses the defect caused by the altered signal sequence and restores export and, in most cases, normal processing of the LamB protein. Three different second-site suppressor mutations have been identified. These suppressors, termed *prl* (*p*rotein *l*ocalization) *A*, *prlB*, and *prlC*, may define cellular components that interact with the signal sequence during the export process (Emr et al. 1981).

One of the suppressors, *prlA*, has been characterized in some detail. It causes phenotypic suppression of all export-defective signal sequence mutations in *lamB*, *malE* (Emr and Bassford 1982), and *phoA* (Michaelis and Beckwith 1982; Michaelis et al. 1983b). In some cases, *prlA* restores export to levels that are approximately 85% of that seen in wild-type strains. Despite this powerful suppression, *prlA* causes no growth defects nor does it alter normal protein export. Since the suppressors restore export of several different proteins with defective signal sequences, the cellular component altered by the *prlA* mutation probably interacts with this sequence during the export process.

Genetic mapping shows that the gene altered by the suppressor mutation is a component of the *spc* operon (Emr et al. 1981). This operon, which maps at 72 minutes on the *E. coli* chromosome, has been shown to specify ten different ribosomal proteins (Jaskunas et al. 1977a,b; Post et al. 1980). Using a combination of genetic, biochemical, and recombinant DNA techniques, we have shown that the *prlA* mutations lie at the extreme promoter-distal end of the *spc* operon and do not alter any of the previously known genes. The protein specified by this heretofore unidentified gene appears to be the cellular component altered by the *prlA* mutations. We propose that this new gene be termed *prlA* (Shultz et al. 1982).

prlA restores not only export but also processing of every tested signal sequence mutation except one. This finding demonstrates that signal sequence mutations that prevent export do not prevent processing. If the mutant protein can be exported, it can be processed. Thus, the processing enzyme must be located outside the cytoplasm. Indeed, considering the proteins associated with cellular fractionation techniques, signal sequence processing may constitute a more reliable test for export from the cytoplasm. The only export-defective signal sequence mutation that is suppressed by *prlA* without concomitant processing is *lamB*S60. This mutation is an in-frame deletion that removes 12 amino acids from the LamB signal sequence. The fact that export is restored without processing in this case suggests that the action of signal peptidase is not required for either the export or function of LamB.

Several other lines of evidence support our conclusion that PrlA is a component of the cellular protein export machinery. First, as described above, Ito and co-workers have isolated a temperature-sensitive lethal mutation using techniques of local mutagenesis. At the nonpermissive temperature, the export and processing of several different proteins is drastically reduced. Although the nature of this mutation is not known, expression of the ribosomal protein L15 has been shown to be greatly reduced at high temperature (Ito et al. 1983). This mutation may also decrease expression of PrlA, because the *prlA* gene is located adjacent to and promoter-distal from the gene specifying L15. Thus in the absence of PrlA, protein export may be blocked at an early step.

A second line of evidence indicating that PrlA is a component of the export machinery comes from studies with the temperature-sensitive *secA* (*secA*[ts]). One might anticipate that second-site mutations in genes coding for proteins that interact with SecA could compensate for the export defect. To test this possibility, pseudorevertants of a *secA*[ts] strain that are no longer temperature-sensitive have been isolated. These pseudorevertants no longer exhibit an export-defective phenotype. One of these second-site mutations appears to map in *prlA* (E. Brickman et al., pers. comm.). Thus, it seems likely that both SecA and PrlA are components of the export machinery. Furthermore, these two proteins may interact at some point during the export process.

The function of the PrlA protein remains to be elucidated. Although the gene does not appear to code for a known ribosomal protein, its location in an operon that specifies ten other ribosomal proteins suggests that PrlA may be associated with the machinery of protein synthesis. The genetic data that we have obtained suggests that PrlA may be involved in signal sequence recognition. Accordingly, PrlA may function in the coupling of translation and export.

A Mechanism to Couple Protein Synthesis and Export

Studies with eukaryotic cells indicate that a biochemical mechanism couples translation and export. This mechanism involves SRP and docking protein (Fig. 1). There is also evidence that a similar mechanism exists in prokaryotic systems. The first experimental results to support this coupling mechanism came from the analysis of a mutation that alters one of the basic amino acids in the hydrophilic portion of the LamB signal sequence. This mutation was shown by DNA sequence analysis to change the arginine codon at position 6 to a serine codon (Table 2) (Hall et al. 1983). Studies using a *lamB-lacZ* fusion that contains the LamB signal sequence but produces a hybrid protein that remains in the cytoplasm showed that the mutation does not affect translation. However, when the mutation is recombined into a *lamB-lacZ* fusion that specifies an exported hybrid protein, expression is reduced markedly. Moreover, when the mutation is present in an otherwise wild-type *lamB* gene, expression of LamB is reduced as well (Hall and Schwartz 1982). Recently, S. Inouye et al. (1982) constructed similar mutations in Lpp using in vitro techniques. These mutations also appear to affect expression. The correlation between export and synthesis suggests that these two processes may be coupled in vivo.

Further evidence to support the existence of a mechanism that couples synthesis and export comes from studies using the *secA* nonsense mutation. When this mutation is present in a strain carrying a temperature-sensitive nonsense suppressor, expression of SecA can be controlled conditionally. Oliver and Beckwith (1982b) showed that synthesis of MalE is greatly reduced when this strain is incubated at high temperature. This result should be contrasted to results described previously with *prlA*. In this case, the absence of the gene product appears to prevent export but does not prevent synthesis, since precursor accumulation can be demonstrated.

We speculate that SecA performs a function in *E. coli* that is analogous to that of the docking protein, whereas PrlA may be a component of SRP. If SecA is not present, the translation block cannot be removed. Conversely, if

PrlA is absent, the block cannot occur, thus preventing export and processing. The fact that SecA is associated with the membrane provides yet another similarity. Although the evidence is as yet incomplete, we posit that synthesis and export may be tightly coupled in *E. coli*.

If a mechanism to couple synthesis and export exists, then biochemical experiments that use pulse-labeling techniques must be viewed with caution. Such experiments are based on the assumption that translation proceeds at a constant rate, and this assumption may not be valid. Similarly, genetic selections for export-defective mutants that require synthesis of the exported protein may not yield the desired mutation. Indeed, the selection technique for the isolation of *secA*^{ts} uses a *malE-lacZ* fusion and requires synthesis of the hybrid protein (Lac$^+$). Mutations in *secA* that destroy function cannot be isolated in this manner, because such a mutation would prevent synthesis of β-galactosidase activity. The *secA*^{ts} allele must produce an *altered* gene product.

CONCLUSIONS

There is a striking concordance of predictions of the signal hypothesis with many aspects of the protein localization process in both eukaryotic and prokaryotic systems. Until recently, nearly all of the features of this model have been derived from work on eukaryotic systems and thus are untested in vivo. Genetic analysis in bacteria has served to validate a number of these predictions. Through the isolation of mutants, the position and function of information signals within an exported protein has been determined. In addition, several genes that specify cellular components required for the localization process have been identified; however, numerous questions regarding the mechanisms, the machinery, and the pathway(s) of specific protein sorting to the correct noncytoplasmic location remain to be answered. We believe that answers to such questions will require genetic analysis for which bacteria, particularly *E. coli*, are clearly the organism of choice. The genetic studies described in this review demonstrate the potential of this approach.

ACKNOWLEDGMENTS

Research was sponsored by the National Cancer Institute, Department of Health and Human Services (DHHS), under contract NO1-CO-23909 with Litton Bionetics, Inc. The contents of this paper do not necessarily reflect the views or policies of the DHHS, nor does mention of trade names, commercial products, or organizations imply endorsement by the U.S. Government. S.B. is the recipient of a National Research Service Award from the National Cancer Institute, DHHS, Public Health Service grant 1F32-CAO-6786-01.

REFERENCES

Bassford, P.J., T.J. Silhavy, and J.R. Beckwith. 1979. Use of gene fusion to study secretion of maltose-binding protein into *Escherichia coli* periplasm. *J. Bacteriol.* **139:** 19.

Beck, E. and E. Bremer. 1980. Nucleotide sequence of the gene *ompA* coding for the outer membrane protein II* of *Escherichia coli* K-12. *Nucleic Acids Res.* **8:** 3011.

Bedouelle, H. and M. Hofnung. 1981. Functional implications of secondary structure analysis of wild-type and mutant bacterial signal peptides. *Prog. Clin. Biol. Res.* **63:** 309.

Bedouelle, H., P.J. Bassford, A.V. Fowler, I. Zabin, J. Beckwith, and M. Hofnung. 1980. Mutations which alter the function of the signal sequence of the maltose binding protein of *Escherichia coli. Nature* **285:** 78.

Benson, S.A. and T.J. Silhavy. 1983. Information within the mature LamB protein necessary for localization to the outer membrane of *E. coli* K12. *Cell* **32:** 1325.

Blobel, G. 1980. Intracellular protein topogenesis. *Proc. Natl. Acad. Sci.* **77:** 1496.

Blobel, G. and B. Dobberstein. 1975. Transfer of proteins across membranes. I. Presence of proteolytically processed and unprocessed nascent immunoglobulin light chains on membrane-bound ribosomes of murine myeloma. *J. Cell Biol.* **67:** 835.

Boeke, J.D. and P. Model. 1982. A prokaryotic membrane anchor sequence: Carboxyl terminus of bacteriophage f1 gene *III* protein retains it in the membrane. *Proc. Natl. Acad. Sci.* **79:** 5200.

Braun, V. 1975. Covalent lipoprotein from the outer membrane of *Escherichia coli. Biochim. Biophys. Acta* **415:** 335.

Bremer, E., E. Beck, I. Hindennach, I. Sonntag, and U. Henning. 1980. Cloned structural gene *ompA* for an integral outer membrane protein of *Escherichia coli* K-12 localization on hybrid plasmid PTU-100 and expression of a fragment of the gene. *Mol. Gen. Genet.* **179:** 13.

Bremer, E., S.T. Cole, I. Hindennach, U. Henning, E. Beck, C. Kurz, and H. Schaller. 1982. Export of a protein into the outer membrane of *Escherichia coli* K12. *Eur. J. Biochem.* **122:** 223.

Casadaban, M.J. 1976. Transposition and fusion of the *lac* operon to selected promoters in *E. coli* using bacteriophage λ and Mu. *J. Mol. Biol.* **104:** 541.

Chang, C.N., G. Blobel, and P. Model. 1978. Detection of prokaryotic signal peptidase in an *Escherichia coli* membrane fraction—Endo-proteolytic cleavage of nascent F1 precoat protein. *Proc. Natl. Acad. Sci.* **75:** 361.

Chang, C.N., P. Model, and G. Blobel. 1979. Membrane biogenesis—Cotranslational integration of the bacteriophage F1 coat protein into an *Escherichia coli* membrane-fraction. *Proc. Natl. Acad. Sci.* **76:** 1251.

Chou, P.Y. and G.D. Fasman. 1978. Empirical predictions of protein conformation. *Annu. Rev. Biochem.* **47:** 251.

Chua, N.-H. and G.W. Schmidt. 1978. Posttranslational transport into intact chloroplasts of a precursor to the small subunit of ribulose 1,5-biphosphate carboxylase. *Proc. Natl. Acad. Sci.* **75:** 6110.

Davis, B.D. and P.C. Tai. 1980. Mechanism of protein secretion across membranes. *Nature* **283:** 433.

Di Rienzo, J.M., K. Nakamura, and M. Inouye. 1978. The outer membrane proteins of gram-negative bacteria biosynthesis assembly and functions. *Annu. Rev. Biochem.* **47:** 481.

Dobberstein, D., G. Blobel, and N.-H. Chua. 1977. *In vitro* synthesis and processing of a putative precursor for the small subunit of ribulose-1,5-biphosphate carboxylase of *Chlamydomonas reinhardtii. Proc. Natl. Acad. Sci.* **74:** 1082.

Ehring, R., K. Beyreuther, J.K. Wright, and P. Overath. 1980. *In vitro* and *in vivo* products of *E. coli* lactose permease gene are identical. *Nature* **283:** 537.

Emr, S.D. and P.J. Bassford, Jr. 1982. Localization and processing of outer membrane and periplasmic proteins in *Escherichia coli* strains harboring export-specific suppressor mutations. *J. Biol. Chem.* **257:** 5852.

Emr, S.D. and T.J. Silhavy. 1982. The molecular components of the signal sequence that function in the initiation of protein export. *J. Cell Biol.* **95:** 689.

————. 1983. Genetic evidence for the role of signal sequence secondary structure in protein secretion. *Proc. Natl. Acad. Sci.* **80:** 4599.

Emr, S.D., M.N. Hall, and T.J. Silhavy. 1980a. A mechanism of protein localization: The signal hypothesis and bacteria. *J. Cell Biol.* **86:** 701.

Emr, S.D., S. Hanley-Way, and T.J. Silhavy. 1981. Supressor mutations that restore export of a protein with a defective signal sequence. *Cell* **23:** 79.

Emr, S.D., J. Hedgpeth, J.M. Clement, T.J. Silhavy, and M. Hofnung. 1980b. Sequence analysis of mutations that prevent export of lambda receptor. An *Escherichia coli* outer membrane protein. *Nature* **285:** 82.

Fraser, T.H. and B.J. Bruce. 1978. Chicken ovalbumin is synthesized and secreted by *Escherichia coli*. *Proc. Natl. Acad. Sci.* **75:** 5936.

Hall, M.N. and M. Schwartz. 1982. Reconsidering the early steps of protein secretion. *Ann. Microbiol.* **133:** 123.

Hall, M.N. and T.J. Silhavy. 1981a. Genetic analysis of the major outer membrane proteins of *Escherichia coli*. *Annu. Rev. Genet.* **15:** 91.

————. 1981b. The *ompB* locus and the regulation of the major outer membrane porin proteins of *Escherichia coli* K-12. *J. Mol. Biol.* **146:** 23.

Hall, M.M., M. Schwartz, and T.J. Silhavy. 1982. Sequence information within the *lamB* gene is required for proper routing of the bacteriophage λ receptor protein to the outer membrane of *Escherichia coli* K-12. *J. Mol. Biol.* **156:** 93.

Hall, M.N., J. Gabay, M. Debarbouille, and M. Schwartz. 1983. A role for mRNA secondary structure in the control of translation initiation. *Nature* **295:** 616.

Hedgpeth, J., J.M. Clement, C. Marchal, D. Perrin, and M. Hofnung. 1980. DNA-sequence encoding the NH$_2$-terminal peptide involved in transport of lambda receptor. An *Escherichia coli* secondary protein. *Proc. Natl. Acad. Sci.* **77:** 2621.

Highfield, P.E. and R.J. Ellis. 1978. Synthesis and transport of the small subunit of chloroplast ribulose biophosphate carboxylase. *Nature* **271:** 420.

Hofnung, M., ed. 1982. *The maltose system as a tool in molecular biology.* Masson, Paris. Reprinted from *Ann. Microbiol.* **133:** 1.

Ichihara, S., M. Hussain, and S. Mizushima. 1981. Characterization of new membrane lipoproteins and their precursors of *Escherichia coli*. *J. Biol. Chem.* **256:** 3125.

Inouye, H., W. Barnes, and J. Beckwith. 1982. Signal sequence of alkaline phosphatase of *Escherichia coli*. *J. Bacteriol.* **149:** 434.

Inouye, M. and S. Halegoua. 1979. Secretion and membrane localization of proteins in *Escherichia coli*. *CRC Crit. Rev. Biochem.* **7:** 339.

Inouye, S., S. Wang, J. Sekizawa, S. Halegoua, and M. Inouye. 1977. Amino acid sequence for the peptide extension on the prolipoprotein of the *Escherichia coli* outer membrane. *Proc. Natl. Acad. Sci.* **74:** 1004.

Inouye, S., X. Soberon, T. Franceschini, K. Nakamura, K. Itakura, and M. Inouye. 1982. Role of positive charge on the amino-terminal region of the signal peptide in protein secretion across the membrane. *Proc. Natl. Acad. Sci.* **79:** 3438.

Ito, K. and J.R. Beckwith. 1981. Role of the mature protein sequence of maltose-binding protein in its secretion across the *E. coli* cytoplasmic membrane. *Cell* **25:** 143.

Ito, K., M. Wittekind, M. Nomura, A. Miura, K. Shiba, T. Yura, and H. Nashimoto. 1983. A temperature-sensitive mutant of *E. coli* exhibiting slow processing of exported proteins. *Cell* **32:** 789.

Jaskunas, S.R., A.M. Fallon, and M. Nomura. 1977a. Identification and organization of ribosomal protein genes of *Escherichia coli* carried by λ*fus*2 transducing phage. *J. Biol. Chem.* **252:** 7323.

Jaskunas, S.R., A.M. Fallon, M. Nomura, B.G. Williams, and F.R. Blattner. 1977b. Expression of ribosomal protein genes cloned in Charon vector phages and identification of their promoters. *J. Biol. Chem.* **252:** 7355.

Koshland, D. and D. Botstein. 1980. Secretion of beta-lactamase requires the carboxy end of the protein. *Cell* **20:** 749.

————. 1982. Evidence for posttranslational translocation of B-lactamase across the bacterial inner membrane. *Cell* **30:** 893.

Koshland, D., R.T. Sauer, and D. Botstein. 1982. Diverse effects of mutations in the signal sequence on the secretion of β-lactamase in *Salmonella typhimurium*. *Cell* **30:** 903.

Kreil, G. 1981. Transfer of proteins across membranes. *Annu. Rev. Biochem.* **50:** 317.

Lin, J.J.-C., H. Kanazawa, J. Ozols, and H.C. Wu. 1978. *Escherichia coli* mutant with an amino-acid alteration within signal sequence of outer membrane prolipoprotein. *Proc. Natl. Acad. Sci.* **75:** 4891.

Lodish, H.F. and J.E. Rothman. 1979. The assembly of cell membranes: The two sides of a biological membrane differ in structure and function. *Sci. Am.* **240:** 48.

Meyer, D.I. and B. Dobberstein. 1980. Identification and characterization of a membrane component essential for the translocation of nascent proteins across the membrane of the endoplasmic reticulum. *J. Cell Biol.* **87:** 503.

Meyer, D.I., E. Krause, and B. Dobberstein. 1982. Secretory protein translocating across membranes—The role of the "docking protein." *Nature* **297:** 647.

Michaelis, S. 1982. "Mutations which alter the signal sequence of alkaline phosphatase in *E. coli*." Ph.D. thesis. Harvard University, Cambridge, Massachusetts.

Michaelis, S. and J. Beckwith. 1982. Mechanism of incorporation of cell envelope proteins in *Escherichia coli*. *Annu. Rev. Microbiol.* **36:** 435.

Michaelis, S., L. Guarente, and J. Beckwith. 1983a. *In vitro* construction and characterization of *phoA-lacZ* gene fusions in *Escherichia coli*. *J. Bacteriol.* **154:** 356.

Michaelis, S., H. Inouye, D. Oliver, and J. Beckwith. 1983b. Mutations that alter the signal sequence of alkaline phosphatase in *Escherichia coli*. *J. Bacteriol.* **154:** 366.

Moreno, F., A.V. Fowler, M. Hall, T.J. Silhavy, I. Zabin, and M. Schwartz. 1980. A signal sequence is not sufficient to lead beta-galactosidase out of the cytoplasm. *Nature* **286:** 356.

Müller, M., I. Ibrahimé, C.N. Chang, P. Walter, and G. Blobel. 1982. A bacterial secretory protein requires signal recognition particle for translocation across mammalian endoplasmic reticulum. *J. Biol. Chem.* **257:** 60.

Neupert, W. and G. Schatz. 1981. How proteins are transported into mitochondria. *Trends Biochem. Sci.* **6:** 1.

Nielsen, J., G.H. Flemming, J. Hoppe, P. Friedl, and K. von Meyenburg. 1981. The nucleotide sequence of the *atp* genes coding for the F_0 subunits a,b,c and the F_1 subunit δ of the membrane-bound ATP synthase of *Escherichia coli*. *Mol. Gen. Genet.* **184:** 33.

Oliver, D.B. and J. Beckwith. 1981. *Escherichia coli* mutant pleiotropically defective in the export of secreted proteins. *Cell* **25:** 765.

————. 1982a. Identification of a new gene (*secA*) and gene product involved in the secretion of envelope proteins in *Escherichia coli*. *J. Bacteriol.* **150:** 686.

————. 1982b. Regulation of a membrane component required for protein secretion in *Escherichia coli*. *Cell* **30:** 311.

Oliver, D., C. Kumamoto, M. Quinlan, and J. Beckwith. 1982. Pleiotropic mutants affecting secretory apparatus of *Escherichia coli*. *Ann. Microbiol.* **133:** 105.

Osborn, M.J. and H.C.-P. Wu. 1980. Proteins of the outer membrane of gram-negative bacteria. *Annu. Rev. Microbiol.* **34:** 369.

Palade, G.E. 1975. Intracellular aspects of the process of protein synthesis. *Science* **189:** 347.

Post, L.E., A.E. Arfsten, G.R. Davis, and M. Nomura. 1980. DNA sequence of the promoter region for the A ribosomal protein operon in *Escherichia coli*. *J. Biol. Chem.* **255:** 4653.

Pratt, J.M., I.B. Holland, and B.G. Spratt. 1981. Precursor forms of penicillin-binding proteins 5 and 6 of *E. coli* cytoplasmic membrane. *Nature* **293:** 307.

Rogers, J., P. Early, C. Carter, K. Calame, M. Bond, L. Hood, and R. Wall. 1980. Two mRNAs with different 3′ ends encode membrane-bound and secreted forms of immunoglobulin μ chain. *Cell* **22:** 303.

Roggenkamp, R., B. Kustermann-Kuhn, and C. Hollenberg. 1981. Expression and processing of bacterial β-lactamase in the yeast *Saccharomyces cerevisiae*. *Proc. Natl. Acad. Sci.* **78:** 4466.

Rothman, J.E. and H.F. Lodish. 1977. Synchronized transmembrane insertion and glycosylation of the nascent membrane protein. *Nature* **269:** 775.

Sabatini, D.D., G. Kreibich, T. Morimoto, and M. Adesnik. 1982. Mechanisms for the incorporation of proteins in membranes and organelles. *J. Cell. Biol.* **92:** 1.

Schaller, H., E. Beck, and M. Takanami. 1978. Sequence and regulatory signals of the filamentous phage genome. In *The single-stranded DNA phages* (ed. D. Denhardt et al.), p. 139. Cold Spring Harbor Laboratory, Cold Spring Harbor, New York.

Schedl, P. and P. Primakoff. 1973. Mutants of *Escherichia coli* thermosensitive for the synthesis of transfer RNA. *Proc. Natl. Acad. Sci.* **70:** 2091.

Schmidt, G.W., A. Devillers-Thiery, H. Desruisseaux, G. Blobel, and N.-H. Chua. 1979. NH$_2$-terminal amino acid sequences of precursor and mature forms of the ribulose-1,5-biphosphate carboxylase small subunit from *Chlamydomonas reinhardtii*. *J. Cell Biol.* **83:** 615.

Schwartz, M., M. Roa, and M. Debarbouille. 1981. Mutations that affect *lamB* gene expression at a posttranscriptional level. *Proc. Natl. Acad. Sci.* **78:** 2937.

Shultz, J., T.J. Silhavy, M.L. Berman, N. Fiil, and S.D. Emr. 1982. A previously unidentified gene in the *spc* operon of *Escherichia coli* K-12 specifies a component of the protein export machinery. *Cell* **31:** 227.

Shuman, H.A., T.J. Silhavy, and J.R. Beckwith. 1980. Labeling of proteins with beta-galactosidase by gene fusion—Identification of a cytoplasmic membrane component of the *Escherichia coli* maltose-transport system. *J. Biol. Chem.* **255:** 168.

Silhavy, T.J. and J. Beckwith. 1983. Isolation and characterization of mutants of *Escherichia coli* K-12 affected in protein localization. *Methods Enzymol.* **97:** 11.

Silhavy, T.J., P.J. Bassford, Jr., and J.R. Beckwith. 1979. A genetic approach to the study of protein localization in *Escherichia coli*. In *Bacterial outer membranes: Biogenesis and functions* (ed. M. Inouye), p. 203. Wiley, New York.

Silhavy, T.J., M.J. Casadaban, H.A. Shuman, and J.R. Beckwith. 1976. Conversion of β-galactosidase to a membrane-bound state by gene fusion. *Proc. Natl. Acad. Sci.* **73:** 3423.

Silhavy, T.J., H.A. Shuman, J. Beckwith, and M. Schwartz. 1977. Use of gene fusions to study outer membrane protein localization in *Escherichia coli*. *Proc. Natl. Acad. Sci.* **74:** 5411.

Silver, P., C. Watts, and W. Wickner. 1981. Membrane assembly from purified components. I. Isolated M13 procoat does not require ribosomes or soluble proteins for processing by membranes. *Cell* **25:** 341.

Sugimoto, K., H. Sugisaki, T. Okamoto, and M. Takanami. 1977. Studies on bacteriophage fd DNA. IV. The sequence of messenger RNA for the major coat protein gene. *J. Mol. Biol.* **110:** 487.

Sutcliffe, J.G. 1978. Nucleotide sequence of the ampicillin resistance gene of *Escherichia coli* plasmid pBR322. *Proc. Natl. Acad. Sci.* **75:** 3737.

Talmadge, K., J. Brosius, and W. Gilbert. 1981. An internal signal sequence directs secretion and processing of proinsulin in bacteria. *Nature* **294:** 176.

Talmadge, K., S. Stahl, and W. Gilbert. 1980. Eukaryotic signal sequence transports insulin antigen in *Escherichia coli*. *Proc. Natl. Acad. Sci.* **77:** 3369.

Wandersman, C., F. Moreno, and M. Schwartz. 1980. Pleiotropic mutations rendering *Escherichia coli* K-12 resistant to bacteriophage TP-1. *J. Bacteriol.* **143:** 1374.

Wanner, B.L., A. Sarthy, and J. Beckwith. 1979. *Escherichia coli* pleiotropic mutant that reduces amounts of several periplasmic and outer membrane proteins. *J. Bacteriol.* **140:** 229.

Walter, P. and G. Blobel. 1980. Purification of a membrane-associated protein complex required for protein translocation across the endoplasmic-reticulum. *Proc. Natl. Acad. Sci.* **77:** 7112.

——— . 1981. Translocation of proteins across the endoplasmic reticulum. III. Signal recognition protein (SRP) causes signal sequence-dependent and site-specific arrest of chain elongation that is released by microsomal membranes. *J. Cell Biol.* **91:** 557.

Watts, C., P. Silver, and W. Wickner. 1981. Membrane assembly from purified components. 2. Assembly of M13 procoat into liposomes reconstituted with purified leader peptidase. *Cell* **25:** 347.

White, J. and A. Helenius. 1980. PH-dependent fusion between the Semliki-forest virus membrane and liposomes. *Proc. Natl. Acad. Sci.* **77:** 3273.

Wickner, W. 1979. Assembly of proteins into biological membranes—Membrane trigger hypothesis. *Annu. Rev. Biochem.* **48:** 23.

————— . 1980. Assembly of proteins into membranes. *Science* **210:** 861.

Wu, H.C. and J.J.-C. Lin. 1976. *Escherichia coli* mutants altered in murein lipoprotein. *J. Bacteriol.* **126:** 147.

Zilberstein, A., M.D. Snider, M. Porter, and H.F. Lodish. 1980. Mutants of vesicular stomatitis virus blocked at different stages in maturation of the viral glycoprotein. *Cell* **21:** 417.

Membrane-mediated Regulation of Gene Expression in Bacteria

Wolfgang Epstein
Department of Biochemistry
University of Chicago
Chicago, Illinois 60637

The final step in the regulation of gene expression in prokaryotes is a cytoplasmic event, since transcription and translation occur inside the cells and there is no nuclear membrane. Most examples of genetic regulation involve effector molecules which, after transport from the outside, are sensed inside the cell by soluble cytoplasmic regulatory proteins. However, a number of exceptions to this totally internal and soluble scheme are known in which membrane proteins participate in gene regulation. In certain cases, membranes can participate in regulation in response to effectors in the cytoplasm when one of the regulatory proteins is membrane-bound rather than free. Alternatively, regulation by a membrane protein that spans the cytoplasmic membrane can occur when the external concentration of an effector molecule is sensed to produce an intracellular regulatory signal. Sensing external effector is one way of controlling genes by a compound in the medium, where that compound is also a metabolic intermediate always present in the cell. External sensing of effectors occurs in many chemotactic responses of bacteria, a subject reviewed by Hazelbauer and Parkinson (this volume). Membranes have also been implicated where regulation is in response to water activity as reflected by osmotic forces. In one example, the osmotic pressure of the medium is sensed; in another, it is the turgor pressure, the difference in osmotic pressure between the inside and outside of the cell, that controls expression.

MEMBRANE PROTEINS AS INTERNAL REGULATORY SENSORS

Internal sensing of proline by a membrane-bound protein controls the enzymes of proline degradation in *Salmonella typhimurium*. Proline degradation requires the induction of two activities: a specific transport system coded by *putP* and the *putA*-coded bifunctional membrane-bound proline oxidase, which performs the two steps converting proline to glutamic acid (Menzel and Roth 1981b). These two genes are adjacent but are transcribed in opposite directions from a common controlling region (Ratzkin and Roth 1978; Menzel and Roth 1981a; Maloy and Roth 1982). Proline oxidase is also a repressor of the expression of *putP* and of its own gene *putA* (Menzel and Roth 1981a).

281

Insertions and other mutations in *putA* that presumably abolish all functions of proline oxidase express *putP* constitutively. The enzymatic and regulatory functions of proline oxidase are separable by mutation: Some point mutations retain proline oxidase activity but are consitutive for transport and oxidation, whereas others abolish proline oxidation but retain inducibility of *putP* expression. Constitutivity produced by *putA* mutations is recessive to *putA*$^+$ in *trans*, consistent with the properties of constitutive mutations in a gene coding for a repressor (Menzel and Roth 1981a). Proline oxidase is abundant in proline-grown cells, constituting about 2% of the protein in a strain diploid for the *put* region. This protein is membrane-associated but this association seems relatively weak, since many buffers release up to half of the activity in soluble form. Protein solubilized by nonionic detergents migrates as discrete monomer and dimer bands in electrophoretic gels not containing detergent and as a single band in detergent-free sucrose gradients (Menzel and Roth 1981b). The weak membrane association of proline oxidase suggests that it does not have extensive hydrophobic regions and therefore does not span the membrane to be exposed to the periplasmic space. Proline oxidase has been purified to homogeneity. The active form appears to be a dimer of M_r 130,000 subunits. The purified enzyme from *S. typhimurium* has low affinity for proline, with a K_m of 80 mM or 50 mM, depending on whether measured as a soluble or a membrane-associated enzyme (Menzel and Roth 1981c). In cytoplasmic vesicles of *Escherichia coli* prepared under conditions that stabilize the membrane-bound form, the K_m for proline is much lower, about 3 mM (Abrahamson et al. 1983).

Proline induces the *put* activities inside the cell, after accumulation by the *putP* transport system. The finding that *putP* mutants are not inducible for either *put* activity originally suggested that these mutants defined a regulatory gene. It is now clear that the induction defect is due to loss of transport. When a proline-containing dipeptide is provided, induction of proline oxidase is seen in *putP* mutants (Ratzkin et al. 1978). The peptide is transported by another transport system and cleaved in the cell to liberate proline. Similar results have been reported for the *putP* mutant of *E. coli* (Wood and Zadworny 1979). Apparently, to avoid induction by the low concentrations of proline always present to satisfy the needs of protein synthesis, the *put* system has a repressor with relatively low affinity for proline. Proline oxidase may have been recruited as repressor because its low affinity for proline made it well suited for this role. The affinity measured by biochemical assays of enzymatic activity can be presumed to be the affinity for the regulatory actions of the protein. The basal level of *put* activity could be due in part to the endogenous levels of proline, a possibility testable by measuring *put*-gene expression in a proline auxotroph during proline-limited growth in a chemostat. The *put* genes resemble other inducible catabolic activities in showing a dependence on the cAMP-*crp* system (Ratzkin et al. 1978).

Repression by a protein that is itself induced to very high levels raises a number of questions. Continued induction when proline oxidase levels are high indicates that virtually none of this very abundant protein (1% of protein in a haploid strain, equivalent to ~10,000 molecules/cell) can be in a repressing configuration. The low affinity for proline, on the other hand, implies that even at very high concentrations of proline a significant fraction of proline oxidase is not complexed with proline. There must be an efficient way to make essentially all proline oxidase unavailable to repress even when it is not saturated with proline. A model in which the nonrepressing state is rapidly attained in the presence of proline and converts more slowly to the repressing state in the absence of proline can explain these findings. Conversion to a nonrepressing state has been suggested to occur by sequestration of proline oxidase in the membrane, with only free enzyme able to repress (Menzel and Roth 1981a). This idea is supported by recent findings that oxygen or other electron acceptors are required for maximal induction by $putA^+$ strains but not in $putA$ mutants. Because the enzyme must interact with membrane-bound electron-transport components for function, this requirement suggests that full induction requires a membrane association (Maloy and Roth 1983). An essential feature of the model is slow dissociation of enzyme from the membrane in the absence of proline; if dissociation were rapid and dissociated enzyme repressed, full induction would never occur. Other models, such as ones in which the enzyme is always membrane-bound but converts only slowly from nonrepressing conformation, also fit the observations.

INDUCTION FROM WITHOUT

The best characterized example of the sensing of an external regulatory effector is the induction of Uhp, the hexose phosphate transport system of *E. coli* (for review, see Dietz 1976). Regulation of the analogous system in *S. typhimurium* seems to be very similar (Eidels et al. 1974). The Uhp transport system depends on a set of clustered genes and has transport specificity for hexose phosphates, some other sugar phosphates, and inorganic phosphate (Dietz 1976; Ezzell and Dobrogosz 1978). Induction is highly specific; only glucose-6-P and 2-deoxyglucose-6-P are good inducers. Several lines of evidence have implicated external glucose-6-P as inducer. In a mutant that accumulates high concentrations of glucose-6-P when given glucose, there is no induction of this transport system unless glucose-6-P is added to the medium (Dietz and Heppel 1971). The kinetics of induction indicate that transport is not required for induction (Winkler 1971). The K_m for transport is about 25 μM, much higher than the requirement for induction, which is about 1 μM (Shattuck-Eidens and Kadner 1981). External sensing of inducer was demonstrated in a *uhpT-lacZ* fusion strain where induction could be measured in the absence of transport. Low concentrations of glucose-6-P were more effective in inducing this strain than a similar strain in which a functional *uhpT* gene was also present. The difference can be attributed to rapid reduction in external inducer concentration due to metabolism by the latter strain with a functional transport system (Shattuck-Eidens and Kadner 1981).

Mutants of the system (for review, see Dietz 1976) have been obtained in selections for loss of Uhp activity or in selections for constitutive expression of Uhp. Mutations that abolish activity are of two types, some reverting at low frequency to the wild type and others reverting at higher rates but becoming constitutive for Uhp. A recent analysis of plasmids carrying *uhp* genes and of insertion and point mutations that abolish Uhp activity identifies three genes in the order *uhpT*, *uhpR*, and *uhpA* (Kadner and Shattuck-Eidens 1983; Shattuck-Eidens and Kadner 1983). The *uhpT* gene is probably a structural gene for Uhp transport, because transcriptional *lacZ-uhpT* fusions in a strain also carrying a wild-type *uhp* region in *trans* express β-galactosidase coordinately with Uhp transport activity (Shattuck-Eidens and Kadner 1981). Identification of *uhpT* as a structural gene is consistent with the low reversion rate of *uhpT* mutations, their reversion to the wild-type inducible phenotype, and the loss of the insertion in the infrequent revertants of *uhpT*::Tn*10* insertion mutations. Whether this is the only structural gene for Uhp activity is not known. Mutations in *uhpR* are characterized by high rates of reversion to constitutive Uhp expression. Insertion mutations revert by acquiring a nearby suppressor mutation without losing their insertion, indicating that the requirement for *uhpR* function is readily bypassed by mutation. The *uhpA* gene appears to be essential for Uhp expression; insertion mutations revert only by loss of the insertion, and the revertants are inducible in the majority. The suggestion that *uhpA* codes for a positive regulator is consistent with constitutive expression of Uhp from plasmid and chromosomal *uhp* regions in strains with a multicopy plasmid carrying all of the *uhp* genes. Expression is inducible if the plasmid carries other *uhp* genes, except *uhpA*. This result, attributable to a gene dosage effect for *uhpA*, contrasts with the virtual absence of a gene dosage effect for *uhpT*. Gene dosage effects are commonly seen for transport genes (Rhoads et al. 1976; Teather et al. 1980), so the lack of such an effect for *uhpT* is most plausibly ascribed to some sort of autoregulation.

Consideration of models for Uhp regulation is somewhat premature, because the transcriptional organization of these genes has not been delineated, and the sites of mutations that produce constitutive expression have not been determined. These mutations, whether isolated directly from the wild type or as revertants of *uhpR* mutations, could define yet more *uhp* genes. None of the models for this type of control fit Uhp well, but one suggested by Dills et al. (1980) is a useful point of departure. This model postulates a membrane-spanning protein (M), which changes the conformation of its cytoplasmic portion in response to binding inducer outside the cell. This conformational change is postulated to alter binding of a second protein (Reg), which regulates expression at the DNA level. The amount of M is in excess of the amount of Reg, so that all Reg in the cell can be sequestered by binding to M. For a system where Reg is a positive regulatory protein, M binds Reg tightly in the absence of inducer and is released when inducer binds. For a system under negative control the situation is reversed, with M gaining high affinity to sequester Reg only when inducer is present. The properties of the *uhpA* mutants and effects when the gene is on a multicopy plasmid suggest that *uhpA* product is a positive regulator, the Reg component in this model. This leaves the *uhpR* gene to code for the M component. However, the properties of insertion mutations in this gene do not readily fit the expectations of this model. In a system with a positive regulator, null mutations in the gene for M should generally be constitutive, yet insertion mutations in *uhpR* are obtained rather frequently in a selection for the Uhp⁻ phenotype. The genetic analysis suggests that regulation is more complex than this simple model proposes, perhaps including both positive and negative control proteins. Additional information, including mapping of the mutations that produce constitutive expression and determination of which gene codes for the inducer-sensing function, will be useful in constructing a more specific model for the regulation of this system.

OSMOTIC CONTROL OF GENE EXPRESSION

The relative amounts of two major outer membrane proteins of *E. coli* K12 are regulated by medium osmolarity. These proteins, coded by the *ompC* and *ompF* genes, are called porins because they form aqueous pores allowing polar molecules to cross the outer membrane barrier (for reviews, see Inouye 1979; Nikaido and Nakae 1979; Osborn and Wu 1980). In media of low osmolarity, such as nutrient broth, the *ompF* porin predominates; in media of high osmolarity, such as Luria broth and tryptone soy broth (both contain ~0.1 M NaCl in addition to nutrients), or when osmolarity is increased by the addition of sugars, the *ompC* porin is more abundant (Hasegawa et al. 1976; Lugtenberg et al. 1976; Bassford et al. 1977; van Alphen and Lugtenberg 1977). Although the relative amounts of these proteins vary, their sum remains relatively constant. These two porins form pores of similar diameter that exclude molecules larger than about 600 daltons, but the *ompF* porin appears to mediate much higher rates of transport (Nikaido 1981). Thus, loss of the *ompF* porin results in a markedly reduced rate of outer membrane transport for nucleotides (Beacham et al. 1977; van Alphen et al. 1978a) and for a number of antibiotics (Bavoil and Nikaido 1977; Chopra and Eccles 1978; Pugsley and Schnaitman 1978; van Alphen et al. 1978b). Each of these two porins also serves as the receptor for specific bacteriophages (Datta et al. 1977; Hantke 1978). Selections for resistance to the appropriate phage or antibiotic yield mutations in the structural genes for these porins and in a regulatory locus called *ompB*.

The *ompB* locus is at minute 74, far from the structural gene for *ompC* at minute 47 or that for *ompF* at minute 21 of the *E. coli* map (Bachmann and Low 1980). Mutations in the *ompB* region were shown by Hall and Silhavy

(1981a) to define two genes called *ompR* and *envZ*. Both genes have been cloned, their products identified, and their sequences determined (Taylor et al. 1981; Mizuno et al. 1982a,b; Wurtzel et al. 1982). The genes constitute an operon with transcription proceeding from *ompR* to *envZ*. The product of *ompR* is a 32,500-dalton soluble protein; the *envZ* product is a membrane-bound protein of 44,000 daltons. The difficulty in identifying the *envZ* product (Mizuno et al. 1982b) suggests *envZ* is expressed at a much lower level than is *ompR*.

The *ompR*- and *envZ*-gene products regulate porin synthesis at the level of transcription as shown by studies of strains with transcriptional *omp::lacZ* fusions (Hall and Silhavy 1979, 1981a). Regulatory mutations with all possible phenotypes have been obtained. Mutations that do not make significant amounts of either porin (C^-F^-), as well as ones leading to synthesis of only the *ompF* porin (C^-F^+) are found in *ompR* (Hall and Silhavy 1981b). Mutations in *envZ* are associated with the C^+F^-, C^-F^+, or C^-F^- phenotypes (Hall and Silhavy 1981b; Garrett et al. 1983; Taylor et al. 1983). Insertion mutations in *ompR* and nonsense mutations in *ompR* or in *envZ* are associated with the C^-F^- phenotype, indicating that this is probably the null phenotype and that both gene products are required for expression of both porin genes (Garrett et al. 1983). Polar effects of nonsense and insertion mutations in *ompR* or *envZ* may contribute to the phenotype of such mutations. Mutations in *envZ* do not bypass the requirement for *ompR*; a double mutant combining *envZ3* (C^-F^+) and an *ompR* mutation has the C^-F^- phenotype (Taylor et al. 1983).

The sole effect of null mutations in *envZ* appears to be the reduction of *ompC* and *ompF* expression to low basal levels (Garrett et al. 1983), but other mutations in or near *envZ* that generally have a C^+F^- phenotype reduce expression of a number of other outer membrane and periplasmic proteins (Wanner et al. 1979; Wandersman et al. 1980; Lundrigen and Earhart 1981; Cavard et al. 1982). Procaine and phenethyl alcohol alter expression of a number of envelope proteins (Gayda et al. 1979; Halegoua and Inouye 1979; Lazdunski et al. 1979; Pugsley et al. 1980; Pages and Lazdunski 1981), an effect dependent on osmolarity (Pages and Lazdunski 1982) and apparently mediated by the *envZ* product (Taylor et al. 1983). Regulation by the *ompB* proteins generally appears to be at the level of transcription, but posttranscriptional control was reported for the effect of *perA*, a presumed *envZ* allele, on alkaline phosphatase (Wanner et al. 1979). Posttranscriptional control has been reported as the mechanism underlying the dependence of *ompC*, *ompF*, and *lamB* porin synthesis on concomitant lipid synthesis (Bocquet-Pages et al. 1981; Murgier et al. 1982). Whether this control is related to *ompB*-mediated osmotic regulation of porins is not known.

The regulation of porins may be related to that for other envelope components, which varies with the osmolarity of the medium. High osmolarity is associated with a modest reduction of lipopolysaccharide content on a dry weight basis (van Alphen and Lugtenberg 1977) and a marked reduction in the content of MDO, an anionic oligosaccharide associated with the cell envelope (Kennedy 1982). MDO regulation differs from that of the porins. Recent work indicates that *envZ* mutations do not alter MDO regulation and that control of MDO by osmolarity is a direct effect on enzyme activity (E.P. Kennedy, pers. comm.) Earlier results implied that osmolarity controlled the amount of enzyme(s) present (Kennedy 1982).

Steady-state measurements of the relationship of osmolarity to porin synthesis indicate that osmotic pressure exerted across the total cell envelope (inner and outer membranes) is much more effective than is osmotic pressure exerted only across the inner membrane. Kawaji et al. (1979) showed that sugars too large to cross the outer membrane are about three times as potent on a molar basis as are sugars that cross the outer membrane. Kinetic analysis of porin synthesis after a shift to a medium of different osmolarity showed a transient response in which one porin was made at a high rate and the other was apparently not synthesized at all; after about 1½ generations, under the particular conditions of the study, synthesis shifted to that characteristic of the steady state in the new medium (van Alphen and Lugtenberg 1977). These results show that the cells do not simply sense the osmolarity; they must be monitoring their outer membrane composition as well and are much more responsive to osmotic pressure across the outer membrane.

There is as yet no known adaptive advantage in osmotic regulation of porin synthesis to guide ideas about the mechanism of such control. The increased MDO synthesis in dilute media was suggested to increase osmolarity in the periplasmic space because the high concentration of fixed anions would generate a large Donnan potential and attract many counter cations (Kennedy 1982). Under these conditions the excess of cations over fixed anions contributes significantly to osmolarity; the ions associated with fixed anions have little effect because their osmotic effects are small (Alexandrowicz 1962).

The specific roles of the products of the *ompB* operon in porin regulation and how the regulatory determinants are sensed can only be vaguely guessed at. The genetic analysis of Hall and Silhavy (1981b) led them to suggest that regulation involved different forms of the *ompF* protein, perhaps a change from monomer to oligomer, and that this change was controlled by the *envZ* protein. This model readily explains the phenotypes of *ompR* mutations but needs to be modified to fit the more recent finding that expression of both porins requires the *envZ* product (T.J. Silhavy, pers. comm.). The *envZ* product must modify the *ompR* protein for the latter to be effective at all, with the extent of such modification or the relative amount of modified *ompR* product determining the relative amount of synthesis of each porin. The more difficult aspects—what is sensed and how—suggest questions rather than models at this point. Do *ompB* mutations that exhibit discernible osmotic regulation retain the high sensitivity to osmotic pressure across the outer membrane (Kawaji et al. 1979) and the unusual kinetics of porin synthesis upon a change in osmolarity (van Alphen and Lugtenberg 1977)? What physiological differences in cells growing in media of low, as compared to high, osmolarity can suggest an adaptive value to the shift in porins?

REGULATION BY TURGOR PRESSURE

A high-affinity potassium (K^+) transport system of *E. coli* called Kdp serves to scavenge this ion from media containing little K^+ (Rhoads et al. 1976). In media containing ample amounts of K^+, a separate, constitutive K^+ transport system also present in wild-type cells is adequate, and the Kdp system is not made. A study of such regulation, which at first glance appears to be repression by high concentrations of external K^+, showed that the determining factor was not the concentration of K^+ in the medium or in the cell but, rather, the ability of the cell to satisfy its need for K^+ (Laimins et al. 1981). In a transcriptional *kdp-lacZ* fusion strain with a wild-type constitutive K^+ transport system, expression of Kdp did not occur above about 3 mM K^+. When activity of the constitutive system in this strain was markedly reduced by mutations, Kdp expression occurred in media containing as much as 40 mM K^+. If the cells do not sense K^+ directly, they must sense it as satisfying some cellular role. K^+ is a major intracellular cation that contributes to internal osmolarity and is believed to regulate internal osmotic pressure to maintain a relatively fixed turgor pressure (Epstein and Schultz 1968). Thus, the need for K^+ could be sensed as a modest reduction in turgor pressure. This interpretation of Kdp regulation was supported by showing that reducing turgor pressure by a sudden increase in external osmolarity produced transient expression of β-galactosidase in the fusion strain (Laimins et al. 1981).

The Kdp system depends on the expression of the *kdp* operon, which codes for three inner membrane proteins that make up this transport system (Laimins et al. 1978; Rhoads et al. 1978). Adjacent to the operon is the *kdpD* gene, which codes for a positive regulator of the operon. These factors suggest a simple model in which a single protein, the *kdpD* product, senses turgor pressure and interacts with the promoter of the *kdp* operon. The tentative identification of the *kdpD* product as a 95,000-dalton membrane protein is consistent with this model. Recent work suggests that there may be an additional gene at the *kdpD* end of the region involved in regulating Kdp expression (J.E. Hesse, unpubl.). Better definition of the regulatory genes for Kdp expression and characterization of their products is needed to define the regulatory elements of the turgor pressure control of the Kdp system.

SUMMARY

The role of membranes in regulation is less an established fact than a promising area for investigation. The number of situations where membranes are suggested to mediate genetic regulation is large, but detailed investigations are few. In only a few cases, of which those considered above may be representative, is participation by membranes established or highly plausible. Where a membrane protein senses a small molecule effector, the sensing interaction is probably analogous to those mediated by soluble proteins. Sensing of osmotic forces is poorly understood; neither the location of such sensing, the protein involved, nor the mechanism of sensing are established. The unraveling of the molecular mechanisms underlying this regulatory role of membrane proteins promises to document one more example of the remarkable diversity of function found in organisms once thought to be simple.

ACKNOWLEDGMENTS

I thank Robert Kadner and Thomas Silhavy for providing preprints and unpublished data.

REFERENCES

Abrahamson, J.L.A., L.G. Baker, J.T. Stephenson, and J.M. Wood. 1983. Proline dehydrogenase from *Escherichia coli* K12. Properties of the membrane-associated enzyme. *Eur. J. Biochem.* **134:** 77.

Alexandrowicz, S. 1962. Osmotic and Donnan equilibriums in poly(acrylic acid)-sodium bromide solutions. *J. Polymer Sci.* **56:** 115.

Bachmann, G.J. and K.B. Low. 1980. Linkage map of *Escherichia coli* K-12, Edition 6. *Microbiol. Rev.* **44:** 1.

Bassford, P.J., Jr., D.L. Diedrich, C.A. Schnaitman, and P. Reeves. 1977. Outer membrane proteins of *Escherichia coli*. VI. Protein alteration in bacteriophage-resistant mutants. *J. Bacteriol.* **131:** 608.

Bavoil, P. and H. Nikaido. 1977. Pleiotropic transport mutants of *Escherichia coli* lack porin, a major outer membrane protein. *Mol. Gen. Genet.* **158:** 23.

Beacham, I.R., D. Haas, and E. Yagil. 1977. Mutants of *Escherichia coli* "cryptic" for certain periplasmic enzymes: Evidence for an alteration of the outer membrane. *J. Bacteriol.* **129:** 1034.

Bocquet-Pages, C., C. Lazdunski, and A. Lazdunski. 1981. Lipid-synthesis-dependent biosynthesis (or assembly) of major outer-membrane proteins of *Escherichia coli*. *Eur. J. Biochem.* **118:** 105.

Cavard, D., J.M. Pages, and C.J. Lazdunski. 1982. A protease as a possible sensor of environmental conditions in *E. coli* outer membrane. *Mol. Gen. Genet.* **188:** 508.

Chopra, I. and S.J. Eccles. 1978. Diffusion of tetracycline across the outer membrane of *Escherichia coli*: Involvement of protein 1a. *Biochem. Biophys. Res. Commun.* **58:** 550.

Datta, D.B., B. Arden, and U. Henning. 1977. Major proteins of the *Escherichia coli* outer cell envelope membrane as bacteriophage receptors. *J. Bacteriol.* **131:** 821.

Dietz, G.W., Jr. 1976. The hexose phosphate transport system of *Escherichia coli*. *Adv. Enzymol.* **46:** 237.

Dietz, G.W. and L.A. Heppel. 1971. Studies in the uptake of hexose phosphates. II. The induction of the glucose-6-phosphate transport system by exogenous but not by endogenously formed glucose-6-phosphate. *J. Biol. Chem.* **246:** 2885.

Dills, S.S., A. Apperson, M.R. Schmidt, and M.H. Saier, Jr. 1980. Carbohydrate transport in bacteria. *Microbiol. Rev.* **44:** 385.

Eidels, L., P.D. Rick, N.P. Stimler, and M.J. Osborn. 1974. Transport of D-arabinose-5-phosphate and D-seduloheptulose-7-phosphate by the hexose phosphate transport system of *Salmonella typhimurium*. *J. Bacteriol.* **119:** 138.

Epstein, W. and S.G. Schultz. 1968. Ion transport and osmoregulation in bacteria. In *Microbial protoplasts, spheroplasts and L-forms* (ed. L.B. Guze), p. 186. Williams and Wilkins, Baltimore, Maryland.

Ezzell, J.W. and W.J. Dobrogosz. 1978. Cyclic AMP regulation of the hexose phosphate transport system in *Escherichia coli*. *J. Bacteriol.* **133:** 1047.

Garrett, S., R.K. Taylor, and T.J. Silhavy. 1983. Isolation and characterization of chain-terminating nonsense mutations in a porin regulator gene, *envZ*. *J. Bacteriol.* **156:** 62.

Gayda, R.C., G.W. Henderson, and A. Markovitz. 1979. Neuroactive drugs inhibit trypsin and outer membrane protein processing in *Escherichia coli* K-12. *Proc. Natl. Acad. Sci.* **76:** 2138.

Halegoua, S. and M. Inouye. 1979. Translocation and assembly of outer membrane proteins of *Escherichia coli*. Selective accumulation of precursors and novel assembly of intermediates caused by phenethyl alcohol. *J. Mol. Biol.* **130:** 39.

Hall, M.N. and T.J. Silhavy. 1979. Transcriptional regulation of *Escherichia coli* K-12 major outer membrane protein 1b. *J. Bacteriol.* **140:** 342.

————— . 1981a. The *ompB* locus and the regulation of the major outer membrane porin proteins of *Escherichia coli*. *J. Mol. Biol.* **146:** 23.

————— . 1981b. Genetic analysis of the *ompB* locus in *Escherichia coli* K-12. *J. Mol. Biol.* **151:** 1.

Hantke, K. 1978. Major outer membrane proteins of *Escherichia coli* K-12 serve as receptors for the phage T2 (protein 1a) and 434 (protein 1b). *Mol. Gen. Genet.* **164:** 131.

Hasegawa, Y., H. Yamada, and S. Mizushima. 1976. Interactions of outer membrane proteins O-8 and O-9 with the peptidoglycan sacculus of *Escherichia coli* K-12. *J. Biochem.* **80:** 1401.

Inouye, M., ed. 1979. Bacterial outer membranes; biogenesis and function. John Wiley and Sons, New York.

Kadner, R.J. and D.M. Shattuck-Eidens. 1983. Genetic control of the hexose phosphate transport system of *Escherichia coli*: Mapping of deletion and insertion mutations in the *uhp* region. *J. Bacteriol.* **155:** 1052.

Kawaji, H., T. Mizuno, and S. Mizushima. 1979. Influence of molecular size and osmolarity of sugars and dextrans on the synthesis of outer membrane proteins O-8 and O-9 of *Escherichia coli* K-12. *J. Bacteriol.* **140:** 843.

Kennedy, E.P. 1982. Osmotic regulation and the biosynthesis of membrane-derived oligosaccharides in *Escherichia coli*. *Proc. Natl. Acad. Sci.* **79:** 1092.

Laimins, L.A., D.B. Rhoads, and W. Epstein. 1981. Osmotic control of *kdp* operon expression in *Escherichia coli*. *Proc. Natl. Acad. Sci.* **78:** 464.

Laimins, L.A., D.B. Rhoads, K. Altendorf, and W. Epstein. 1978. Identification of the structural proteins of an ATP-driven potassium transport system in *Escherichia coli*. *Proc. Natl. Acad. Sci.* **75:** 3216.

Lazdunski, C., D. Baty, and J.M. Pages. 1979. Procaine, a local anesthetic interacting with the cell membrane, inhibits the processing of precursor forms of periplasmic proteins in *Escherichia coli*. *Eur. J. Biochem.* **96:** 49.

Lugtenberg, B., R. Peters, H. Bernheimer, and W. Berendsen. 1976. Influence of cultural conditions and mutations on the composition of the outer membrane proteins of *Escherichia coli*. *Mol. Gen. Genet.* **147:** 251.

Lundrigan, M. and C.F. Earhart. 1981. Reduction in three iron-regulated outer membrane proteins and protein a by the *Escherichia coli* K-12 *perA* mutation. *J. Bacteriol.* **146:** 804.

Maloy, S. and J. Roth. 1983. Regulation of proline utilization in *Salmonella typhimurium*. Characterization of *put*::Mu d(Ap, *lac*) operon fusions. *J. Bacteriol.* **154:** 561.

Menzel, R. and J. Roth. 1981a. Regulation of the genes for proline utilization in *Salmonella typhimurium*: Autogenous repression by the *putA* gene product. *J. Mol. Biol.* **148:** 21.

————— . 1981b. Purification of the *putA* gene product. A bifunctional membrane-bound protein from *Salmonella typhimurium* responsible for the two-step oxidation of proline to glutamate. *J. Biol. Chem.* **256:** 9755.

—————— . 1981c. Enzymatic properties of the purified *putA* protein from *Salmonella typhimurium*. *J. Biol. Chem.* **256:** 9762.

Mizuno, T., E.T. Wurtzel, and M. Inouye. 1982a. Cloning of the regulatory genes (*ompR* and *envZ*) for the matrix proteins of the *Escherichia coli* outer membrane. *J. Bacteriol.* **150:** 1462.

—————— . 1982b. Osmoregulation of gene expression. II. DNA sequence of the *envZ* gene of the *ompB* operon of *Escherichia coli* and characterization of its gene product. *J. Biol. Chem.* **257:** 13692.

Murgier, M., C. Pages, C. Lazdunski, and A. Lazdunski. 1982. Translational control of *ompF*, *ompC* and *lamB* genetic expression during lipid synthesis inhibition of *Escherichia coli*. *FEMS Microbiol. Lett.* **13:** 307.

Nikaido, H. 1981. Outer membrane permeability of bacteria: Resistance and accessibility of targets. In *Beta-lactam antibiotics* (ed. M.R.J. Salton and G.D. Shockman), p. 249. Academic Press, New York.

Nikaido, H. and T. Nakae. 1979. The outer membrane of Gram-negative bacteria. *Adv. Microb. Physiol.* **20:** 163.

Osborn, M.J. and H.C.P. Wu. 1980. Proteins of the outer membrane of Gram-negative bacteria. *Annu. Rev. Microbiol.* **34:** 369.

Pages, J.-M. and C. Lazdunski. 1981. Action of phenyl alcohol on the processing of precursor forms of periplasmic proteins in *Escherichia coli*. *FEMS Mibrobiol. Lett.* **12:** 65.

—————— . 1982. Transcriptional regulation of *ompF* and *lamB* genetic expression by local anesthetics. *FEMS Microbiol. Lett.* **15:** 153.

Pugsley, A.P. and C.A. Schnaitman. 1978. Outer membrane proteins of *Escherichia coli*. VII. Evidence that bacteriophage-directed protein 2 functions as a pore. *J. Bacteriol.* **133:** 1181.

Pugsley, A.P., D.J. Conrad, C.A. Schnaitman, and T.I. Gregg. 1980. In vivo effects of local anesthetics on the production of major outer membrane proteins by *Escherichia coli*. *Biochim. Biophys. Acta* **599:** 1.

Ratzkin, B. and J. Roth. 1978. Cluster of genes controlling proline degradation in *Salmonella typhimurium*. *J. Bacteriol.* **133:** 744.

Ratzkin, B., M. Grabnar, and J. Roth. 1978. Regulation of the major proline permease gene of *Salmonella typhimurium*. *J. Bacteriol.* **133:** 737.

Rhoads, D.B., L. Laimins, and W. Epstein. 1978. Functional organization of the *kdp* genes of *Escherichia coli* K-12. *J. Bacteriol.* **135:** 445.

Rhoads, D.B., F.B. Waters, and W. Epstein. 1976. Cation transport in *Escherichia coli*. VIII. Potassium transport mutants. *J. Gen. Physiol.* **67:** 325.

Shattuck-Eidens, D.M. and R.J. Kadner. 1981. Exogenous induction of the *Escherichia coli* hexose phosphate transport system defined by *uhp-lac* operon fusions. *J. Bacteriol.* **148:** 203.

—————— . 1983. Molecular cloning of the *uhp* region and evidence for a positive activator for expression of the hexose phosphate transport system of *Escherichia coli*. *J. Bacteriol.* **155:** 1062.

Taylor, R.K., M.N. Hall, and T.J. Silhavy. 1983. Isolation and characterization of mutations altering expression of the major outer membrane porin proteins using the local anaesthetic procaine. *J. Mol. Biol.* **166:** 273.

Taylor, R.K., M.N. Hall, L. Enquist, and T.J. Silhavy. 1981. Identification of OmpR: A positive regulatory protein controlling expression of the major outer membrane matrix porin proteins of *Escherichia coli* K-12. *J. Bacteriol.* **147:** 255.

Teather, R.M., J. Bramhall, I. Riede, J.K. Wright, G. Aichele, U. Wilhelm, and P. Overath. 1980. Lactose carrier protein of *Escherichia coli*. Structure and expression of plasmids carrying the Y gene of the *lac* operon. *Eur. J. Biochem.* **108:** 223.

van Alphen, W. and B. Lugtenberg. 1977. Influence of osmolarity of the growth medium on the outer membrane protein pattern of *Escherichia coli*. *J. Bacteriol.* **131:** 623.

van Alphen, W., N. van Selm, and B. Lugtenberg. 1978a. Pores in the outer membrane of

Escherichia coli K-12. Involvement of proteins b and e in the functioning of pores for nucleotides. *Mol. Gen. Genet.* **159:** 75.

van Alphen, W., R. van Boxtel, N. van Selm, and B. Lugtenberg. 1978b. Pores in the outer membrane of *Escherichia coli* K-12. Involvement of proteins b and c in the permeation of cephaloridine and ampicillin. *FEMS Microbiol. Lett.* **3:** 103.

Wandersman, C., F. Moreno, and M. Schwartz. 1980. Pleiotropic mutations rendering *Escherichia coli* K-12 resistant to bacteriophage TP1. *J. Bacteriol.* **143:** 1374.

Wanner, B.L., A. Sarthy, and J. Beckwith. 1979. *Escherichia coli* pleiotropic mutant that reduces amounts of several periplasmic and outer membrane proteins. *J. Bacteriol.* **140:** 229.

Winkler, H.H. 1971. Kinetics of exogenous induction of the hexose-6-phosphate transport system of *Escherichia coli*. *J. Bacteriol.* **107:** 74.

Wood, J.M. and D. Zadworny. 1979. Characterization of an inducible porter required for L-proline catabolism by *Escherichia coli* K12. *Can. J. Biochem.* **57:** 1191.

Wurtzel, E.T., M.-Y. Chou, and M. Inouye. 1982. Osmoregulation of gene expression. I. DNA sequence of the *ompR* gene of the *ompB* operon of *Escherichia coli* and characterization of its gene product. *J. Biol. Chem.* **257:** 13685.

Bacterial Chemotaxis: Molecular Genetics of Sensory Transduction and Chemotactic Gene Expression

John S. Parkinson
Biology Department
University of Utah
Salt Lake City, Utah 84112

Gerald L. Hazelbauer
Biochemistry/Biophysics Program
Washington State University
Pullman, Washington 99164

INTRODUCTION

Motile bacteria exhibit locomoter responses to a variety of environmental stimuli. Chemotaxis in the enteric bacteria *Escherichia coli* and *Salmonella typhimurium* is the best-studied of these behaviors and is an excellent model system for investigating a wide range of problems in the areas of molecular genetics and sensory transduction. The many individual components of the chemotaxis machinery are produced and assembled in a controlled manner to create an integrated system that functions through a complex set of interactions. These features pose challenging problems in the realm of gene expression and function. Conversely, the existence of a defined sensory-motor system in organisms that are amenable to detailed genetic analysis provides a powerful approach to the study of sensory mechanisms. In this paper we consider some specific aspects of the study of motility and chemotaxis that illustrate these features. Bacterial chemotaxis is a popular subject for reviews, so the interested reader can find additional information and alternative viewpoints (Ordal 1980; Koshland 1981; Taylor and Lazlo 1981) or emphases on flagella and motility (Macnab 1980; Silverman 1980), genetics (Parkinson 1981), chemoreception (Hazelbauer and Parkinson 1977), or transducers and methylation (Springer et al. 1979; Boyd and Simon 1982; Hazelbauer and Harayama 1983).

CHEMOTACTIC BEHAVIOR

In uniform chemical environments, bacteria swim about in a three-dimensional random walk, produced by periods of smooth swimming punctuated by tumbling episodes that reorient the cell in a new, randomly chosen direction (Berg and Brown 1972). In a spatial gradient of an attractant or a repellent compound, the cell biases its random walk to achieve net migration in the favored direction (Berg and Brown 1972; Dalquist et al. 1972). To do this the organism monitors the concentrations of chemicals in its environment and alters tumbling probability whenever a temporal change in concentration is perceived (Berg and Brown 1972; Macnab and Koshland 1972). Favorable changes (attractant increases, repellent decreases) inhibit tumbling, and unfavorable ones enhance tumbling. Thus, chemotactic movements are carried out by effectively increasing the swimming time in favorable directions and decreasing the time spent heading in unfavorable directions.

Swimming and tumbling are produced by movements of the bacterial flagella, the structure and function of which are biologically unique and quite different from those of eukaryotic flagella. The bacterial flagellum consists of a long filament attached by a hook to a basal complex embedded in the cell membrane (DePamphilis and Adler 1971a,b). The filament is a moderately rigid left-handed helix of identical protein subunits, whereas the hook is thought to be much more flexible. The basal end comprises a rotary motor that turns the filament like a propeller (Berg and Anderson 1973; Silverman and Simon 1974a). Rotation is powered by proton motive force, not by ATP (Larsen et al. 1974a). Counterclockwise rotation of the left-handed helices results in a concerted pushing force by the several flagella of the cell producing smooth swimming (Macnab 1977). Clockwise rotation results in deformation of the filaments and uncoordinated pulling forces, producing tumbling (Macnab and Ornston 1977). Thus, the behavioral pattern of swimming and tumbling reflects an alternation between counterclockwise and clockwise rotation of the flagellar rotary motor.

The response of a bacterium to a change in its chemical environment is transient and occurs in two stages, termed excitation and adaptation. Excitation occurs upon detection of a change in the concentration of an active compound and results in a shift in the balance between counterclockwise and clockwise rotation of the flagellar motor (Larsen et al. 1974b). For example, addition of an attractant to a suspension of motile cells results in an immediate suppression of tumbling in response to the temporal gradient (Macnab and Koshland 1972; Segall et al. 1982). However, after a time ranging from seconds to several minutes, depending upon the compound and the magnitude of the gradient, the cells resume their original pattern of swimming and tumbling, even though the stimulating chemical is still present (Berg and Tedesco 1975; Spudich and Koshland 1975). Adaptation is defined as the reestablishment of the original pattern of flagellar rotation following a temporal stimulus. Thus, bacteria, like many sensory cells in higher organisms, adapt to the continued presence of a stimulus. The occurrence of adaptation ensures that the sensory system is functionally sensitive only to changes in the chemical environment and not to the absolute concentrations of attractant and repellent compounds.

CHEMOTAXIS MACHINERY

The chemotaxis machinery of *E. coli* comprises a network of signaling elements through which sensory information about the chemical environment is transmitted to the flagella. The complexity of this sensory system is indicated by the existence of approximately 50 genetic loci required either for motility or for chemotaxis in *E. coli*. Mutational alterations in these genes can lead to four basic types of chemotaxis defects, which presumably reflect the positions and roles of the gene products in the information pathway (Fig. 1).

Mutants that have normal motility and lack responses to only a few structurally related compounds define the sensory devices or chemoreceptors (Adler 1969) responsible for measuring chemical concentrations in the environment (Hazelbauer and Parkinson 1977). Mutants with specific defects in galactose (Hazelbauer and Adler 1971), maltose (Hazelbauer 1975), or ribose (Aksamit and Koshland 1974) taxis allowed identification of the periplasmic binding proteins for those three sugars as specific chemoreceptors. These binding proteins serve as the primary recognition devices both for uptake and for detection of chemotactic gradients. However, transport and metabolism are not essential for eliciting chemotactic responses to these sugars. It is the concentration of the sugar molecules themselves, not the energy they yield, that the chemoreceptors detect.

Mutants with normal motility that lack responses handled by a number of different receptors are considered defective in signaling or transducer function. Four transducer genes (*tar*, *tsr*, *tap*, and *trg*) have been identified in *E. coli*. Their products are integral membrane proteins that can be reversibly methylated by soluble chemotaxis-specific enzymes located inside the cell. The *tsr*- and *tar*-gene products also serve as the chemoreceptors for serine and aspartate, respectively, and are presumed to bind those compounds at the outer face of the cytoplasmic membrane. Thus, the transducers are transmembrane signal-

CHEMORECEPTION ---->	SENSORY TRANSDUCTION ---->	SWIMMING PATTERN ---->	MOTILITY
gatA (galactitol)		cheA	
guta (glucitol)		cheB	fla (24 genes)
malE (maltose)	tar	(cheC)	
mglB (galactose)	tsr	(cheD)	flb (4 genes)
mtlA (mannitol)	trg	cheR	
ptsF (fructose)	[tap]	(cheV)	motA, motB
ptsG (glucose)		cheW	
ptsM (mannose)		cheX	hag
rbsP (ribose)		cheY	
		cheZ	

Figure 1 Pathway of information processing by the chemotaxis machinery. ~50 different gene products are involved in chemotactic behavior. Alteration or loss of these components can result in any of four basic mutant phenotypes (see text). The severity of these defects reflects the position of the gene product in the information-handling pathway.

ing proteins. They play central roles in both the excitation and adaptation processes and are discussed in detail in the next section.

Generally, nonchemotactic or *che* mutants are unable to carry out chemotactic responses to any stimuli (Armstrong et al. 1967). Although motile, *che* mutants have aberrant swimming patterns due to an excessive counterclockwise or clockwise bias in flagellar rotation, suggesting that the *che* functions are involved in either the generation or regulation of spontaneous flagellar reversals (see Parkinson 1981). Many such mutants are still capable of responding to chemotactic stimuli but have altered response thresholds or adaptation behavior that preclude an effective response in spatial gradients. A few *che* mutants represent special kinds of transducer mutations (*cheD* [Parkinson 1980]) or flagellar mutations (*cheC* and *cheV* [Silverman and Simon 1973; Collins and Stocker 1976; Warrick et al. 1977; Parkinson 1981; Parkinson et al. 1983a]). The *cheR* and *cheB* mutants lack transducer-specific methyltransferase and methylesterase activities, respectively (see below). The other *che* loci (*cheA*, *cheW*, *cheY*, *cheZ*) produce cytoplasmic proteins whose roles are still not well understood (see below).

Mutations in at least 30 different genes can cause motility defects. Although many of these gene products have not yet been identified, it is generally assumed that they are structural components of the flagellar basal complex or proteins involved in the assembly of that complex. The expression of these genes is regulated in a complex manner (see below).

INFORMATION PROCESSING

Excitation

Changes in attractant or repellent concentration elicit flagellar responses in approximately 200 msec (Segall et al. 1982). In the case of serine and aspartate, the two most powerful attractants, the Tsr and Tar transducers serve as the gradient-sensing devices (Hedblom and Adler 1980; Wang and Koshland 1980). It seems likely that occupancy of a serine- or aspartate-binding site in these proteins causes a conformational change in the transducer, which in turn generates an excitatory signal of some sort. That signal affects the "gear shift" of the flagellar rotary motors to initiate an appropriate swimming response. Loss of ligand from occupied transducers upon a reduction in ligand concentration (a negative stimulus) must induce a conformational change that generates an excitatory signal of the opposite polarity to that generated by a positive stimulus. Ligand-occupied molecules of maltose receptor interact with the Tar protein (Koiwai and Hayashi 1979; Richarme 1982), presumably generating an excitatory change similar to that generated by direct binding of aspartate. The galactose and ribosome receptors interact with a different membrane protein, the product of *trg* (Kondoh et al. 1979; Hazelbauer et al. 1980; Koiwai et al. 1980), in an analogous manner. Sugars such as glucose, which are transported into the cell by the the sugar phosphotransferase system (PTS), are detected by the membrane-associated enzymes II of the PTS (Adler and Epstein 1974; Lengler et al. 1981). Metabolism of PTS sugars is not needed to elicit a chemotactic response; however, responses to PTS sugars, to oxygen and other electron acceptors, and to some repellents appear not to be mediated through any of the known transducer molecules, and these stimuli may influence the flagellar motor directly or through a pathway that does not involve a transducer molecule (Stock et al. 1981; Niwano and Taylor 1982; Pecher et al. 1983).

There is no information available about the character of the excitatory alteration undergone by transducers, nor do we understand the nature of the excitatory linkage between transducer and flagellar motor. Analysis of a membrane fraction enriched in areas surrounding the motors revealed no enrichment of transducer proteins, implying that the link is not direct physical interaction of transducer and motor (Engström and Hazelbauer 1982). Instead, it appears that occupied transducers generate an excitatory signal that passes through the cell from transducer to motor. A good candidate for the excitatory signal would be a small molecule whose intracellular concentration is controlled by the transducers. A number of observations suggest that such a compound is related metabolically or structurally to phosphorylated nucleosides (Aswad and Koshland 1974; Springer et al. 1975; Galloway and Taylor 1980; Kondoh 1980; Arai 1981; Stock et al. 1981; Shioi et al. 1982; Springer et al. 1982; Hazelbauer and Harayama 1983).

Adaptation

Adaptation to a stimulus that is processed through one of the known transducers is associated with a net change in the number of methyl groups attached to the transducer proteins engaged in excitatory signaling (Kort et al. 1975; Springer et al. 1977a,b). Adaptation to positive stimuli (attractant increases, repellent decreases) is linked to increased methylation, and adaptation to negative stimuli is linked to demethylation (Goy et al. 1977; Springer et al. 1979). Adaptation proceeds slowly at a constant rate over the time period between excitation and reappearance of the original behavioral pattern in the adapted state (Berg and Tedesco 1975). The time course of methylation changes exhibits the same pattern (Goy et al. 1977). Thus, it appears that the excitatory change induced by ligand or ligand-receptor binding to the transducer is effectively canceled out by the addition or removal of methyl groups on the transducer molecules.

Methyl groups are attached to transducer proteins by a specific methyltransferase, encoded by the *cheR* gene (Springer and Koshland 1977). This enzyme catalyzes transfer of a methyl group from *S*-adenosylmethionine to the carboxyl group of several glutamic acid residues in the transducer molecule. A specific methylesterase encoded by the *cheB* gene (Stock and Koshland 1978) catalyzes demethylation, producing methanol (Toews and Adler 1979) and a regenerated glutamyl residue. Mutations or physiological treatments that prevent either methylation or demethylation, e.g., mutations in *cheR* or *cheB* or starvation for methionine or *S*-adenosylmethionine, lead to adaptation defects (Armstrong 1972; Aswad and Koshland 1975; Springer et al. 1975, 1977a; Goy et al. 1978; Parkinson and Revello 1978; Rubik and Koshland 1978).

The spontaneous flagellar reversals exhibited by wild-type cells after undergoing adaptation or in the absence of chemotactic stimuli are crucial for chemotaxis. The mechanism responsible for generating these reversals may simply involve random fluctuations in the basal level(s) of transducer signal(s); however, the individual flagella on an unstimulated or adapted cell do not appear to reverse coordinately, as might be expected if they were regulated solely by a freely diffusable signal (Macnab 1983; Macnab and Han 1983). It seems likely that the cell possesses some other means for generating spontaneous flagellar reversals, and several of the Che proteins appear to play a role in this process. For example, *cheC* and *cheV* mutations can result in either a counterclockwise or clockwise flagellar bias (Parkinson et al. 1983a). These mutations represent special alterations of the *flaA* and *flaB* genes, respectively, the products of which are probably components of the flagellar switching mechanism. Moreover, reversion studies have demonstrated that both the *cheY*- and *cheZ*-gene products interact, probably directly, with the FlaA and FlaB proteins (Parkinson and Parker 1979; Parkinson 1981; Parkinson et al. 1983a). These interactions have opposing effects on the direction of flagellar rotation, and the competition between the CheY and CheZ proteins may be responsible for triggering spontaneous flagellar reversals.

What determines the steady-state level of transducer methylation? Since permanent changes in methylation state only take place on those transducer molecules engaged in signaling, it seems highly likely that excitation alters the substrate properties of transducers, either exposing or concealing potential methylation sites to the methylation-demethylation enzymes. In addition, however, chemotactic stimuli seem to exert some control over the activities of the enzymes themselves. For example, demethylation of all transducers ceases upon addition of attractant to cells and only resumes when adaptation is complete (Toews and Adler 1979). There is genetic evidence that this control may be mediated by a feedback loop from the flagellum (Parkinson 1981). The complementation behavior of *cheR*, *cheB*, *cheY*, and *cheZ* mutants indicates that CheY may modulate CheR (methyltransferase) activity and that CheZ protein may modulate CheB (methylesterase) activity (Parkinson 1976, 1978; Stock and Koshland 1978; DeFranco et al. 1979). These interactions may enable the cell to regulate methylation-demethylation activity as a function of the direction of flagellar rotation, because the CheY and CheZ proteins also interact with the flagellar switch, as outlined above.

TRANSDUCERS

Transducer molecules are central to the chemotactic mechanism, functioning in both excitation and adaptation. Intensive study of transducers utilizing primarily techniques of molecular genetics and molecular biology has made these proteins the most extensively characterized components of the bacterial sensory system. These studies have revealed biochemical complexities that are neither required nor anticipated by current models of transducer function.

Covalent Modifications of Transducers

The Tsr, Tar, and Trg proteins are subject to two different types of covalent modification, each of which occurs at more than one site on the molecule. One is carboxyl methylation of specific glutamyl residues to form carboxyl methyl esters as outlined previously. The chemical identity of the other modification has not been established directly, but evidence strongly supports the notion that it involves deamidation of glutamines to yield glutamyl residues. Both modifications were detected because they alter migration of the polypeptide chain in SDS-polyacrylamide gel electrophoresis (SDS-PAGE). Transducers with radioactive methyl groups appear on fluorographs of SDS gels as a series of bands in the region of M_r 60,000 (Silverman and Simon 1977a; Springer et al. 1977b). Investigation of the relationship among the multiple bands observed from a single gene product by stimulation-chase experiments (Engström and Hazelbauer 1980) and by analysis of [^{35}S]methionine-labeled protein synthesized from cloned transducer genes (Boyd and Simon 1980; Chelsky and Dahlquist 1980; DeFranco and Koshland 1980) led to the conclusion that the Tsr and Tar proteins were multiply methylated and that increased mobility in SDS-PAGE was correlated with increased numbers of methylated glutamyl residues on the transducer molecules. Later studies revealed the same pattern for the Trg protein (Harayama et al. 1982). The correlation between extent of methylation and increased electrophoretic mobility in SDS was confirmed by diagonal patterns of Tsr and Tar spots (Engström and Hazelbauer 1980; Hazelbauer and Engström 1981) observed on two-dimensional gels (O'Farrell 1975). Increased levels of carboxyl methylation, observed as unit shifts in the pI of a transducer polypeptide to less negative values, corresponded to positions of lower apparent molecular weight.

However, high-resolution, two-dimensional gels revealed that there were many more electrophoretically separable forms of a transducer protein than could be accounted for by multiple methylation (Figs. 2 and 3). In fact, transducers also exhibited multiple forms in methyltransferase mutants, implying that other features in addition to methylation created heterogeneity of the polypeptide chains (Hazelbauer and Engström 1981). Studies using cloned transducer genes and null mutations in the genes for the methyltransferase (*cheR*) and the demethylase (*cheB*) identified the additional feature as an

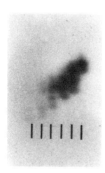

Figure 2   Multiple electrophoretic forms of the Tsr transducer protein visualized on two-dimensional gels. Fluorograph of a two-dimensional gel (isoelectric focusing gradient from basic [*left*] to acidic [*right*] and SDS-PAGE from higher [*top*] to lower [*bottom*] apparent molecular weight) of methyl-^{3}H-labeled Tsr protein. Only the region including the Tsr protein is shown. The apparent molecular weights of Tsr forms range from 55,000 to 65,000, and the pI, from 5.3 to 5.4. The six charge groups of spots focusing at different pIs are indicated by vertical lines. The protein is from unstimulated wild-type cells, and thus the most methylated forms of the protein (least acidic pI) are present in only minor amounts.

irreversible covalent modification that was dependent on an active *cheB* gene (Rollins and Dahlquist 1981; Sherris and Parkinson 1981). The modification affected the charge and electrophoretic migration of transducer polypeptides in the same manner as demethylation, creating a more negative pI and a higher apparent molecular weight, even though the modification occurred on proteins that were totally unmethylated. Analysis by high-performance liquid chromatography of a tryptic peptide of the Trg protein that contains *cheB*-dependent modification sites strongly suggests a glutamine-to-glutamate reaction (M. Kehry et al., in prep.). The modified residue has a pK between pH 2.2 and pH 5.8, and the relative elution position of the modified and unmodified peptides

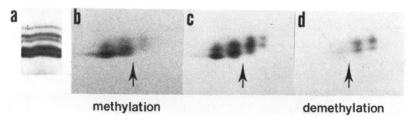

methylation   demethylation

Figure 3   Multiple electrophoretic forms of the Tar transducer protein. (*a*) Pattern of bands on a one-dimensional SDS-PAGE (11% low cross-linking gel) containing methyl-^{3}H-labeled Tar protein. The sample was methyl-labeled cells containing a pBR322 hybrid plasmid carrying the *tar* gene (Harayama et al. 1982). (*b*–*d*) Pattern of spots on two-dimensional gels after adaptation to positive or negative stimuli. These are fluorographs as in Fig. 2, except that the region shown is the one including the Tar protein. The Tar forms focus at pIs between 5.3 and 5.4. (↑) The center of the four charge groups. The attractant stimulus (*b*) was the addition of α-methylaspartate at a concentration (10 mM) sufficient to saturate the Tar receptor site. The repellent stimulus (*d*) was the addition of a maximally stimulating concentration (5 mM) of NiCl. The fluorograph in *b* is a shorter exposure than the other two to allow resolution of individual spots. With equal exposures, the total intensity of spots for the attractant-stimulated condition (*b*) is ~50% greater than for the unstimulated condition (*c*). (Adapted from Hazelbauer et al. 1982.)

at the two pH values mimics the relative positions of phenylthiohydantoin derivatives of glutamic acid and glutamine. The identity of the modification as deamidation of glutamine is further supported by studying a tryptic peptide, called K1, of the Tsr protein which, like the Trg peptide, contains sites for both carboxyl methylation and *cheB*-dependent modification (Kehry and Dahlquist 1982a,b). Three positions in the 23-residue peptide were identified as methyl-accepting sites (Kehry et al. 1983). This methionine- and lysine-containing peptide could be identified in the amino acid sequence of the Tsr protein deduced from the nucleotide sequence of the gene (Boyd et al. 1983a,b) (Fig. 4). The codon for the amino acid at one methyl-accepting position specifies glutamic acid, as would be expected, but the codons for the other two positions specify glutamine. If the two glutamines are converted to methyl-accepting glutamyl residues by the *cheB*-dependent modification, this accounts for the observation that only one position in the methionine-lysine peptide is methylated in a *cheB* mutant (Kehry and Dahlquist 1982b). Although it has not been shown directly that the demethylase encoded by *cheB* actually catalyzes the *cheB*-dependent modification, it is plausible that the single enzyme is both a deamidase and a methylesterase, since a variety of glutaminases are known to have significant methylesterase activity (Hartman 1971).

Transducer Structure

The regions of the transducer proteins that contain the methyl-accepting residues exhibit some tantalizing features (Boyd et al. 1983a,b). The methyl-accepting glutamyl is always the second of two adjacent glutamines. The three methyl-accepting sites identified on the 23-amino acid K1 peptide of the Tsr and Tar proteins are spaced 7 amino acids apart. With the exception of a glutamine following the first Glu-Glu pair and the final lysine, all the other

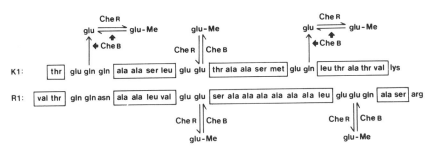

Figure 4   Amino acid sequences of Tsr tryptic peptides that contain methyl-accepting sites. The amino acid sequences were deduced from the nucleotide sequence of *tsr* (Boyd et al. 1983a,b). Hydrophobic residues are enclosed in boxes. The sites of *cheB*-dependent deamidation and methylation are indicated. See the text for a discussion of the evidence for those assignments.

amino acids of the K1 tryptic peptide are hydrophobic. If the region were an α-helix, then the modified residues would all extend on the same side of the helix along a line at a 20° angle to the axis of the helix. This structure might enable that portion of the transducer molecule to orient itself parallel to the face of the cytoplasmic membrane, with the glutamyl residues extending out of the inner face of the membrane and the rest of the helix embedded in the lipid matrix. The arginine-containing R1 peptide, which has one methyl-accepting site in Tar and two in Tsr, could assume a similar structure, although a Glu-Glu pair is one residue displaced from the in-line spacing.

The sequence of the 535 amino acids of the Tsr transducer, as deduced from the nucleotide sequence of the gene, suggests a model for the insertion of the protein in the cytoplasmic membrane (Boyd et al. 1983a,b). Good candidates for membrane-spanning sequences occur very near the amino terminus and in the region of residue 200. The K1 and R1 peptide sequences are in the regions of residues 300 and 500, respectively. These positions imply that the residues between the putative membrane-spanning region constitute an extracytoplasmic domain that forms the ligand-binding site, whereas the residues subsequent to the second membrane-spanning region form an intracellular domain containing sites for covalent modifications and presumably for generation of the excitatory change as well (Fig. 5).

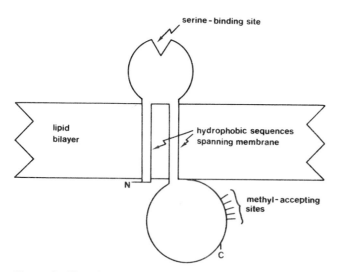

Figure 5 Hypothetical organization of the Tsr protein across the cytoplasmic membrane. The diagram is based on data and suggestions from Boyd et al. (1983a,b). It is not yet known whether the hydrophobic signal sequence at the aminoterminal end of the protein is present in the mature Tsr product; thus, it is possible that the amino end is not embedded in the membrane. See text for discussion.

Genetics of Transducers

Only a limited number of mutations in transducer genes have been characterized, yet there are indications of mutationally separable functions that may parallel the pattern of structural domains in the transducer proteins. Some mutations appear to affect only ligand recognition and thus are likely to be located in the aminoterminal portion of the gene coding for the extracytoplasmic domain. The Tsr protein is excited by changes in occupancy of a high-affinity ($K_d = 5$ μM) serine site (see above) but is also responsive to occupancy at a low-affinity ($K_d = 300$ μM) serine site (Hedblom and Adler 1980), to titration by hydrogen ion of an ionized group on a cytoplasmic domain of the protein (Kihara and Macnab 1981; Repaske and Adler 1981) and to changes in temperature that apparently influence the equilibrium between excited and adapted conformations of the transducer (Maeda and Imae 1979). Selection for mutants unable to respond to low concentrations of serine, but still responsive to repellents that perturb internal pH, produced two phenotypic classes: One type retained the ability to respond to high concentrations of serine and also had the corresponding membrane-associated low-affinity serine-binding sites (Hedblom and Adler 1980). Both classes were still sensitive to pH-perturbing repellents and to temperature (Y. Imae, pers. comm.), and adaptation to such stimuli was correlated with the usual changes in the methylation state of the Tsr protein. These phenotypes can be viewed, respectively, as a specific defect in the high-affinity serine recognition site and as a general defect in the structure of the extracytoplasmic ligand-binding domain. Mutations that appear to affect specific ligand-binding sites have also been identified in the other transducer genes. Some *tar* mutants are insensitive to maltose but respond normally to aspartate (M. Manson, pers. comm.). One *trg* mutant eliminates response to galactose but not ribose, whereas another *trg* mutant has the converse phenotype (Ball 1979), implying that the sites on the Trg protein for the two sugar receptors must be at least partially separate. Both types of mutants exhibit changes in Trg methylation state in response to the Trg-linked stimuli to which the cells are still sensitive (Ball 1979).

Among a collection of presumed missense mutations in *tsr* or *tar* (Parkinson et al. 1983b), approximately half contain mutant transducers that are not modified by either the CheB-dependent reaction or by the methyltransferase. These proteins are likely to have grossly altered conformations. Other *tsr* and *tar* mutants (Parkinson et al. 1983b), as well as some *trg* mutants (Ball 1979), contain transducers inactive in mediating tactic responses but able to serve as substrates for the two modification reactions. Some of these mutants may synthesize transducers that are defective specifically in the excitatory link between the extracytoplasmic and intracytoplasmic domains or in the excitatory change itself. A special class of mutations in *tsr* appears to create a transducer that is locked in the excitatory conformation independent of its degree of receptor site occupancy or methylation (Parkinson 1980). These dominant mutations create cells in which the flagellar motors rotate exclusively in the counterclockwise, smooth swimming direction, and thus no tactic response to any stimulus is observed. Because of this generally nonchemotactic phenotype, these mutants were named *cheD*. Surprisingly, all of the transducer molecules in *cheD* mutants appear to have abnormally high methylation levels (A. Boyd, pers. comm.; A. Callahan and J. Parkinson, unpubl.). If it is assumed that the Tsr protein in *cheD* mutants is locked in the excitatory (i.e., counterclockwise) conformation, the persistence of the excitatory signal might result in extensive methylation of all transducer molecules through feedback control from the flagella in a futile attempt to cancel out the excitation.

This model of *cheD* action is supported by the findings from reversion analyses (Parkinson 1981 and unpubl.). Mutations in the *cheB*, *cheZ*, *flaA*, or *flaB* genes, all of which increase clockwise rotational bias, can partially suppress *cheD* defects. This suppression is not allele-specific but rather seems to be due to restoration of a more normal swimming pattern through the opposing effects of the counterclockwise (*cheD*) and clockwise (suppressor) mutations. Thus, mutations that introduce a clockwise bias to the flagellar control system overcome the counterclockwise effects of the persistent excitatory input from Tsr and thereby allow stimuli handled by other transducers to once more elicit changes in flagellar rotation.

A Cryptic Transducer?

Analysis of *E. coli* DNA carrying the *tar* region has revealed a previously undetected gene, named *tap*, located between the *tar* and *cheR* loci (Boyd et al. 1981; Wang et al. 1982; Slocum and Parkinson 1983). Hybridization studies indicated considerable homology between the *tap* and *tsr* loci, and recent sequence data reveal a close relationship between *tap* and both *tsr* and *tar* (Boyd et al. 1983b). The *tap* locus codes for a protein of M_r 60,000 (Boyd et al. 1981), which appears as multiple bands on an SDS gel (Boyd et al. 1981; Slocum and Parkinson 1983), is methylated (Wang et al. 1982; Slocum and Parkinson 1983), and also contains sites for the CheB-dependent modification (Parkinson et al. 1983b). Thus, *tap* appears to code for a fourth transducer. However, it is not clear what stimulus responses, if any, are mediated by the Tap protein. On the one hand, strains of *E. coli* carrying nonpolar deletions in *tap* exhibit tactic responses indistinguishable from wild-type cells (Parkinson et al. 1983b; Slocum and Parkinson 1983). Moreover, *S. typhimurium*, whose chemotaxis machinery is otherwise very similar to that of *E. coli*, appears to lack a gene homologous to *tap* (D. Koshland, pers. comm.). On the other hand, Wang et al. (1982) have reported that a *tar-tap* deletion mutant containing a multicopy *tap* plasmid exhibited slight responses to aspartate and maltose, stimuli usually mediated by Tar, implying that Tap can be used to process Tar-type stimuli. However, other studies of *tap* function do not support this contention. For example, mutants containing nonpolar *tar* deletions (and therefore *tap*⁺) are unable to respond to aspartate or maltose stimuli (Slocum and Parkinson 1983). Furthermore *tar-tap* gene fusions, apparently generated by recombination between homologous sequences, exhibit varying degrees of chemotaxis to aspartate and maltose, depending upon the amount of *tar* deleted (Parkinson et al. 1983b). Finally, there is no more sequence homology between *tap* and *tar* than there is between any other transducer pairs (Boyd et al. 1983b). It seems that at present there is no compelling evidence for a functional role for the Tap protein in taxis by *E. coli*. The possibility remains, however, that this protein mediates responses to stimuli that have not yet been investigated.

EVOLUTIONARY CONSIDERATIONS

Receptors and Transducers

The genes for the galactose-, maltose-, and ribose-binding proteins are located in operons that contain other genes involved in the specific transport of each of these sugars. Expression of these operons is induced by their respective sugar substrates, and none are linked either physically or by the cascade control scheme to any flagellar or chemotaxis-related genes. Thus, these sugar-binding proteins appear to have originated as recognition components for their respective transport systems. Subsequently, they were recruited as chemoreceptors by the chemotaxis machinery, presumably through modification of existing transducer proteins to permit interaction with the periplasmic sugar-binding proteins. It is surprising that none of the other numerous periplasmic binding proteins, all of which function as transport components in an analogous manner, have been utilized as chemoreceptors.

This limited pattern of receptor recruitment contrasts sharply with the case for sugars transported by the PTS, in which all of the PTS substrates elicit chemotactic responses. In this case, however, tactic responses are dependent on the internal phosphate-transferring proteins, as well as the specific membrane-spanning enzymes II (Adler and Epstein 1974; Lengler et al. 1981), implying that the enzymes II are not chemoreceptors in the conventional sense but, rather, that the phosphorylation of the specific sugar substrate that occurs concomitant with transport perturbs pools of phosphorylated compounds, which are in turn related to the excitatory pathway between transducers and flagella (Lengler et al. 1981; Hazelbauer and Harayama 1983; Pecher et al. 1983).

Unlike the sugar receptors, the genes for the serine and aspartate receptors are clearly integral members of the chemotaxis and motility system. The Tsr and Tar proteins have no known role independent of chemotaxis, and deletion mutants lacking one or both of these proteins are wild type in all other respects examined (Parkinson et al. 1983b). Thus, these proteins may have evolved principally to serve as receptors for the amino acid attractants. The carboxyterminal two thirds of the *tsr*, *tar*, and *tap* gene sequences are strongly homologous to one another in a pattern that does not suggest sequential appearance of the three genes. Rather, they appear to have arisen from a common ancestral gene and to have evolved either through progressive divergence or gene fusions (Boyd et al. 1983b). The place of *trg* in this scheme is not clear at present. There is insufficient homology between *trg* and the other transducer genes to allow detectable hybridization in experiments in which 30% mismatch would be tolerated (J. Bollinger and G. Hazelbauer, unpubl.). However, an antiserum raised to Trg protein purified from cells wholly lacking the other transducers precipitates the Tsr and Tar proteins, as well as Trg (D. Nowlin et al., unpubl.). Perhaps the regions of the Trg protein that contain the sites for covalent modification are homologous in primary or three-dimensional structure to the analogous regions of other transducers (Kehry et al. 1983), and this common structure is recognized by the antiserum.

Recruitment of the maltose-binding protein as a chemoreceptor is easy to envision as resulting from an alteration that allowed interaction of ligand-occupied binding protein and the Tar transducer, which already functioned in the tactic system as a receptor for aspartate. Development of the linkage of galactose- and ribose-binding protein to Trg is more difficult to imagine, since there is no known receptor function performed directly by the Trg transducer, and thus there would be no functional reason for the transducer to exist prior to the creation of the linkage to binding protein. Perhaps the Trg protein originally had a receptor site that was lost.

Transducer Modification Reactions

It appears that transducer molecules can exist in either of two discrete signaling modes, which presumably reflect alternative conformational states, corresponding to counterclockwise and clockwise flagellar rotation. The rotational behavior of the flagella is thought to be determined by the relative numbers of transducer molecules in these two signaling states. Changes in transducer signaling properties can be effected by changes in receptor occupancy (during excitation) and by changes in methylation state (during adaptation). Although the effects of methylation-demethylation on transducer signaling are not yet understood in molecular terms, both physiological and genetic studies indicate that the addition of methyl groups favors clockwise rotation, whereas the removal of methyl groups favors counterclockwise rotation. Thus, the ratio of charged (i.e., unmethylated) to uncharged (i.e., methylated) residues at critical positions in the transducer molecules seems to be important in determining their signaling properties.

The recently discovered CheB-dependent modification reaction appears to mimic both the biochemical and functional consequences of transducer demethylation. As discussed earlier, CheB modification of transducer molecules most likely involves a deamidation of glutamine residues and results in the creation of new methyl-accepting glutamyl sites. Mutants defective in CheR (methyltransferase) function synthesize transducer molecules in which none of the potential methyl-accepting sites are methylated. Some of those sites are introduced directly as glutamate residues during transducer synthesis, whereas others are initially in the form of glutamines, which must undergo CheB modification. The signaling properties of unmethylated transducers are clearly altered by CheB modification. Deletion mutants lacking both CheR and CheB function exhibit a nearly normal balance between counterclockwise and clockwise rotational episodes, whereas *cheR* mutants, which are able to carry out CheB modification, exhibit a very severe counterclockwise rotational bias. Thus, CheB modification, like demethylation, generates charged sites in transducer molecules and favors the counterclockwise signaling mode.

The physiological role of CheB modification is puzzling, but its similarity to demethylation suggests that CheB-catalyzed transducer deamidation may have served as a rudimentary adaptation mechanism prior to the development of the methylation-demethylation cycle. We suggest that the present sensory system could have evolved in several stages, as outlined below.

The "primitive" sensory system probably enabled cells to detect and respond to chemical stimuli but with no provision for adaptation. An analogous situation is seen in *cheR-cheB* deletion mutants, which can respond to both positive and negative stimuli but continue to show stimulus-induced swimming behavior for as long as the stimulating chemical is present. Such cells have a limited, but nevertheless discernible, ability to move along spatial gradients, probably because their sensory system creates a bias toward persistence in favorable directions and causes reorientation upon entry into unfavorable regions, even though the lack of adaptation greatly limits the efficiency of migration. (For a discussion of chemotaxis without adaptation, see Block et al. [1982].)

The next step in the evolutionary process may have been the acquisition of an adaptation mechanism. Sensory adaptation would have enabled cells with a primitive sensory system to respond more efficiently to spatial gradients. Mutants lacking CheR function, but with CheB function, have a rudimentary adaptation system and illustrate how an adaptation mechanism might have been introduced into the bacterial sensory system. These cells increase their tumbling (clockwise) frequency in response to negative stimuli and then exhibit a gradual decrease in tumbling frequency that appears to be a type of adaptation (Parkinson and Revello 1978; Stock et al. 1981). Since no methyl esters exist on transducers in *cheR* cells, this adaptation cannot be due to demethylation. Rather, it appears that this adaptation could be mediated by CheB-dependent deamidation, whose rate is controlled by tactic stimuli in a manner analogous to the control of demethylation (Rollins and Dahlquist 1981; Sherris and Parkinson 1981). This adaptation mechanism is quite inefficient, as judged by the time course of recovery following stimulation. Moreover, since deamidation is irreversible, cells using such an adaptation system would have to replace deamidated transducers continually with newly synthesized, unmodified molecules to maintain their adaptation capability. An even more serious limitation is the inability of such cells to deal with—other than through new transducer synthesis—stimuli that suppress clockwise rotation. Deamidation favors counterclockwise rotation and could not be used to adapt to counterclockwise-enhancing stimuli.

The limitations of using deamidation as an adaptation mechanism could be circumvented by making the reaction reversible. Although amidation of glutamyl residues is not known to occur, methylation could have the same biochemical effect. Methylation of the glutamyl residue eliminates a negative charge just as addition of an amide would. Thus, it is plausible that development of a specific methyltransferase created functional reversibility of the CheB-mediated adaptation mechanism. Since deaminases are known to demethylate methyl esters effectively (Hartman 1971), the initial transducer-modifying enzyme (CheB) would then have been capable of serving not only to create methyl-accepting sites through deamidation of glutamine residues but also to regenerate glutamyl groups through demethylation. The advent of a methyltransferase activity represents the last and potentially the most important step in the evolution of the present adaptation system in bacterial chemotaxis. By creating a mechanism for reversibly modulating the signaling properties of the transducers, the cell became able to cope with both positive and negative stimuli, over large concentration ranges and under adverse physiological conditions that would preclude new transducer synthesis.

cheA LOCUS: OVERLAPPING GENES IN BACTERIA

Mutants defective in *cheA* function are motile but generally nonchemotactic due to an extreme counterclockwise bias in flagellar rotation (Parkinson 1976, 1978). However, they do show weak clockwise responses to strong repellent stimuli, such as a decrease in internal pH (Kihara and Macnab 1981). Moreover, many *cheA* mutants can be phenotypically suppressed by flagellar mutations that cause a clockwise rotational bias (Smith 1981). These observations indicate that CheA function is not essential for clockwise rotation but, rather, may be involved in facilitating clockwise rotation or in setting the cell's threshold for clockwise rotation.

The *cheA* locus is part of an operon containing several other chemotaxis-related genes (see Fig. 6): *cheW*, whose mutant phenotype is similar to that of *cheA*, and *motA* and *motB*, in which mutations lead to flagellar paralysis. The Mot proteins are generally believed to have a role in coupling proton motive force to flagellar rotation, and it is possible that *cheA* and *cheW* serve a similar function specifically for clockwise rotation. It is noteworthy that the threshold membrane potential required for counterclockwise rotation is considerably less than for clockwise rotation, implying that the two rotational modes are mechanistically different (Khan and Macnab 1980).

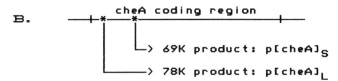

Figure 6 Genetic organization of the *cheA* locus. (*A*) The "mocha" operon (Silverman and Simon 1976). (*B*) The *cheA* locus contains two in-phase translational start sites (*) that are responsible for the synthesis of two similar proteins.

The ability of hybrid λ phages (Silverman and Simon 1977b; Silverman et al. 1977) and recombinant plasmids (Matsumura et al. 1977) carrying *E. coli* genetic material to complement *cheA* defects is correlated with the synthesis of two different proteins, designated p(*cheA*)$_L$ and p(*cheA*)$_S$, which have molecular weights in SDS-PAGE of 78,000 and 69,000, respectively (Smith and Parkinson 1980). Limited proteolysis studies demonstrated that the two proteins were quite similar in primary sequence, except that p(*cheA*)$_L$ yielded a few peptides not found in the smaller protein (Matsumura et al. 1977). Both *cheA* proteins have also been detected by two-dimensional gel analysis of total cell protein from strains devoid of plasmids and prophages (Smith 1981). In all cases, the two products seem to be present in roughly equal amounts. The *cheA* locus of *S. typhimurium* (originally called *cheP*) is functionally homologous to that of *E. coli* and exhibits essentially identical features (DeFranco et al. 1979; DeFranco and Koshland 1981).

The mechanism of expression of the two *cheA* proteins in *E. coli* has been deduced by examining the polypeptide products of different *cheA* nonsense mutations (Smith and Parkinson 1980). Nonsense mutations located in the promoter-distal three fourths of the *cheA*-coding sequence generally made shortened forms of both proteins, demonstrating that both are made from the same coding sequence and in the same reading frame. The size difference between the two nonsense fragments made by mutants of this sort was consistently about M_r 9000, the same size difference observed between p(*cheA*)$_L$ and p(*cheA*)$_S$ in wild-type cells. Thus, the size difference between the two *cheA* proteins must reside largely or entirely at their aminoterminal ends.

If the *cheA* products were related by some type of posttranslational processing, it seems unlikely that nonsense fragments of many different lengths would all be suitable substrates for the processing system. This explanation was convincingly laid to rest by studies of two nonsense mutations located in the promoter-proximal end of the *cheA*-coding sequence. Those mutations made a full-sized p($cheA$)$_S$, but no detectable p($cheA$)$_L$, indicating that p($cheA$)$_L$ and p($cheA$)$_S$ are translated from different initiation sites. Nonsense mutations capable of making p($cheA$)$_S$ presumably map upstream from the p($cheA$)$_S$ initiation site but downstream from the p($cheA$)$_L$ initiation site (see Fig. 6). Nonsense mutations that affect both proteins must map downstream from the p($cheA$)$_S$ startpoint, in the coding sequence common to both proteins.

Recent work with $cheA$-$lacZ$ gene fusions has lent additional support to the two-start model of $cheA$ translation (J. Parkinson and P. Talbert, unpubl.). Fusion of the carboxyterminal $cheA$-coding sequence to the aminoterminal-coding sequence of β-galactosidase leads to the synthesis of two hybrid proteins, differing by approximately 9000 in molecular weight. Nonsense mutations at the beginning of the $cheA$-coding region eliminate only the larger of the two fusion products, whereas nonsense mutations farther downstream in $cheA$ truncate both products.

Complementation studies indicate that there may be two different functions associated with the $cheA$ locus. Missense and nonsense mutations in the aminoterminal-coding sequence unique to p($cheA$)$_L$ complement missense mutations in the carboxyterminal-coding region common to both proteins (Smith and Parkinson 1980). Nonsense mutations in the common coding sequences are unable to complement any $cheA$ mutation. One explanation for this complementation pattern is that p($cheA$)$_L$ may be a bifunctional protein, having one activity associated with its amino terminus and another with its carboxyl terminus. Since this protein is clearly essential for chemotaxis and contains within it the entire primary sequence of p($cheA$)$_S$, the smaller $cheA$ product might not be essential for $cheA$ function. However, nonsense mutations that make only p($cheA$)$_S$ are able to complement some other $cheA$ defects, which implies that, at least under these conditions, the smaller protein can contribute some function. Although definitive evidence is still lacking, it seems probable that both $cheA$ proteins play a role in chemotaxis.

The $cheA$ locus is the first example of overlapping genes in bacteria. Although bacterial and animal viruses have devised a variety of strategies for producing functionally distinct proteins from the same coding sequence, those cases of overlapping genes are generally assumed to be a ploy for enhancing effective coding capacity in spite of the virion-limited constraints on genome size. The $cheA$ system should not be subject to stringent limits on overall genome size and must have evolved for different reasons. What those selective forces may have been remains very much a mystery.

REGULATION OF FLAGELLAR GENE EXPRESSION

Approximately 30 genes are involved in the synthesis, assembly, or function of flagella (see Fig. 1). Mutations in the fla and flb genes generally lead to a nonflagellate phenotype, although some fla and flb mutants do synthesize incomplete basal body structures (Suzuki et al. 1978). Many of these gene products have not yet been identified but are presumed to be required for flagellar assembly or to be essential flagellar structural components located in the basal body or surrounding membrane. Mutants defective in *mot* function possess morphologically normal flagella that are unable to rotate in either sense. The *motA* and *motB* products are inner membrane proteins that may play a role in coupling proton-motive force to power rotation of the flagellar motor. The flagellar hook is specified by the *flaK* gene, and the flagellar filament, by the *hag* gene. As an organelle, the bacterial flagellum rivals the ribosome in complexity, and, like the ribosome, flagellar synthesis is regulated in an elaborate manner. Although the molecular details of the regulation of synthesis and assembly of flagellar components are still poorly understood, this system promises to be a fertile source of novel and sophisticated control strategies, such as the one involved in flagellar phase variation discussed in the paper by Simon and Silverman (this volume). Below, we consider some of the other intriguing aspects of flagellar regulation.

Catabolite and Cell-cycle Control

The synthesis of flagella is strongly repressed by growth on glucose and similar sugars; addition of cAMP to the growth medium overcomes this catabolite repression. Mutants defective in the adenylcyclase (*cya*) or in the cAMP receptor protein (*cap*) gene are unable to synthesize flagella and contain no internal pools of hook protein or flagellin subunits. Silverman and Simon (1974b) isolated motile pseudorevertants from a *cap* mutant and showed that they contained suppressor mutations linked to the *flbB-flaI* operon. These *cfs* (constitutive *flagellar synthesis*) mutants were dominant in complementation tests but required a functional *flaI* gene *cis* to the *cfs* mutation to get flagellar synthesis under catabolite repression conditions or in *cya* or *cap* mutants. These findings indicate that the *flaI* and *flbB* genes may be the only flagellar genes under direct catabolite repression control and that they, in turn, could serve as a positive effector for the expression of all other flagellar genes. According to this model, *cfs* mutations would most likely represent cAMP- and cAMP receptor protein (CRP)-independent alteration of the *flbB-flaI* promoter.

  E. coli typically possesses five to eight flagella distributed randomly over the surface of the cell. Both the number and distribution of these flagella are maintained over a wide range in growth rates, implying that flagellar synthesis and assembly are somehow coupled to the cell-division cycle. At present, little is known about this coupling. Outer membrane structure may be involved because *galU* mutants, which synthesize an incomplete lipopolysaccharide, are defective in flagellar synthesis (Komeda et al. 1977). These mutants were also shown to synthesize reduced amounts of flagellin mRNA, implying that this control is mediated at the transcripional level. Motile pseudorevertants of *galU* strains carry suppressor mutations at the *flaH* locus, which surprisingly also seemed to have *cfs*-like effects on sensitivity to catabolite repression (Komeda et al. 1977).

Flagellar synthesis in *E. coli* is also blocked by growth at high temperatures. Motile strains that were able to form flagella at 42°C contained mutations linked to either the *flaI* or *flaD* locus. The *flaI* class may actually contain *cfs* mutations, because such mutants are somewhat motile at high temperature. However, the existence of temperature-insensitive *flaD* strains could mean that temperature sensitivity is the consequence of an interaction between the *flaD*- and *flaI*-gene products (Silverman and Simon 1977c).

Cascade Control of Flagellar Genes

All flagellar and chemotaxis-related genes are ultimately dependent on *flbB* and *flaI* function for expression. In addition, however, many other *fla* and *flb* mutants fail to synthesize immunologically detectable flagellin subunits, indicating that there are additional controls over *hag* gene expression. Komeda and Iino (1979) used transcriptional fusions of the *lacZ* gene to the *hag* promoter to investigate *hag* expression in various flagellar mutants. They found β-galactosidase activity, i.e., *hag* expression in *flaS*, *flaT*, *flaV*, and *flbC* mutants but no evidence of *hag* transcription in any other *fla* or *flb* mutant backgrounds.

Komeda (1982) has recently extended the *lacZ* fusion approach to study transcription from other *fla* and *flb* promoters. His findings demonstrate a hierarchy of transcriptional controls among the flagellar genes, as summarized in Figure 7. The "executive" genes *flaI* and *flbB* must function to express genes in groups 2 and 3. Group-3 genes are, in turn, needed for the expression of genes in groups 4 and 5. This scheme reveals a striking correspondence between the position of a gene product in the morphogenetic pathway and its

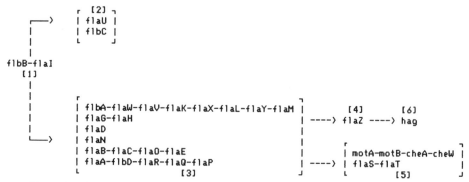

Figure 7   Cascade control of flagellar gene expression. This scheme is based on the results of Komeda (1982). The arrows connecting groups of flagellar genes indicate those functions that are needed for expression of other genes. For example, *hag* expression is dependent on *flaZ*, which is in turn dependent on all of the group-3 genes; these are in turn dependent on *flbB-flaI*. (Group numbers are indicated in brackets.)

position in the control hierarchy. For example, flagellin is the last component needed to complete the flagellum, and its structural gene, *hag*, is only expressed if the cell is able to make and assemble all components of the basal body-hook complex. Although the *flaE* gene is in group 3 (see Fig. 7), it is not in fact required for the expression of genes in groups 4, 5, or 6. It was placed in group 3 because the promoter for the *flaE* operon is only dependent on the genes in group 1 for expression. Unlike the other members of the operon, mutants defective in *flaE* function can assemble basal bodies but are unable to correctly terminate hook polymerization.

The type of cascade regulation observed in the flagellar system probably depends on both positive and negative control mechanisms. For example, to coordinate gene expression and morphogenesis, it seems likely that one or more of the basal body structural components also functions as a repressor of other flagellar promoters. Defects in basal body assembly would lead to an increased level of such a protein in the cytoplasm and consequent repression of subsequent genes in the morphogenetic pathway. This control strategy would only require that one or a few proteins serve as repressors. The bulk of the flagellar structural components would not need to have a dual role, because any assembly defect that led to a buildup of free basal body components, and thus to an accumulation of the critical protein, would exert the same regulatory effect. The group-2 gene products are obvious repressor candidates, because most other flagellar components can still be synthesized in their absence (see Fig. 7).

The executive genes *flbB* and *flaI* obviously serve as positive effectors of flagellar gene expression, but it is not yet clear whether they function directly at all flagellar promoters or whether in some cases they act indirectly through the cascade control network. For example, *flbB-flaI* function may directly activate group-2 and group-3 promoters, which in turn might allow expression of a gene product needed to activate promoters in group 4 and group 5. The *flaD*- and *flaG-flaH*-gene products seem likely candidates for secondary positive effectors ("junior executives") because of their involvement in catabolite, temperature, and *galU* effects.

CONCLUSIONS

The chemotaxis system of *E. coli* has, in recent years, afforded an intriguing view of the molecular events involved in stimulus detection, signaling, flagellar rotation, and sensory adaptation. In this paper we focused on only a few of these topics, primarily the role of the transmembrane proteins that constitute the heart (brain?) of the transduction system. These proteins are the prokaryotic counterparts of the acetylcholine receptor, and their study may prove just as important to our general understanding of sensory transduction processes. A different and somewhat neglected aspect of bacterial chemotaxis concerns the regulatory strategies needed to coordinate the synthesis, assembly, and interaction of over 50 different gene products, the stoichiometry of which must be carefully balanced to ensure optimum efficiency. We discussed only two of those control mechanisms, overlapping genes and cascade regulation, but others, particularly those operating at the translational level, are used as well. Clearly, the chemotaxis system in bacteria has just begun to yield its secrets. Its very complexity promises many more surprises over the coming years.

ACKNOWLEDGMENTS

Research performed in our laboratories was supported by grants GM-19559 and GM-28706 (to J.S.P) and GM-29963 (to G.L.H.) from the National Institutes of Health and by a McKnight Neuroscience Development Award (to G.L.H.).

REFERENCES

Adler, J. 1969. Chemoreceptors in bacteria. *Science* **166:** 1588.

Adler, J. and W. Epstein. 1974. Phosphotransferase-system enzymes as chemoreceptors for certain sugars in *Escherichia coli* chemotaxis. *Proc. Natl. Acad. Sci.* **71:** 2895.

Aksamit, R. and D.E. Koshland, Jr. 1974. Identification of the ribose binding protein as the receptor for ribose chemotaxis in *Salmonella typhimurium*. *Biochemistry* **13:** 4473.

Arai, T. 1981. Effect of arsenate on chemotactic behavior of *Escherichia coli*. *J. Bacteriol.* **145:** 803.

Armstrong, J.B. 1972. An S-adenosylmethionine requirement for chemotaxis in *Escherichia coli*. *Can. J. Microbiol.* **18:** 1695.

Armstrong, J.B., J. Adler, and M.M. Dahl. 1967. Nonchemotactic mutants of *Escherichia coli*. *J. Bacteriol.* **93:** 390.

Aswad, D. and D.E. Koshland, Jr. 1974. Role of methionine in bacterial chemotaxis. *J. Bacteriol.* **118:** 640.

————. 1975. Evidence for an S-adenosylmethionine requirement in the chemotactic behavior of *Salmonella typhimurium*. *J. Mol. Biol.* **97:** 207.

Ball, C.B. 1979. "Sensory transduction in *Escherichia coli*: Role of methyl-accepting proteins in chemotaxis to sugars." Ph.D. thesis, University of Wisconsin, Madison.

Berg, H.C. and R.A. Anderson. 1973. Bacteria swim by rotating their flagellar filaments. *Nature* **245:** 380.

Berg, H.C. and D.A. Brown. 1972. Chemotaxis in *Escherichia coli* analyzed by three-dimensional tracking. *Nature* **239:** 500.

Berg, H.C. and P.M. Tedesco. 1975. Transient response to chemotactic stimuli in *Escherichia coli*. *Proc. Natl. Acad. Sci.* **72:** 3235.

Block, S.M., J.E. Segall, and H.C. Berg. 1982. Impulse responses in bacterial chemotaxis. *Cell* **31:** 215.

Boyd, A. and M.I. Simon. 1980. Stimulus-elicited methylation generates multiple electrophoretic forms of methyl-accepting chemotaxis proteins in *Escherichia coli*. *J. Bacteriol.* **143:** 809.

————. 1982. Bacterial chemotaxis. *Annu. Rev. Physiol.* **44:** 501.

Boyd, A., K. Kendall, and M.I. Simon. 1983a. Structure of the serine chemoreceptor in *Escherichia coli*. *Nature* **301:** 623.

Boyd, A., A. Krikos, and M. Simon. 1981. Sensory transducers of *E. coli* are encoded by homologous genes. *Cell* **26:** 333.

Boyd, A., A. Krikos, N. Mutoh, and M. Simon. 1983b. A family of homologous genes encoding sensory transducers in *E. coli*. In *Mobility and recognition in cell biology* (ed. H. Sund and C. Veeger), p. 551. de Gruyter, Berlin.

Chelsky, D. and F.W. Dahlquist. 1980. Structural studies of methyl-accepting chemotaxis proteins of *Escherichia coli*: Evidence for multiple methylation sites. *Proc. Natl. Acad. Sci.* **77**: 2434.

Collins, A.L. and B.A.D. Stocker. 1976. *Salmonella typhimurium* mutants generally defective in chemotaxis. *J. Bacteriol.* **128**: 754.

Dahlquist, F.W., P. Lovely, and D.E. Koshland, Jr. 1972. Quantitative analysis of bacterial migration in chemotaxis. *Nat. New Biol.* **236**: 120.

DeFranco, A.L. and D.E. Koshland, Jr. 1980. Multiple methylation in the processing of sensory signals during bacterial chemotaxis. *Proc. Natl. Acad. Sci.* **77**: 2429.

—————. 1981. Molecular cloning of chemotaxis genes and overproduction of gene products in the bacterial sensing system. *J. Bacteriol.* **147**: 390.

DeFranco, A.L., J.S. Parkinson, and D.E. Koshland, Jr. 1979. Functional homology of chemotaxis genes in *Escherichia coli* and *Salmonella typhimurium*. *J. Bacteriol.* **139**: 107.

DePamphilis, M.L. and J. Adler. 1971a. Fine structure and isolation of the hook-basal body complex of flagella from *Escherichia coli* and *Bacillus subtilis*. *J. Bacteriol.* **105**: 384.

—————. 1971b. Attachment of flagellar basal bodies to the cell envelope: Specific attachment to the outer, lipopolysaccharide membrane and the cytoplasmic membrane. *J. Bacteriol.* **105**: 396.

Engström, P. and G.L. Hazelbauer. 1980. Multiple methylation of methyl-accepting chemotaxis proteins during adaptation of *E. coli* to chemical stimuli. *Cell* **20**: 165.

—————. 1982. Methyl-accepting chemotaxis proteins are distributed in the membrane independently from basal ends of bacterial flagella. *Biochim. Biophys. Acta* **686**: 19.

Galloway, R.J. and B.L. Taylor. 1980. Histidine starvation and adenosine 5'-triphosphate depletion in chemotaxis of *Salmonella typhimurium*. *J. Bacteriol.* **144**: 1068.

Goy, M.F., M.S. Springer, and J. Adler. 1977. Sensory transduction in *Escherichia coli*: Role of a protein methylation reaction in sensory adaptation. *Proc. Natl. Acad. Sci.* **74**: 4964.

—————. 1978. Failure of sensory adaptation in bacterial mutants that are defective in a protein methylation reaction. *Cell* **15**: 1231.

Harayama, S., P. Engstrom, H. Wolf-Watz, T. Iino, and G.L. Hazelbauer. 1982. Cloning of *trg*, a gene for a sensory transducer in *Escherichia coli*. *J. Bacteriol.* **152**: 372.

Hartman, S.C. 1971. Glutaminases and γ-glutamyltransferases. In *The enzymes* (ed. P.D. Boyer), p. 79. Academic Press, New York.

Hazelbauer, G.L. 1975. The maltose chemoreceptor of *Escherichia coli*. *J. Bacteriol.* **122**: 206.

Hazelbauer, G.L. and J. Adler. 1971. Role of the galactose binding protein in chemotaxis of *Escherichia coli* toward galactose. *Nat. New Biol.* **230**: 101.

Hazelbauer, G.L. and P. Engström. 1981. Multiple forms of methyl-accepting chemotaxis protein distinguished by a factor in addition to multiple methylation. *J. Bacteriol.* **145**: 35.

Hazelbauer, G.L. and S. Harayama. 1983. Sensory transduction in bacterial chemotaxis. *Int. Rev. Cytol.* **81**: 33.

Hazelbauer, G.L. and J.S. Parkinson. 1977. Bacterial chemotaxis. In *Receptors and recognition: Microbial interactions* (ed. J.L. Reissig), vol. 3, p. 60. Chapman and Hill, New York.

Hazelbauer, G.L. P. Engström, and S. Harayama. 1980. Methyl-accepting chemotaxis protein III and transducer gene *trg*. *J. Bacteriol.* **145**: 43.

—————. 1982. The complex nature of methyl-accepting chemotaxis proteins of enteric bacteria. In *Transmethylation* (ed. R.T. Borchardt, et al.), p. 83. Macmillan Press, London.

Hedblom, M.L. and J. Adler. 1980. Genetic and biochemical properties of *Escherichia coli* mutants with defects in serine chemotaxis. *J. Bacteriol.* **144**: 1048.

Kehry, M.R. and F.W. Dahlquist. 1982a. The methyl-accepting chemotaxis proteins of *E. coli*: Identification of the multiple methylation sites on MCPI. *J. Biol. Chem.* **257**: 10378.

—————. 1982b. Adaptation in bacterial chemotaxis: CheB-dependent modification permits additional methylations of sensory transducer proteins. *Cell* **29**: 761.

Kehry, M.R., F.W. Dahlquist, and M.W. Bond. 1983. Bacterial chemotaxis: The chemical properties of the cheB-dependent modification. In *Mobility and recognition in cell biology* (ed. H. Sund and C. Veeger), p.533. de Gruyter, Berlin.

Khan, S. and R.M. Macnab. 1980. The steady-state counterclockwise/clockwise ratio of bacterial flagellar motors is regulated by proton motive force. *J. Mol. Biol.* **138:** 563.

Kihara, M. and R.M. Macnab. 1981. Cytoplasmic pH taxis and weak-acid repellent taxis of bacteria. *J. Bacteriol.* **145:** 1209.

Koiwai, O. and H. Hayashi. 1979. Studies on bacterial chemotaxis. IV. Interaction of maltose receptor with a membrane-bound chemosensing component. *J. Biochem.* **86:** 27.

Koiwai, O., S. Minoshima, and H. Hayashi. 1980. Studies on bacterial chemotaxis. V. Possible involvement of four species of the methyl-accepting chemotaxis protein in chemotaxis of *Escherichia coli*. *J. Biochem.* **87:** 1365.

Komeda, Y. 1982. Fusion of flagellar operons to lactose genes on a Mu*lac* bacteriophage. *J. Bacteriol.* **150:** 16.

Komeda, Y. and T. Iino. 1979. Regulation of the expression of the flagellin gene (*hag*) in *Escherichia coli* K-12: Analysis of *hag-lac* gene fusions. *J. Bacteriol.* **139:** 721.

Komeda, Y., T. Icho, and T. Iino. 1977. Effects of *galU* mutation on flagellar formation in *Escherichia coli*. *J. Bacteriol.* **129:** 908.

Kondoh, H. 1980. Tumbling chemotaxis mutants of *Escherichia coli*: Possible gene-dependent effect of methionine starvation. *J. Bacteriol.* **142:** 527.

Kondoh, H., C.B. Ball, and J. Adler. 1979. Identification of a methyl-accepting chemotaxis protein for the ribose and galactose chemoreceptors of *Escherichia coli*. *Proc. Natl. Acad. Sci.* **76:** 260.

Kort, E.N., M.F. Goy, S.H. Larsen, and J. Adler. 1975. Methylation of a protein involved in bacterial chemotaxis. *Proc. Natl. Acad. Sci.* **72:** 3939.

Koshland, D.E., Jr. 1981. Biochemistry of sensing and adaptation in a simple bacterial system. *Annu. Rev. Biochem.* **50:** 765.

Larsen, S.H., J. Adler, J.J. Gargus, and R.W. Hogg. 1974a. Chemomechanical coupling without ATP: The source of energy for motility and chemotaxis in bacteria. *Proc. Natl. Acad. Sci.* **71:** 1239.

Larsen, S.H., R.W. Reader, E.N. Kort, W.-W. Tso, and J. Adler. 1974b. Change in direction of flagellar rotation is the basis of the chemotactic response in *Escherichia coli*. *Nature* **249:** 74.

Lengler, J., A.-M. Auburger. R. Mayer, and A. Pecher. 1981. The phosphoenolpyruvate-dependent carbohydrate: Phosphotransferase system enzymes II as chemoreceptors in chemotaxis of *Escherichia coli* K12. *Mol. Gen. Genet.* **183:** 163.

Macnab, R.M. 1977. Bacterial flagella rotating in bundles: A study in helical geometry. *Proc. Natl. Acad. Sci.* **74:** 221.

_____. 1980. Sensing the environment: Bacterial chemotaxis. In *Biological regulation and development* (ed. R. Goldberger), vol. 2, p. 377. Plenum Press, New York.

_____. 1983. Bacterial motility: Energization and switching of the flagellar motor. In *Mobility and recognition in cell biology* (ed. H. Sund and C. Veeger), p. 499. de Gruyter, Berlin.

Macnab, R.M. and D.P. Han. 1983. Asynchronous switching of flagellar motors on a single bacterial cell. *Cell* **32:** 109.

Macnab, R.M. and D.E. Koshland, Jr. 1972. The gradient-sensing mechanism in bacterial chemotaxis. *Proc. Natl. Acad. Sci.* **69:** 2509.

Macnab, R.M. and M.K. Ornston. 1977. Normal-to-curly flagellar transitions and their role in bacterial tumbling. Stabilization of an alternative quaternary structure by mechanical force. *J. Mol. Biol.* **112:** 1.

Maeda, K. and Y. Imae. 1979. Thermosensory transduction in *Escherichia coli*: Inhibition of the thermoresponse by L-serine. *Proc. Natl. Acad. Sci.* **76:** 91.

Matsumura, P., M. Silverman, and M. Simon. 1977. Synthesis of *mot* and *che* gene products of *Escherichia coli* programmed by hybrid ColE1 plasmids in minicells. *J. Bacteriol.* **132:** 996.

Niwano, M. and B.L. Taylor. 1982. Novel sensory adaptation mechanism in bacterial chemotaxis to oxygen and phosphotransferase substrates. *Proc. Natl. Acad. Sci.* **79:** 11.

O'Farrell, P.H. 1975. High resolution two-dimensional electrophoresis of proteins. *J. Biol. Chem.* **250:** 4007.

Ordal, G.W. 1980. Bacterial chemotaxis: A primitive sensory system. *Bioscience* **30:** 408.

Parkinson, J.S. 1976. *cheA*, *cheB* and *cheC* genes of *Escherichia coli* and their role in chemotaxis. *J. Bacteriol.* **126:** 758.,

―――――――. 1978. Complementation analysis and deletion mapping of *Escherichia coli* mutants defective in chemotaxis. *J. Bacteriol.* **135:** 45.

―――――――. 1980. Novel mutations affecting a signaling component for chemotaxis of *Escherichia coli*. *J. Bacteriol.* **142:** 953.

―――――――. 1981. Genetics of bacterial chemotaxis. *Symp. Soc. Gen. Microbiol.* **37:** 265.

Parkinson, J.S. and S.R. Parker. 1979. Interaction of the *cheC* and *cheZ* gene products is required for chemotactic behavior in *Escherichia coli*. *Proc. Natl. Acad. Sci.* **76:** 2390.

Parkinson, J.S. and P.T. Revello. 1978. Sensory adaptation mutants of *E. coli*. *Cell* **15:** 1221.

Parkinson, J.S., S.R. Parker. P.B. Talbert, and S.E. Houts. 1983a. Interactions between chemotaxis genes and flagellar genes in *Escherichia coli*. *J. Bacteriol.* **155:** 265.

Parkinson, J.S., M.K. Slocum, A.M. Callahan, D. Sherris, and S.E. Houts. 1983b. Genetics of transmembrane signaling proteins in *E. coli*. In *Mobility and recognition in cell biology* (ed. H. Sund and C. Veeger), p. 563. de Gruyter, Berlin.

Pecher, A., I. Renner, and J. Lengler. 1983. The phosphoenolpyruvate-dependent carbohydrate phosphotransferase system enzymes II, a new class of chemosensors in bacterial chemotaxis. In *Mobility and recognition in cell biology* (ed. H. Sund and C. Veeger), p. 517. de Gruyter, Berlin.

Repaske, D. and J. Adler. 1981. Change in intracellular pH of *Escherichia coli* mediates the chemotactic response to certain attractants and repellents. *J. Bacteriol.* **145:** 1196.

Richarme, G. 1982. Interaction of the maltose-binding protein with membrane vesicles of *Escherichia coli*. *J. Bacteriol.* **149:** 662.

Rollins, C. and F.W. Dahlquist. 1981. The methyl-accepting chemotaxis proteins of *E. coli*: A repellent-stimulated, covalent modification, distinct from methylation. *Cell* **25:** 333.

Rubik, B.A. and D.E. Koshland. Jr. 1978. Potentiation, desensitization, and inversion of response in bacterial sensing of chemical stimuli. *Proc. Natl. Acad. Sci.* **75:** 2820.

Segall, J.E., M.D. Manson, and H.C. Berg. 1982. Signal processing times in bacterial chemotaxis. *Nature* **296:** 855.

Sherris, D. and J.S. Parkinson. 1981. Posttranslational processing of methyl-accepting chemotaxis proteins in *Escherichia coli*. *Proc. Natl. Acad. Sci.* **78:** 6051.

Shioi, J.-I., R.J. Galloway, M. Niwano, R.E. Chinnock, and B.L. Taylor. 1982. Requirement of ATP in bacterial chemotaxis. *J. Biol. Chem.* **257:** 7969.

Silverman, M. 1980. Building bacterial flagella. *Q. Rev. Biol.* **55:** 395.

Silverman, M. and M. Simon. 1973. Genetic analysis of bacteriophage Mu-induced flagellar mutants in *Escherichia coli*. *J. Bacteriol.* **116:** 116.

―――――――. 1974a. Flagellar rotation and the mechanism of bacterial motility. *Nature* **259:** 73.

―――――――. 1974b. Characterization of *Escherichia coli* flagellar mutants that are insensitive to catabolite repression. *J. Bacteriol.* **120:** 1196.

―――――――. 1976. Operon controlling motility and chemotaxis in *E. coli*. *Nature* **264:** 577.

―――――――. 1977a. Chemotaxis in *Escherichia coli*: Methylation of *che* gene products. *Proc. Natl. Acad. Sci.* **74:** 3317.

―――――――. 1977b. Identification of polypeptides necessary for chemotaxis in *Escherichia coli*. *J. Bacteriol.* **130:** 1317.

―――――――. 1977c. Bacterial flagella. *Annu. Rev. Microbiol.* **31:** 397.

Silverman, M., P. Matsumura, M Hilmen, and M. Simon. 1977. Characterization of lambda-*Escherichia coli* hybrids carrying chemotaxis genes. *J. Bacteriol.* **130:** 877.

Slocum, M.K. and J.S. Parkinson. 1983. Genetics of methyl-accepting chemotaxis proteins in *Escherichia coli*: Organization of the *tar* region. *J. Bacteriol.* **155**: 565.

Smith, R.A. 1981. "Detailed analysis of a genetic locus that contains a pair of overlapping genes and is involved in bacterial chemotaxis." Ph.D. thesis, University of Utah, Salt Lake City.

Smith, R.A. and J.S. Parkinson. 1980. Overlapping genes at the *cheA* locus of *Escherichia coli*. *Proc. Natl. Acad. Sci.* **77**: 5370.

Springer, M., M.F. Goy, and J. Adler. 1977a. Sensory transduction in *Escherichia coli*: A requirement for methionine in sensory adaptation. *Proc. Natl. Acad. Sci.* **74**: 183.

—————— . 1977b. Sensory transduction in *Escherichia coli*: Two complementary pathways of information processing that involve methylated proteins. *Proc. Natl. Acad. Sci.* **74**: 3312.

—————— . 1979. Protein methylation in behavioural control mechanisms and in signal transduction. *Nature* **280**: 279.

Springer, M.S., B. Zenolari, and P.A. Pierzchala. 1982. Ordered methylation of the methyl-accepting chemotaxis proteins of *Escherichia coli*. *J. Biol. Chem.* **257**: 6861.

Springer, M.S., E.N. Kort, S.H. Larsen, G.O. Ordal, R.W. Reader, and J. Adler. 1975. Role of methionine in bacterial chemotaxis: Requirement for tumbling and involvement in information processing. *Proc. Natl. Acad. Sci.* **72**: 4640.

Springer, W.R. and D.E. Koshland, Jr. 1977. Identification of a protein methyltransferase as the *cheR* gene product in the bacterial sensing system. *Proc. Natl. Acad. Sci.* **74**: 533.

Spudich, J.L. and D.E. Koshland, Jr. 1975. Quantitation of the sensory response in bacterial chemotaxis. *Proc. Natl. Acad. Sci.* **72**: 710.

Stock, J.B. and D.E. Koshland, Jr. 1978. A protein methylesterase involved in bacterial sensing. *Proc. Natl. Acad. Sci.* **75**: 3659.

Stock, J., A.M. Maderis, and D.E. Koshland, Jr. 1981. Bacterial chemotaxis in the absence of receptor carboxylmethylation. *Cell* **27**: 37.

Suzuki, T., T. Iino, T. Horiguchi, and S. Yamaguchi. 1978. Incomplete flagellar structures in nonflagellate mutants of *Salmonella typhimurium*. *J. Bacteriol.* **33**: 904.

Taylor, B.L. and D.J. Lazlo. 1981. The role of proteins in chemical perception in bacteria. In *The perception of behavioral chemicals* (ed. D.M. Norris), p. 2. Elsevier/North-Holland, Amsterdam.

Toews, M.L. and J. Adler. 1979. Methanol formation *in vivo* from methylated chemotaxis proteins in *Escherichia coli*. *J. Biol. Chem.* **254**: 1761.

Wang, E.A. and D.E. Koshland, Jr. 1980. Receptor structure in the bacterial sensing system. *Proc. Natl. Acad. Sci.* **77**: 7157.

Wang, E.A., K.L. Mowry, D.O. Clegg, and D.E. Koshland, Jr. 1982. Tandem duplication and multiple functions of a receptor gene in bacterial chemotaxis. *J. Biol. Chem.* **257**: 4673.

Warrick, H.M., B.L. Taylor, and D.E. Koshland, Jr. 1977. The chemotactic mechanism of *Salmonella typhimurium*: Preliminary mapping and characterization of mutants. *J. Bacteriol.* **130**: 223.

Subject Index

319

Bacteriophage λ *(continued)*
N protein, 135
nut site, 148
receptor, 257–258
retroregulation in, 150–152
sib, 150
transcription termination sites, 127–132, 144
Bacteriophage M13, 193
Bacteriophage MS2, 25, 27, 40, 42, 185–187
frameshift deletions and lysis gene, 52
Bacteriophage Mu. *See also* Gene fusions, Mud*lac* and
clonal polymorphism, 213
G-loop inversion, 218–219, 220–221
gin, 218–221
Bacteriophage P1
C-loop inversion, 218–219, 221
cin, site-specific recombination function, 218–221
clonal polymorphism in, 213
recombinase, 222
Bacteriophage Qβ, 185–187
role of terminator readthrough, 48
Bacteriophage T4
autogenous control of *regA*, 199–201, 204
DNA replication, 203
DNA replication and *regA*, 201
DNA replication and transcription, 200
genes of, 199–200
genes controlled by *regA*, 199
gene-*32* protein, 187–194
rate of RNA synthesis, 199
regA gene, 199–201, 204
rIIB frameshift mutants, 51
bgl operon, 223
Bla protein export, 257–258, 261–262, 267
Borrelia surface antigen and clonal polymorphism, 211–213

cAMP, 163–164, 166, 286, 311
Carbomycin, 102, 110
Cascade control of flagellar genes, 312–313
Catabolite repression, 163–164
flagella protein synthesis and, 311–312
put genes, 282
Celesticetin, 95
Cell division, arrest during DNA repair, 171–172
Che. *See* Chemotaxis
che loci, 295–299, 301, 303, 304, 308–310
genetics of, 308–310
Chemoreceptors, 295
Chemotaxis, 293–318. *See also* Flagella
adaptation by deamidation, 307–308
adaptation step, 294, 297–298, 306–308
behavior, 293–294
che A locus and overlapping genes, 308–310
CheA proteins and flagella motion,

308–309
CheB-dependent modification, 306–308
cryptic transducer, 304
evolution of sensory system, 305–307
excitation step, 294–297, 306
flagellar reversals and, 298
information processing and, 296–297
machinery, 295–296
methylation and adaptation, 297–298, 306–308
methylesterase and adaptation, 298
methyltransferase and adaptation, 297–298
methyltransferase, mutants affecting, 307
Mot proteins and flagellar action, 308–309
mutants affecting, 295–296, 298, 303–304, 307–310
periplasmic binding proteins and, 295
swimming patterns of mutants, 296
Tar protein, 295–296, 299–300, 304–306
transducer genes, 295
transducer genetics, 303–304
transducer protein modification, 299–301, 306–308
transducer proteins, 295–299, 300, 306–307
transducer protein structure, 301–302
transducers, genetics of, 303–304
Tsr protein, 295–297, 299–303
Cirramycin, 101, 103
cis-active sites and transposable elements, 234–236
Citrobacter freundii, clonal polymorphism, 213, 221
Clindamycin, 95, 109, 116
Clonal polymorphism, 211–227. *See also* Recombinational regulation
bacteriophage Mu, 213, 218–221
bacteriophage P1, 213, 218–220, 222
Borrelia surface antigen, 211–213
Citrobacter freundii Vi antigen, 213, 221
Escherichia coli pilin, 213
Neisseria gonococcus, 213, 223
Rhodopseudomonas spheriodes, 213
Serratia marcescens, 213
Streptomyces reticuli, 213
Vibrio harveyi, 211, 213
Clostridium welchii, 100
Coding. *See* Translation
Codon-anticodon pairs, 25, 27–30
Codon frequency
translation efficiency, 55–56
tRNA abundance, 56
Codon preference and protein abundance, 55–56
Codons, 23, 25, 27–29, 35, 39, 40–42
Coevolution, 203
Cointegrates, 221
Cooperativity of protein-nucleic acid binding, 188–190
Corynebacterium diphtheriae, 100